Molecular Aspects in Catalytic Materials for Pollution Elimination and Green Chemistry

Molecular Aspects in Catalytic Materials for Pollution Elimination and Green Chemistry

Editor

Junjiang Zhu

MDPI • Basel • Beijing • Wuhan • Barcelona • Belgrade • Manchester • Tokyo • Cluj • Tianjin

Editor
Junjiang Zhu
College of Chemistry and
Chemical Engineering
Wuhan Textile University
Wuhan
China

Editorial Office
MDPI
St. Alban-Anlage 66
4052 Basel, Switzerland

This is a reprint of articles from the Special Issue published online in the open access journal *International Journal of Molecular Sciences* (ISSN 1422-0067) (available at: www.mdpi.com/journal/ijms/special_issues/Catalytic_Material).

For citation purposes, cite each article independently as indicated on the article page online and as indicated below:

LastName, A.A.; LastName, B.B.; LastName, C.C. Article Title. *Journal Name* **Year**, *Volume Number*, Page Range.

ISBN 978-3-0365-7411-0 (Hbk)
ISBN 978-3-0365-7410-3 (PDF)

© 2023 by the authors. Articles in this book are Open Access and distributed under the Creative Commons Attribution (CC BY) license, which allows users to download, copy and build upon published articles, as long as the author and publisher are properly credited, which ensures maximum dissemination and a wider impact of our publications.

The book as a whole is distributed by MDPI under the terms and conditions of the Creative Commons license CC BY-NC-ND.

Contents

About the Editor vii

Junjiang Zhu
Editorial to the Special Issue "Molecular Aspects in Catalytic Materials for Pollution Elimination and Green Chemistry"
Reprinted from: *Int. J. Mol. Sci.* **2023**, 24, 7033, doi:10.3390/ijms24087033 1

Xiao Xu, Xianglong Yang, Yunlong Tao, Wen Zhu, Xing Ding and Junjiang Zhu et al.
Enhanced Exciton Effect and Singlet Oxygen Generation Triggered by Tunable Oxygen Vacancies on Bi_2MoO_6 for Efficient Photocatalytic Degradation of Sodium Pentachlorophenol
Reprinted from: *Int. J. Mol. Sci.* **2022**, 23, 15221, doi:10.3390/ijms232315221 5

Mikhail Lyulyukin, Nikita Kovalevskiy, Andrey Bukhtiyarov, Denis Kozlov and Dmitry Selishchev
Kinetic Aspects of Benzene Degradation over TiO_2-N and Composite $Fe/Bi_2WO_6/TiO_2$-N Photocatalysts under Irradiation with Visible Light
Reprinted from: *Int. J. Mol. Sci.* **2023**, 24, 5693, doi:10.3390/ijms24065693 17

Syed Taj Ud Din, Wan-Feng Xie and Woochul Yang
Synthesis of Co_3O_4 Nanoparticles-Decorated $Bi_{12}O_{17}Cl_2$ Hierarchical Microspheres for Enhanced Photocatalytic Degradation of RhB and BPA
Reprinted from: *Int. J. Mol. Sci.* **2022**, 23, 15028, doi:10.3390/ijms232315028 35

Xiaorong Cai, Yaning Wang, Shuting Tang, Liuye Mo, Zhe Leng and Yixian Zang et al.
Rhombohedral/Cubic In_2O_3 Phase Junction Hybridized with Polymeric Carbon Nitride for Photodegradation of Organic Pollutants
Reprinted from: *Int. J. Mol. Sci.* **2022**, 23, 14293, doi:10.3390/ijms232214293 55

Xue Bai, Wanyu Chen, Bao Wang, Tianxiao Sun, Bin Wu and Yuheng Wang
Photocatalytic Degradation of Some Typical Antibiotics: Recent Advances and Future Outlooks
Reprinted from: *Int. J. Mol. Sci.* **2022**, 23, 8130, doi:10.3390/ijms23158130 69

Sanxiu Li, Yufei Kang, Chenyang Mo, Yage Peng, Haijun Ma and Juan Peng
Nitrogen-Doped Bismuth Nanosheet as an Efficient Electrocatalyst to CO_2 Reduction for Production of Formate
Reprinted from: *Int. J. Mol. Sci.* **2022**, 23, 14485, doi:10.3390/ijms232214485 89

Wei Long, Zhilong Chen, Yinfei Huang and Xinping Kang
Selective Oxidation of Toluene to Benzaldehyde Using Co-ZIF Nano-Catalyst
Reprinted from: *Int. J. Mol. Sci.* **2022**, 23, 12881, doi:10.3390/ijms232112881 101

Ning Zhang, Yiou Shan, Jiaxin Song, Xiaoqiang Fan, Lian Kong and Xia Xiao et al.
Dendritic Mesoporous Silica Nanoparticle Supported PtSn Catalysts for Propane Dehydrogenation
Reprinted from: *Int. J. Mol. Sci.* **2022**, 23, 12724, doi:10.3390/ijms232112724 117

Wei Gao, Sai Tang, Ting Wu, Jianhong Wu, Kai Cheng and Minggui Xia
Solid Fe Resources Separated from Rolling Oil Sludge for CO Oxidation
Reprinted from: *Int. J. Mol. Sci.* **2022**, 23, 12134, doi:10.3390/ijms232012134 129

Hao Lin, Yao Xiao, Aixia Geng, Huiting Bi, Xiao Xu and Xuelian Xu et al.
Research Progress on Graphitic Carbon Nitride/Metal Oxide Composites: Synthesis and Photocatalytic Applications
Reprinted from: *Int. J. Mol. Sci.* **2022**, 23, 12979, doi:10.3390/ijms232112979 143

About the Editor

Junjiang Zhu

Prof. Junjiang Zhu received his Ph.D. degree in Physical Chemistry in 2005 from the Changchun Institute of Applied Chemistry, Chinese Academy of Sciences. Thereafter, he has worked at Fuzhou University, University of Porto, Technical University of Berlin, South-Central University for Nationalities, and Shenyang Normal University as a post-doctor, visiting scholar, and teaching staff. In June 2018, he joined the College of Chemistry and Chemical Engineering of Wuhan Textile University. He was appointed as the Chutian Distinguished Professor of Hubei Province in 2018 and as the Dean of the College of Chemistry and Chemical Engineering in 2022.

Editorial

Editorial to the Special Issue "Molecular Aspects in Catalytic Materials for Pollution Elimination and Green Chemistry"

Junjiang Zhu

Hubei Key Laboratory of Biomass Fibers and Eco-dyeing & Finishing, College of Chemistry and Chemical Engineering, Wuhan Textile University, Wuhan 430200, China; jjzhu@wtu.edu.cn

The Special Issue on "Molecular Aspects in Catalytic Materials for Pollution Elimination and Green Chemistry" encompasses two aims: one is to remove the pollutants produced in the downstream, and the other is to synthesize chemicals by a green route, avoiding the production of pollutants. The aim of this Special Issue is to explore catalysis technology to resolve current environmental problems using photocatalysis, electrocatalysis, and thermocatalysis. This Special Issue also describes research works related to the catalytic removal of pollutants, or the catalytic synthesis of chemicals by a green route.

Xu et al. [1] reported a method of constructing tunable oxygen vacancies (OVs) to accelerate molecular oxygen activation for boosting photocatalytic performance. In this paper, the in situ introduction of OVs on Bi_2MoO_6 was accomplished by using a calcination treatment in an H_2/Ar atmosphere. The introduced OVs not only facilitated carrier separation, but also strengthened the exciton effect, which accelerates singlet oxygen generation through the energy transfer process. Superior carrier separation and abundant singlet oxygen favor photocatalytic NaPCP degradation. The optimal BMO-001-300 sample exhibited the fastest NaPCP degradation rate of 0.033 min^{-1}, which is 3.8 times higher than that of the pristine Bi_2MoO_6. NaPCP was effectively degraded and mineralized through dechlorination, dihydroxylation, and benzene ring opening. The present work provided a novel insight into ROS-mediated photocatalytic degradation.

Lyulyukin et al. [2] synthesized composite materials based on nanocrystalline anatase TiO_2 doped with nitrogen and bismuth tungstate by using a hydrothermal method. To determine the correlations between their physicochemical characteristics and photocatalytic activity, the kinetic aspects were studied both in batch and continuous-flow reactors, using ethanol and benzene as test compounds. The researchers found that the Bi_2WO_6/TiO_2-N heterostructure enhanced with Fe species efficiently utilized visible light in the blue region, and exhibited much higher activity in the degradation of ethanol vapor compared to pristine TiO_2-N. However, the increased activity of Fe/Bi_2WO_6/TiO_2-N had an adverse effect on the degradation of benzene vapor. This result might be due to the fast accumulation of non-volatile intermediates on its surface at high benzene concentrations, which caused a temporary deactivation of the photocatalyst.

Din et al. [3] synthesized three-dimensional (3D) hierarchical $Bi_{12}O_{17}Cl_2$ (BOC) microsphere via a facile solvothermal method by using a binary solvent, which was applied for the photocatalytic degradation of Rhodamine-B (RhB) and Bisphenol-A (BPA). Subsequently, they prepared Co_3O_4 nanoparticles (NPs)-decorated BOC (Co_3O_4/BOC) heterostructures to further enhance their photocatalytic performances. The BOC microspheres, composed of thin (~20 nm thick) nanosheets, showed a 3D hierarchical morphology and a high surface area. Compared to the BOC, the 20-Co_3O_4/BOC heterostructure exhibited enhanced degradation efficiency of RhB (97.4%) and BPA (88.4%). The authors hypothesized that the high surface area, the extension of absorption to visible light region, and the suppression of photoexcited electron-hole recombination are main factors accounting for the better photocatalytic performances. This work provides a new vista to construct high-performance heterostructures for photocatalytic applications.

Citation: Zhu, J. Editorial to the Special Issue "Molecular Aspects in Catalytic Materials for Pollution Elimination and Green Chemistry". *Int. J. Mol. Sci.* **2023**, *24*, 7033. https://doi.org/10.3390/ijms24087033

Received: 31 March 2023
Accepted: 7 April 2023
Published: 11 April 2023

Copyright: © 2023 by the author. Licensee MDPI, Basel, Switzerland. This article is an open access article distributed under the terms and conditions of the Creative Commons Attribution (CC BY) license (https://creativecommons.org/licenses/by/4.0/).

Phase junctions constructed as photocatalysts show great prospects for organic degradation with visible light. Cai et al. [4] synthesized an elaborate rhombohedral corundum/cubic In_2O_3 phase junction (named MIO) combined with polymeric carbon nitride (PCN) by using an in situ calcination method. The authors attributed the excellent performance of MIO/PCN to the intimate interface contact between MIO and PCN, which provides a reliable charge transmission channel, thereby improving the separation efficiency of charge carriers. This work illustrates that MOF-modified materials have a great potential for solving environmental pollution without generating secondary pollution.

The need for effective and rapid processes for photocatalytic degradation has led to an increased interest in finding more sustainable catalysts for antibiotic degradation. Bai et al. [5] provided an overview on the removal of pharmaceutical antibiotics through photocatalysis, and the recent progress using different nanostructure-based photocatalysts. They also reviewed the possible sources of antibiotic pollutants released through the ecological chain, and the consequences and damages caused by antibiotics in wastewater on the environment and human health. The fundamental dynamic processes of nanomaterials and the degradation mechanisms of antibiotics were discussed, and recent studies regarding different photocatalytic materials for the degradation of some typical and commonly used antibiotics were summarized. Finally, the challenges and future opportunities for photocatalytic degradation of commonly used antibiotics were highlighted.

Formate is a desirable product of an electrochemical CO_2 reduction (CO_2RR) and has great economic value. Li et al. [6] reported a facile one-step method to synthesize nitrogen-doped bismuth nanosheets (N-BiNSs), as electrocatalysts, for CO_2RR to produce formate. The N-BiNSs exhibited a high formate Faradic efficiency with a stable current density. Moreover, the N-BiNSs for CO_2RR yielded a large current density for formate production in a flow-cell measurement, achieving the commercial requirement. The authors found that nitrogen doping could induce charge transfer from the N atom to the Bi atom, thus modulating the electronic structure of N-Bi nanosheets. Density Functional Theory results demonstrated that the N-BiNSs reduced the adsorption energy of the *OCHO intermediate and promoted the mass transfer of charges. This study provides a valuable strategy to enhance the catalytic performance of bismuth-based catalysts for CO_2RR by using a nitrogen-doping strategy.

Long et al. [7] synthesized a nanometer-size Co-ZIF (zeolitic imidazolate frameworks) catalyst for selective oxidation of toluene to benzaldehyde under mild conditions, which showed a BET surface area of 924.25 m^2/g. The authors investigated the effect of reaction temperatures, oxygen pressure, mass amount of N-hydroxyphthalimide (NHPI), and reaction time on the catalytic activity. The results showed that the Co-ZIF catalyst gave the best result of 92.30% toluene conversion and 91.31% selectivity to benzaldehyde under 0.12 MPa and 313 K. This new nanometer-size Co-ZIF catalyst exhibits good application prospects in the selective oxidation of toluene to benzaldehyde.

Zhang et al. [8] prepared the PtSn catalysts for propane dehydrogenation by incipient-wetness impregnation using a dendritic mesoporous silica support. The authors found that changing the Pt/Sn ratios influenced the interaction between Pt and Sn. The best catalytic performance was obtained from the Pt_1Sn_2/DMSN catalyst, with an initial propane conversion of 34.9%. They proposed that the good catalytic performance of this catalyst is a result of the small nanoparticle size of PtSn and the favorable chemical state and dispersion degree of Pt and Sn species.

The efficient recycling of valuable resources from rolling oil sludge (ROS) is meaningful. Gao et al. [9] reported the recycling of solid Fe resources from ROS by a catalytic hydrogenation technique, and discussed its catalytic performance for CO oxidation. The solid Fe resources calcinated in air (Fe_2O_3-H) exhibited comparable activity for CO oxidation to those prepared by the calcinations of ferric nitrate (Fe_2O_3-C). Following this, the authors supported the Fe resources on 13X zeolite, and pretreated the sample with CO atmosphere, which led to a complete CO conversion at 250 °C on the 20 wt.% Fe_2O_3-H/13X sample.

Graphitic carbon nitride (g-C_3N_4) is thoroughly studied owing to its remarkable properties exhibited in photocatalysis. Lin et al. [10] summarized the research progress on the synthesis of g-C_3N_4 and its coupling with single- or multiple-metal oxides, and its photocatalytic applications in energy production and environmental protection, including the splitting of water to hydrogen, the reduction of CO_2 to valuable fuels, the degradation of organic pollutants, and the disinfection of bacteria. At the end, challenges and prospects in the synthesis and photocatalytic applications of g-C_3N_4-based composites were proposed, and an outlook was provided.

Conflicts of Interest: The author declares no conflict of interest.

References

1. Xu, X.; Yang, X.; Tao, Y.; Zhu, W.; Ding, X.; Zhu, J.; Chen, H. Enhanced Exciton Effect and Singlet Oxygen Generation Triggered by Tunable Oxygen Vacancies on Bi_2MoO_6 for Efficient Photocatalytic Degradation of Sodium Pentachlorophenol. *Int. J. Mol. Sci.* **2022**, *23*, 15221. [CrossRef] [PubMed]
2. Lyulyukin, M.; Kovalevskiy, N.; Bukhtiyarov, A.; Kozlov, D.; Selishchev, D. Kinetic Aspects of Benzene Degradation over TiO_2-N and Composite Fe/Bi_2WO_6/TiO_2-N Photocatalysts under Irradiation with Visible Light. *Int. J. Mol. Sci.* **2023**, *24*, 5693. [CrossRef] [PubMed]
3. Din, S.T.; Xie, W.-F.; Yang, W. Synthesis of Co_3O_4 Nanoparticles-Decorated $Bi_{12}O_{17}Cl_2$ Hierarchical Microspheres for Enhanced Photocatalytic Degradation of RhB and BPA. *Int. J. Mol. Sci.* **2022**, *23*, 15028. [CrossRef] [PubMed]
4. Cai, X.; Wang, Y.; Tang, S.; Mo, L.; Leng, Z.; Zang, Y.; Jing, F.; Zang, S. Rhombohedral/Cubic In_2O_3 Phase Junction Hybridized with Polymeric Carbon Nitride for Photodegradation of Organic Pollutants. *Int. J. Mol. Sci.* **2022**, *23*, 14293. [CrossRef] [PubMed]
5. Bai, X.; Chen, W.; Wang, B.; Sun, T.; Wu, B.; Wang, Y. Photocatalytic Degradation of Some Typical Antibiotics: Recent Advances and Future Outlooks. *Int. J. Mol. Sci.* **2022**, *23*, 8130. [CrossRef] [PubMed]
6. Li, S.; Kang, Y.; Mo, C.; Peng, Y.; Ma, H.; Peng, J. Nitrogen-Doped Bismuth Nanosheet as an Efficient Electrocatalyst to CO_2 Reduction for Production of Formate. *Int. J. Mol. Sci.* **2022**, *23*, 14485. [CrossRef] [PubMed]
7. Long, W.; Chen, Z.; Huang, Y.; Kang, X. Selective Oxidation of Toluene to Benzaldehyde Using Co-ZIF Nano-Catalyst. *Int. J. Mol. Sci.* **2022**, *23*, 12881. [CrossRef] [PubMed]
8. Zhang, N.; Shan, Y.; Song, J.; Fan, X.; Kong, L.; Xiao, X.; Xie, Z.; Zhao, Z. Dendritic Mesoporous Silica Nanoparticle Supported PtSn Catalysts for Propane Dehydrogenation. *Int. J. Mol. Sci.* **2022**, *23*, 12724. [CrossRef] [PubMed]
9. Gao, W.; Tang, S.; Wu, T.; Wu, J.; Cheng, K.; Xia, M. Solid Fe Resources Separated from Rolling Oil Sludge for CO Oxidation. *Int. J. Mol. Sci.* **2022**, *23*, 12134. [CrossRef] [PubMed]
10. Lin, H.; Xiao, Y.; Geng, A.; Bi, H.; Xu, X.; Xu, X.; Zhu, J. Research Progress on Graphitic Carbon Nitride/Metal Oxide Composites: Synthesis and Photocatalytic Applications. *Int. J. Mol. Sci.* **2022**, *23*, 12979. [CrossRef] [PubMed]

Disclaimer/Publisher's Note: The statements, opinions and data contained in all publications are solely those of the individual author(s) and contributor(s) and not of MDPI and/or the editor(s). MDPI and/or the editor(s) disclaim responsibility for any injury to people or property resulting from any ideas, methods, instructions or products referred to in the content.

Article

Enhanced Exciton Effect and Singlet Oxygen Generation Triggered by Tunable Oxygen Vacancies on Bi_2MoO_6 for Efficient Photocatalytic Degradation of Sodium Pentachlorophenol

Xiao Xu [1,2,†], Xianglong Yang [2,3,†], Yunlong Tao [2], Wen Zhu [1], Xing Ding [2], Junjiang Zhu [1,*] and Hao Chen [2,*]

[1] Hubei Key Laboratory of Biomass Fibers and Eco-Dyeing & Finishing, College of Chemistry and Chemical Engineering, Wuhan Textile University, Wuhan 430200, China
[2] College of Science, Huazhong Agricultural University, Wuhan 430070, China
[3] National Reference Laboratory for Agricultural Testing (Biotoxin), Laboratory of Quality and Safety Risk Assessment for Oilseed Products (Wuhan), Key Laboratory of Detection for Mycotoxins, Quality Inspection and Test Center for Oilseed Products, Ministry of Agriculture and Rural Affairs, Oil Crops Research Institute, Chinese Academy of Agricultural Sciences, Wuhan 430062, China
* Correspondence: jjzhu@wtu.edu.cn (J.Z.); hchenhao@mail.hzau.edu.cn (H.C.)
† These authors contributed equally to this work.

Abstract: Construction of the tunable oxygen vacancies (OVs) is widely utilized to accelerate molecular oxygen activation for boosting photocatalytic performance. Herein, the in-situ introduction of OVs on Bi_2MoO_6 was accomplished using a calcination treatment in an H_2/Ar atmosphere. The introduced OVs can not only facilitate carrier separation, but also strengthen the exciton effect, which accelerates singlet oxygen generation through the energy transfer process. Superior carrier separation and abundant singlet oxygen played a crucial role in favoring photocatalytic NaPCP degradation. The optimal BMO-001-300 sample exhibited the fastest NaPCP degradation rate of 0.033 min^{-1}, about 3.8 times higher than that of the pristine Bi_2MoO_6. NaPCP was effectively degraded and mineralized mainly through dechlorination, dehydroxylation and benzene ring opening. The present work will shed light on the construction and roles of OVs in semiconductor-based photocatalysis and provide a novel insight into ROS-mediated photocatalytic degradation.

Keywords: oxygen vacancy; exciton effect; Bi_2MoO_6; sodium pentachlorophenate; photocatalysis

1. Introduction

The excessive discharge of pesticide pollutants produced in agricultural activity has brought serious environmental pollution and caused a heavy burden on the ecological system [1–3]. Among all pesticide pollutants, sodium pentachlorophenate (NaPCP) has been listed as a priority contaminant by the U.S. Environmental Protection Agency due to its bio-refractory and high biological toxicity [4]. It is of great urgency and significance to develop novel technology for efficient NaPCP removal. Compared with other advanced oxidation processes for NaPCP degradation, photocatalytic technology is regarded as one of the most potentially effective methods of pollution control and environmental remediation. It works by utilizing light-induced reactive oxygen species (ROS) to degrade organic contaminants [5–8].

Until now, numerous semiconductors including metal oxides (TiO_2, BiOBr and Bi_2MoO_6, etc.) [9–11] and organic polymers (C_3N_4, MOF and COF, etc.) [12–14] have been chosen as the candidates for NaPCP degradation. However, photocatalytic NaPCP degradation still suffers from a lack of active species and unsatisfying reaction efficiency [15,16]. To improve the reactivity for NaPCP degradation, multifarious modification strategies have been explored in

Citation: Xu, X.; Yang, X.; Tao, Y.; Zhu, W.; Ding, X.; Zhu, J.; Chen, H. Enhanced Exciton Effect and Singlet Oxygen Generation Triggered by Tunable Oxygen Vacancies on Bi_2MoO_6 for Efficient Photocatalytic Degradation of Sodium Pentachlorophenol. *Int. J. Mol. Sci.* 2022, 23, 15221. https://doi.org/10.3390/ijms232315221

Academic Editor: Bogusław Buszewski

Received: 9 October 2022
Accepted: 29 November 2022
Published: 2 December 2022

Publisher's Note: MDPI stays neutral with regard to jurisdictional claims in published maps and institutional affiliations.

Copyright: © 2022 by the authors. Licensee MDPI, Basel, Switzerland. This article is an open access article distributed under the terms and conditions of the Creative Commons Attribution (CC BY) license (https://creativecommons.org/licenses/by/4.0/).

recent years to strengthen the photocatalytic activity of semiconductors, such as heterojunction construction, co-catalyst modification and defect engineering [17–20].

Oxygen vacancies (OVs), the typical anion defects, have been regarded as an effective modification to enhance the photocatalytic reactivity [21,22]. The introduced OVs are capable of extending the light absorption region of semiconductors and accelerating the carrier migration and separation of charge carriers [23,24]. Wei et al. employed transient absorption spectroscopy to confirm that OVs on the WO_3 surface were enabled to induce electron trapping states and inhibit the direct recombination of photogenerated carriers [25]. More importantly, OV-mediated electrons in localized states led to more photogenerated electrons gathering, serving as the active sites for molecular oxygen and reactive substrate activation [26]. Li et al. systematically explored the impacts of OVs on the optical absorption of BiOCl and molecular oxygen activation [27]. The results indicated that the OVs on the catalyst surface can selectively activate oxygen to $•O_2^-$, which dominated the dechlorination process of NaPCP, thus achieving the excellent photocatalytic NaPCP degradation [27]. Their subsequent research found that different facet-dependent OVs exhibited different water adsorption configurations in the form of chemical adsorption or dissociation adsorption, which determined the ability of adsorbed water molecules to be oxidized by holes [28].

Considerable research efforts have been devoted to investigating the role of OVs in ROS generation from two aspects, including the electron transfer mechanism induced by charge-carrier behavior and the energy transfer mechanism caused by the exciton effect [29–33]. The electron transfer process results from the separated carriers' transfer to molecular oxygen, thus realizing the activation of molecular oxygen to generate ROS (such as $•O_2^-$, H_2O_2 and $•OH$). The energy transfer process originates from the transformation from a high-energy singlet exciton to the triple excited state via the inter-system crossover, accompanying molecular oxygen activation to produce $^1O_2$ [34]. As reported, the OV-mediated electron capture state on the semiconductor surface can activate O_2 to produce $•O_2^-$ and H_2O_2 through the electron transfer process utilizing localized electrons [35]. The introduced OVs also increase the highly delocalized valence electrons at adjacent atoms, which will further weaken the charge shielding effect and enhance the electron-hole interaction, thus endowing the catalyst with increased exciton binding energy [36]. Achieving tunable OVs is of great significance for arranging the exciton or carrier behavior. Although many modification strategies have been developed to introduce tunable OVs, including solvothermal processes and reduction with UV illumination and doping, these are complicated and uncontrollable. There is therefore an urgent need to solve the significant challenge of achieving the accurate and controllable construction of OVs.

Herein, we choose Bi_2MoO_6, a typical layered Aurivillius material, as the photocatalyst model, and realize the mild and controllable construction of OVs on the Bi_2MoO_6 surface via adjusting the calcination temperature under H_2/Ar atmosphere. The tunable OVs introduced here are proven to have several function schemes, not only endowing Bi_2MoO_6 with efficient carrier separation and an optimized band gap via the constructing defects level, but also enabling the weak charge screening effect, which is of great significance for the enhanced exciton effect and $^1O_2$ production through the energy transfer process. Subsequently, the photocatalytic performance of the NaPCP degradation is evaluated to explore the contribution of different active species in the degradation process. Finally, a possible degradation mechanism of NaPCP is proposed with the main degradation step of dechlorination, dehydroxylation and benzene ring oxidation and breakage based on the in-situ DRIFTS experiments.

2. Results and Discussion

2.1. Characterization of the Samples

As shown in Figure 1a, Bi_2MoO_6 with (001) facet exposed was firstly prepared by adjusting the pH of the reaction solution referring to our previous work [37]. Subsequently, OVs on BMO-001 were in-situ constructed by calcination treatment at 300 °C under H_2/Ar

atmosphere to attain BMO-001-300. The as-prepared samples were characterized by XRD to investigate the crystal phase and structure in Figure 1b. All samples exhibited clear diffraction peaks, which were consistent with the characteristic diffraction peaks of tetragonal phase Bi_2MoO_6 (PDF No. 21-0102). This result indicated that pure Bi_2MoO_6 photocatalysts were successfully prepared via the typical hydrothermal method, and the subsequent calcination treatment hardly caused any change in the crystal structure of the Bi_2MoO_6 sample. In addition, the high intensity of the (002) diffraction peak implied the preferential growth and exposure of the (001) facet on Bi_2MoO_6, which was consistent with the result reported in our previous work [16,37]. The micro morphologies of all samples were depicted in Figure S1 with TEM images. The specific surface area of these samples also had no obvious change (Figure S2). The as-prepared samples exhibited similar nanosheet morphology, demonstrating that the calcination treatment process did not cause any destruction to the structure and size of Bi_2MoO_6 nanosheets. As shown in Figure 1c,d, the BMO-001-300 nanosheets exhibited a width of 200 nm and a thickness of 28 nm. HRTEM images of BMO-001-300 displayed distinct orthorhombic lattice fringes with spaces of 0.270 nm and 0.274 nm, corresponding to the interplanar distance of (060) planes and (200) planes, thereby suggesting BMO-001-300 is indeed exposed with the (001) facets (Figure 1e). Identical conclusions were obtained in the selected area electron diffraction (SAED) patterns of BMO-001-300 (Figure 1f).

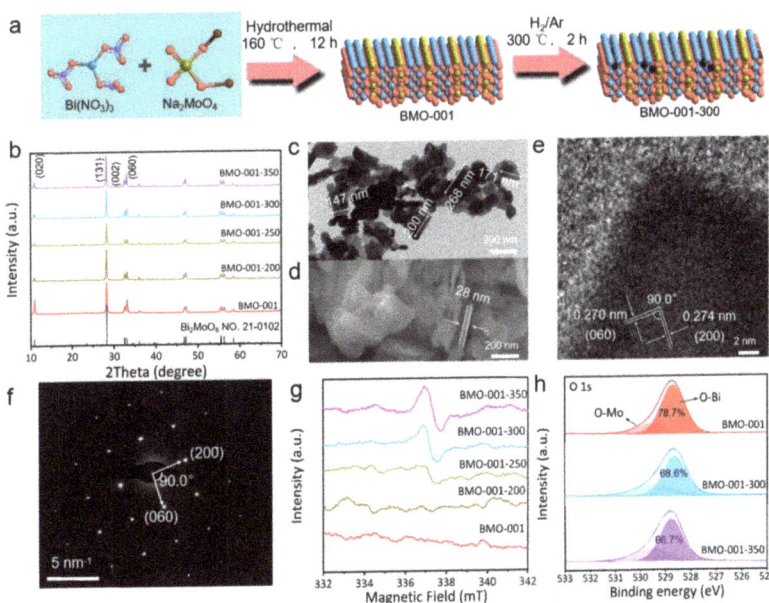

Figure 1. (**a**) Schematic illustration of BMO-001-300 preparation with calcination treatment; (**b**) XRD patterns of as-prepared Bi_2MoO_6 samples; TEM (**c**), SEM (**d**), HRTEM (**e**) and SAED (**f**) images of BMO-001-300; (**g**) EPR spectra of oxygen vacancies on as-prepared Bi_2MoO_6 samples; (**h**) high-resolution XPS spectra of O 1s for BMO-001, BMO-001-300, BMO-001-350.

To confirm the existence of OVs on BMO-001-300, the EPR test was performed by monitoring the single–electron-induced paramagnetic signal of the as-prepared samples. As observed in Figure 1g, a continuously enhanced paramagnetic signal at g = 2.003, corresponding to the OVs signal, gradually appeared with the increase of the calcination temperature during the synthesis process. This result conveyed the fact that the strategy for in-situ construction of OVs in this work was practicable. The XPS test was conducted to analyze the surface electronic states and chemical compositions of samples (Figures S3 and 1h).

The high-resolution XPS spectra of Bi 4f and Mo 3d over BMO-001 indicated that no change occurred on BMO-001-300 and BMO-001-350 (Figure S3b,c). The high-resolution XPS spectra of O 1s were divided into two peaks corresponding to the Bi-O (528.7 eV) and Mo-O bands (529.6 eV) (Figure 1h). Notably, the peak area of the Bi-O signals over BMO-001-300 and BMO-001-350 was much weaker than that of BMO-001, thereby implying that the OVs of BMO-001-300 and BMO-001-350 resulted from the escape of O atoms connected with Bi atoms.

2.2. Photocatalytic Activity

As reported numerous times in the literature, OVs played a vital role in improving photocatalytic performances by inducing coordination unsaturation and electron localization [16,19,30,38]. Thus, photocatalytic degradation of NaPCP with visible light irradiation was conducted to evaluate the roles of surface OVs. As exhibited in Figure 2a, all catalysts treated by calcination performed superior photocatalytic activities for NaPCP degradation, implying that the OVs induced by the calcination treatment used on BMO-001-X (X = 200, 250, 300, 350) could be a great impetus for boosting the photocatalytic performance. Compared with the reported catalysts, the photocatalytic performance of the optimized BMO-001-300 also showed a significant advantage (Table S1). According to the first-order translation kinetics, the rate constant of the reaction was fitted and calculated. Briefly, the reaction kinetic constant of NaPCP degradation on BMO-001-300 was 0.033 min^{-1}, which is slightly higher than BMO-001-350 and 3.8 times that of the pure BMO-001 (0.009 min^{-1}) (Figure 2b). The above results suggest that surface OVs could accelerate the photocatalytic process, but that excessive OVs restrain the degradation process, possibly because excessive OVs act as recombination traps of electron-hole pairs. As shown in Figure 2c, the photocatalytic activity for NaPCP degradation hardly decayed after four recycling runs. Furthermore, no obvious changes were observed in the XRD patterns and the XPS spectra of BMO-001-300 after four runs of photocatalytic tests (Figure S4), indicating that BMO-001-300 kept excellent recycle stability.

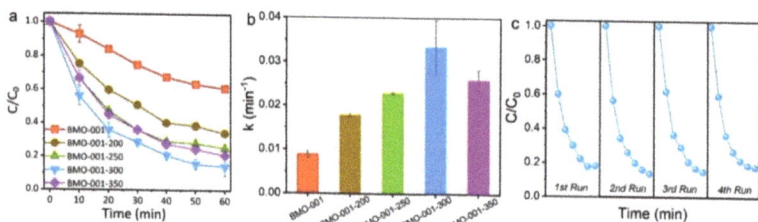

Figure 2. Photocatalytic NaPCP degradation performance curve (**a**) and reaction rate (**b**) on as-prepared Bi$_2$MoO$_6$ samples under visible light irradiation; (**c**) recycle performance of BMO-001-300 for photocatalytic NaPCP degradation.

2.3. Mechanisms of Photocatalytic NaPCP Degradation on BMO-001-300

As shown in Figure S5, these photocatalysts exhibited gradually enhanced light absorption with the OVs increasing, demonstrating that the introduced OVs expanded the light absorption range. As exhibited in the photocurrent response test, the OV-modified BMO-001-300 sample displayed higher photocurrent intensity compared with its counterparts, indicating that the OVs favored the internal charge separation efficiency (Figure 3a). Besides, BMO-001-300 presented a smaller radius of EIS Nyquist plots than BMO-001, suggesting a lower charge transfer resistance and a greater impetus for electron transfer (Figure 3b). It contributed to the accelerated charge transport promoted by the OVs. In addition, the PL behavior was monitored to acquire the related information for charge separation (Figure 3c). The lower PL emission intensity on BMO-001-300 meant a lower recombination probability of the photogenerated carriers, confirming that the existence of OVs hindered the recombination of photogenerated e$^-$/h$^+$ pairs. To reveal the OV-mediated electron transfer process, TR-PL was performed to test the carrier lifetime. As shown in Figure 3d, the electron lifetime of the samples followed the order of BMO-001-300 > BMO-001-350 > BMO-001. The result indicated that the appropriate OVs could favor effective electron separation, greatly extending the electron lifetime, while excessive OVs accelerated carrier recombination. Interestingly, when oxygen in the environment was isolated, the electron lifetime was significantly prolonged, implying that photoelectron transfer was conducted with oxygen in the form of electron transfer or exciton recombination in the photocatalytic process.

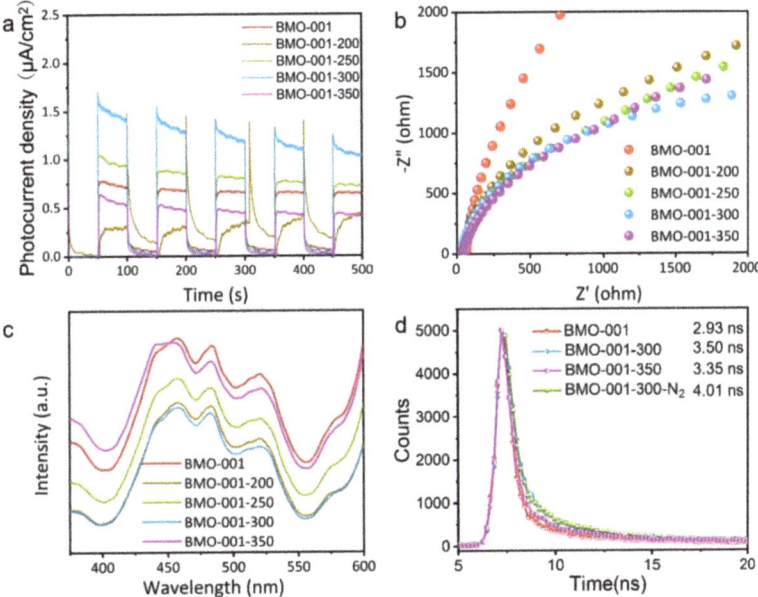

Figure 3. Photocurrent tests (**a**), EIS (**b**) and steady-state photoluminescence spectra (**c**) and time-resolved transient photoluminescence decay spectra (**d**) of as-prepared Bi_2MoO_6 samples.

Photocatalytic oxygen or water molecule activation produced abundant active species with excellent reactivity, displaying a great impetus for photocatalytic reaction. To explore the dominant roles of these active species in the reaction, the photocatalytic NaPCP degradation of BMO-001 and BMO-001-300 were evaluated with isopropanol (IPA), superoxide dismutase (SOD), natural β-carotene (Caro), and triethanolamine (TEOA) as the •OH, •O_2^-, $^1O_2$ and holes scavengers, respectively (Figure 4). As for both BMO-001 and BMO-001-300 photocatalysts, the degradation rate of NaPCP significantly decreased with TEOA capturing the holes, suggesting that the holes became the most important active species by activating NaPCP and lowering the reaction barrier. As reported in several works [1,16,39], •OH and •O_2^- also played an indispensable role in the degradation process. The contribution rates of •OH and •O_2^- in the photocatalytic system of BMO-001-300 reached 70% and 81%, respectively, which demonstrated few differences with that of BMO-001, suggesting that the OVs did not possess a positive tendency in the degradation process triggered by •OH and •O_2^-. It is noteworthy that the absence of $^1O_2$ inhibited 82% of NaPCP degradation for BMO-001-300, which was much higher than the 28% of BMO-001, revealing that $^1O_2$ was the critical factor to improve photocatalytic reactivity.

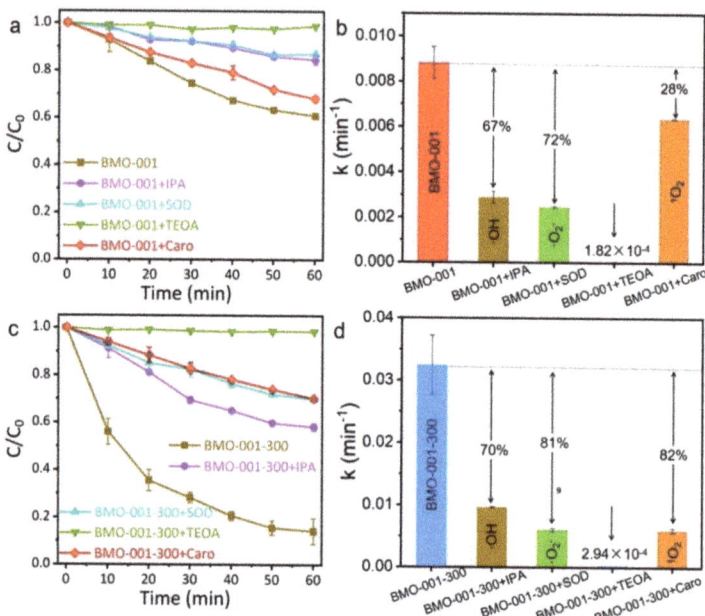

Figure 4. Photocatalytic NaPCP degradation performance curve and reaction rate on BMO-001 (**a**,**b**) and BMO-001-300 (**c**,**d**) with different scavengers (TEOA→h^+, IPA→•OH, SOD→•O_2^-, caro→$^1O_2$) under visible light irradiation, respectively.

Subsequently, we utilized EPR spin-trapping tests to detect the photogenerated ROS, including •OH, •O_2^- and $^1O_2$. As shown in Figure 5a,b, slight enhancement was observed in the quadruple peak with the intensity ratio of 1:2:2:1 and the six-fold peak corresponding to the addition products of DMPO-•OH and DMPO-•O_2^- over BMO-001-300, which proved that OVs played an insignificant role in accelerating •OH and •O_2^- production by activating oxygen or water molecules through the electron transfer process. Interestingly, when TEMP was added to capture $^1O_2$, a clear triple peak signal with the intensity ratio of 1:1:1 was generated. This TEMP-$^1O_2$ signal of BMO-001-300 was much stronger than that of BMO-001 and BMO-001-350 (Figure 5c). The above results indicated that an appropriate amount of OVs can promote the production of $^1O_2$, which may be attributed to the fact that the OV-modified BMO-001-300 material was prone to induce excitonic enhancement. This EPR test result also supported the scavenger experiment for NaPCP degradation, confirming the crucial role of multiple ROS, especially $^1O_2$, in the photocatalytic degradation process.

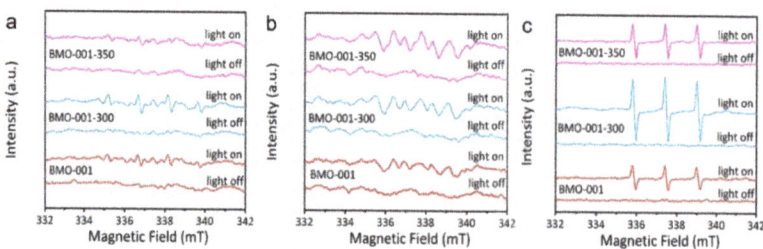

Figure 5. EPR signal of DMPO-•OH (**a**), DMPO-•O_2^- (**b**) and TEMP-$^1O_2$ (**c**) for BMO-001, BMO-001-300 and BMO-001-350 under visible light irradiation.

To verify the feasibility of the OVs on the Bi_2MoO_6 surface to activate oxygen and water molecules to generate ROS, we tested and calculated the band positions of the samples through a series of characterizations. Based on the diffuse reflectance ultraviolet-visible spectra, the plots $(\alpha h\nu)^{1/2}$ versus the energy of absorbed light were estimated to calculate the bandgap energies of the samples (Figure 6a). The sample with more oxygen vacancies obtained a narrower band gap, which should be ascribed to the OV-induced defect levels [18]. The Mott–Schottky test was employed to determine the flat band potentials of samples (Figure 6b). The flat band potentials of BMO-001, BMO-001-300 and BMO-001-350 were −0.29, −0.30 and −0.34 V (vs. Ag/AgCl), corresponding to 0.32, 0.31 and 0.27 V (vs. NHE), which were nearly equal to their Fermi level for n-type semiconductors. Figure 6c illustrated the VB-XPS spectra of all samples, indicating that the energy gaps between their Fermi level (E_f) and VB were 1.66, 2.07 and 2.16 eV for BMO-001, BMO-001-300 and BMO-001-350, respectively. The above results enabled us to pinpoint the VB position of the sample directly and acquire the CB position combined with the band gap values (Table S2). For BMO-001, its CB potential was lower than −0.33 V vs. NHE for the conversion from O_2 to $\bullet O_2^-$, which was not negative enough to activate oxygen. Once OVs were introduced, the obtained BMO-001-300 and BMO-001-350 not only possessed more activation sites, but also exhibited more negative CB positions, which facilitated the transfer of photogenerated electrons to oxygen and may activate molecular oxygen (Figure 6d). Considering the unique electron trap effect of OVs, the OV concentration has a significant influence on the regulation of exciton and carrier behavior [36]. When appropriate OVs were introduced using a calcination treatment at 300 °C, BMO-001-300 maintained a notable Coulomb-interaction-mediated excitonic effect, which resulted from the attenuated dielectric screening induced by the OVs. Excess OVs provided more electron localization centers, enabled the acceleration of the electron transfer behavior of carrier separation, and correspondingly weakened the energy transfer process induced by the excitons, finally resulting in the production of increased $\bullet O_2^-$ and decreased $^1O_2$.

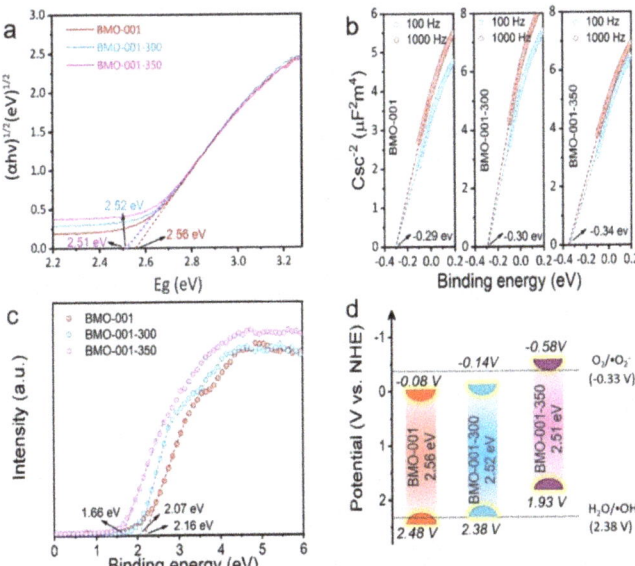

Figure 6. The plots of $(\alpha h\nu)^{1/2}$ versus photon energy ($h\nu$) (**a**), Mott–Schottky plots (**b**), VB-XPS (**c**) and a schematic illustration of the band structure (**d**) of BMO-001, BMO-001-300 and BMO-001-350.

In-situ DRIFTS experiments were conducted to dynamically detect the reaction intermediates and track the possible photodegradation route. In general, the negative bands and positive bands in in-situ DRIFTS spectra represented the depletion of reactants and the accumulation of products during the reaction, respectively. Both negative and positive peaks in the DRIFTS spectra of the two systems were observed in Figure 7. These increased negative bands at 1450–1600 cm^{-1} were attributed to the vibration peaks of the consumed benzene ring skeleton, which suggested that the carbon skeleton of NaPCP was continuously attacked and decomposed under visible light irradiation. Besides, the negative peaks at 996 cm^{-1} and 1214 cm^{-1} were observed in Figure 7a, corresponding to the stretching vibration of the aromatic C-Cl bond and C-O bond within NaPCP molecules, respectively. These results imply that the C-Cl bond and C-O bond of adsorbed NaPCP were consumed over BMO-001-300 under illumination. As described in Figure 7b, it was obvious that there were three positive bands located in the ranges of 2760–2950 cm^{-1} and the absorption intensity increased over time, which corresponded to the vibrations of the boosted C-H bond.

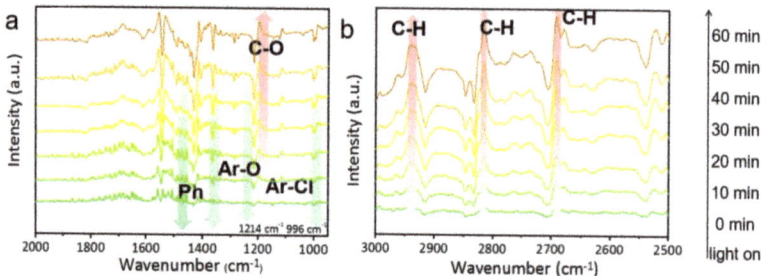

Figure 7. In situ DRIFTS spectra of photocatalytic NaPCP degradation in the range of 2000–1000 cm^{-1} (**a**) and 3000–2500 cm^{-1} (**b**) with BMO-001-300 under irradiation.

Based on the above in-situ DRIFTS analyses, the mechanism of photocatalytic NaPCP degradation over OV-modified BMO-001-300 is summarized. Considering the great contribution of abundant active species, especially holes and $^1O_2$, the adsorbed NaPCP is easier to degrade and mineralize. NaPCP was decomposed mainly through the following three paths, including dehydrogenation, dechlorination and benzene ring decomposition. Meanwhile, the significantly enhanced C-H peaks of the in-situ DRIFTS spectrum indicated that the chemically stable NaPCP was gradually degraded and transformed into non-toxic organic small molecules, and finally completely mineralized.

2.4. Conclusions

In summary, tunable OV construction on the Bi_2MoO_6 surface was completed through the modulation of the temperature in the calcination treatment. The introduced OVs can not only promote carrier separation and optimize the band gap structure, but also greatly drive the exciton effect, thus generating plentiful $^1O_2$. These sufficient active species, especially $^1O_2$, played a crucial role in photocatalytic NaPCP degradation. NaPCP was effectively degraded mainly through dechlorination, dehydroxylation and benzene ring opening. The present work will provide a novel strategy for the tunable introduction of OVs on semiconductors and insights into the efficient degradation of NaPCP.

3. Materials and Methods

3.1. Sample Preparation

All reagents used in this work were of analytical grade and used without further purification.

BMO-001 was fabricated via hydrothermal synthesis according to our previous work. In detail, 0.242 g of $Na_2MoO_4 \cdot 2H_2O$ and 0.970 g of $Bi(NO_3)_3 \cdot 5H_2O$ were dissolved and dispersed in 60 mL of deionized water while stirring continuously. After that, the above suspension was adjusted to pH = 6 using ammonia and then heated at 160 °C for 12 h in the Teflon-lined autoclave. The as-prepared sample was purged with distilled water and ethanol 3 times and finally dried at 60 °C for 12 h.

Bi_2MoO_6 with tunable oxygen vacancies was synthesized by calcination treatment. Specifically, a 300 mg BMO-001 sample was placed in a tubular furnace filled with an H_2/Ar atmosphere, then heated to x °C (x = 200, 250, 300, 350) at 10 °C/min and kept for 2 h, and finally cooled naturally to room temperature. The as-prepared Bi_2MoO_6 with oxygen vacancies were named BMO-001-200, BMO-001-250, BMO-001-300, and BMO-001-350.

3.2. Sample Characterization

The X-ray diffraction (XRD) patterns of the as-prepared samples were measured on a Bruker D8 Advance X-ray diffractometer with Cu Kα radiation. A JEOL 6700-F scanning electron microscope (SEM) and FEI TALOS F200 transmission electron microscope (TEM) were utilized to characterize the morphologies and microstructures of these samples. Diffuse reflection spectra (DRS) were acquired using a PerkinElmer Lambda 650s UV/vis spectrometer. The surface electronic states of these photocatalysts were investigated on a Thermo Fisher ESCALAB 250Xi X-ray photoelectron spectrometer (XPS) with all the binding energies referenced to the C 1s peak at 284.6 eV of the surface amorphous carbon. A Bruker MS-5000 electron paramagnetic resonance (EPR) test was performed to detect the OVs and photogenerated reactive oxygen species. A PerkinElmer LS55 was used with steady-state photoluminescence spectra (PL) on the fluorescence spectrometer with an excitation wavelength of 325 nm. An Edinburgh FLS980 was used with time-resolved fluorescence decay spectra (TR-PL) to characterize photoelectron lifetime with a picosecond diode laser excitation wavelength of 375 nm and emission wavelength of 470 nm.

3.3. Photocatalytic Activity Test

The photocatalytic degradation of NaPCP was conducted under the irradiation of the 300 W Xe lamp (PLS-SXE300D, Beijing Perfectlight, $\lambda \geq 400$ nm). In detail, the 50 mg photocatalyst was dispersed in a 50 mL 50 mg/L NaPCP solution in a cylindrical reactor equipped with a cooling system using magnetic stirring. The suspension was first stirred in the dark for 60 min to reach adsorption-desorption equilibrium. During the illumination, about 2 mL aliquot was sampled and filtrated for High-Performance Liquid Chromatography (HPLC) analysis with a UV detector at 227 nm (Agilent 1260). The photocatalytic NaPCP degradation efficiency (η) was calculated using the formula $\eta(\%) = (1 - C/C_0) \times 100\%$, where C and C_0 were the NaPCP concentrations in the reaction's aqueous solution.

As for the scavenger system of NaPCP degradation, Triethanolamine (TEOA, 1 mL), isopropanol (IPA, 1 mL), superoxide dismutase (SOD, 1 mg) and natural β-carotene (caro, 1 mg) served as the hole, while $\bullet OH$, $\bullet O_2^-$ and $^1O_2$ scavengers were added to investigate the corresponding active species, respectively.

3.4. In-Situ DRIFTS Experiments

To further interpret the degradation mechanisms of NaPCP, in-situ DRIFTS experiments were conducted with solid samples on a Thermo Fisher Nicolet iS50FT-IR spectrometer equipped with a designed reaction cell and a liquid nitrogen-cooled HgCdTe (MCT) detector. The mixture of 0.1 g BMO-001-300 and 0.01 g NaPCP was finely ground for 15 min and then deposited on the substrate in the center of the cell. Then, Ar (50 mL/min) was used to purify the cells at 120 °C for 1 h to ensure these impurities on the sample surface dislodged. Subsequently, the mixture of air (30 mL/min) and trace water vapor was introduced into the reactor with the infrared signal in-situ collected by the MCT detector.

Supplementary Materials: The following supporting information can be downloaded at: https://www.mdpi.com/article/10.3390/ijms232315221/s1. References [40–45] are cited in the Supplementary Materials.

Author Contributions: Conceptualization, X.X.; Methodology, Y.T.; Formal analysis, X.Y. and W.Z.; Investigation, Y.T.; Data curation, X.X. and Y.T.; Writing—original draft, X.X.; Writing—review & editing, X.Y., X.D., J.Z. and H.C.; Supervision, J.Z. and H.C.; Project administration, X.D. All authors have read and agreed to the published version of the manuscript.

Funding: This research was funded by the National Natural Science Foundation of China (51872107, 21976141, 42277485), the Department of Science and Technology of Hubei Province (2021CFA034), the Department of Education of Hubei Province (T2020011).

Conflicts of Interest: The authors declare no conflict of interest.

References

1. Ding, X.; Zhao, K.; Zhang, L. Enhanced photocatalytic removal of sodium pentachlorophenate with self-doped Bi_2WO_6 under visible light by generating more superoxide ions. *Environ. Sci. Technol.* **2014**, *48*, 5823–5831. [CrossRef] [PubMed]
2. Weon, S.; He, F.; Choi, W. Status and challenges in photocatalytic nanotechnology for cleaning air polluted with volatile organic compounds: Visible light utilization and catalyst deactivation. *Environ. Sci.-Nano* **2019**, *6*, 3185–3214. [CrossRef]
3. Hunge, Y.M.; Uchida, A.; Tominaga, Y.; Fujii, Y.; Yadav, A.A.; Kang, S.-W.; Suzuki, N.; Shitanda, I.; Kondo, T.; Itagaki, M.; et al. Visible Light-Assisted Photocatalysis Using Spherical-Shaped $BiVO_4$ Photocatalyst. *Catalysts* **2021**, *11*, 460. [CrossRef]
4. Ding, X.; Yang, X.; Xiong, Z.; Chen, H.; Zhang, L. Environment Pollutants Removal with Bi-Based Photocatalysts. *Prog. Chem.* **2017**, *29*, 1115–1126.
5. Hodges, B.C.; Cates, E.L.; Kim, J.H. Challenges and prospects of advanced oxidation water treatment processes using catalytic nanomaterials. *Nat. Nanotechnol.* **2018**, *13*, 642–650. [CrossRef]
6. Loeb, S.K.; Alvarez, P.J.J.; Brame, J.A.; Cates, E.L.; Choi, W.; Crittenden, J.; Dionysiou, D.D.; Li, Q.; Li-Puma, G.; Quan, X.; et al. The Technology Horizon for Photocatalytic Water Treatment: Sunrise or Sunset? *Environ. Sci. Technol.* **2019**, *53*, 2937–2947. [CrossRef]
7. Hunge, Y.M.; Yadav, A.A.; Kang, S.-W.; Jun Lim, S.; Kim, H. Visible light activated MoS_2/ZnO composites for photocatalytic degradation of ciprofloxacin antibiotic and hydrogen production. *J. Photochem. Photobiol. A* **2023**, *434*, 114250. [CrossRef]
8. Hunge, Y.M.; Yadav, A.A.; Khan, S.; Takagi, K.; Suzuki, N.; Teshima, K.; Terashima, C.; Fujishima, A. Photocatalytic degradation of bisphenol A using titanium dioxide@nanodiamond composites under UV light illumination. *J. Colloid Interface Sci.* **2021**, *582*, 1058–1066. [CrossRef] [PubMed]
9. Xu, M.; Chen, Y.; Qin, J.; Feng, Y.; Li, W.; Chen, W.; Zhu, J.; Li, H.; Bian, Z. Unveiling the Role of Defects on Oxygen Activation and Photodegradation of Organic Pollutants. *Environ. Sci. Technol.* **2018**, *52*, 13879–13886. [CrossRef]
10. Li, Z.; Meng, X. New Insight into Reactive Oxidation Species (ROS) for Bismuth-based Photocatalysis in Phenol Removal. *J. Hazard. Mater.* **2020**, *399*, 122939. [CrossRef] [PubMed]
11. Yang, X.; Xiang, Y.; Qu, Y.; Ding, X.; Chen, H. Novel in situ fabrication of conjugated microporous poly(benzothiadiazole)–Bi_2MoO_6 Z-scheme heterojunction with enhanced visible light photocatalytic activity. *J. Catal.* **2017**, *345*, 319–328. [CrossRef]
12. Li, S.; Huang, T.; Du, P.; Liu, W.; Hu, J. Photocatalytic transformation fate and toxicity of ciprofloxacin related to dissociation species: Experimental and theoretical evidences. *Water Res.* **2020**, *185*, 116286. [CrossRef]
13. Zeng, L.; Guo, X.; He, C.; Duan, C. Metal–Organic Frameworks: Versatile Materials for Heterogeneous Photocatalysis. *ACS Catal.* **2016**, *6*, 7935–7947. [CrossRef]
14. Qian, Y.; Li, D.; Han, Y.; Jiang, H.-L. Photocatalytic Molecular Oxygen Activation by Regulating Excitonic Effects in Covalent Organic Frameworks. *J. Am. Chem. Soc.* **2020**, *142*, 20763–20771. [CrossRef]
15. Liu, Y.; Liu, L.; Wang, Y. A Critical Review on Removal of Gaseous Pollutants Using Sulfate Radical-based Advanced Oxidation Technologies. *Environ. Sci. Technol.* **2021**, *55*, 9691–9710. [CrossRef]
16. Xu, X.; Wang, J.; Chen, T.; Yang, N.; Wang, S.; Ding, X.; Chen, H. Deep insight into ROS mediated direct and hydroxylated dichlorination process for efficient photocatalytic sodium pentachlorophenate mineralization. *Appl. Catal. B Environ.* **2021**, *296*, 120352. [CrossRef]
17. Xu, Q.; Zhang, L.; Cheng, B.; Fan, J.; Yu, J. S-Scheme Heterojunction Photocatalyst. *Chem* **2020**, *6*, 1543–1559. [CrossRef]
18. Qiu, B.; Du, M.; Ma, Y.; Zhu, Q.; Xing, M.; Zhang, J. Integration of redox cocatalysts for artificial photosynthesis. *Energy Environ. Sci.* **2021**, *14*, 5260–5288. [CrossRef]
19. Yang, X.; Wang, S.; Yang, N.; Zhou, W.; Wang, P.; Jiang, K.; Li, S.; Song, H.; Ding, X.; Chen, H.; et al. Oxygen vacancies induced special CO_2 adsorption modes on Bi_2MoO_6 for highly selective conversion to CH_4. *Appl. Catal. B-Environ.* **2019**, *259*, 118088–118095. [CrossRef]
20. Xu, X.; Liu, H.; Wang, J.; Chen, T.; Ding, X.; Chen, H. Insight into surface hydroxyl groups for environmental purification: Characterizations, applications and advances. *Surf. Interfaces* **2021**, *25*, 101272. [CrossRef]
21. Huang, Y.; Yu, Y.; Yu, Y.; Zhang, B. Oxygen Vacancy Engineering in Photocatalysis. *Sol. RRL* **2020**, *4*, 2000037. [CrossRef]
22. Wu, X.; Li, J.; Xie, S.; Duan, P.; Zhang, H.; Feng, J.; Zhang, Q.; Cheng, J.; Wang, Y. Selectivity Control in Photocatalytic Valorization of Biomass-Derived Platform Compounds by Surface Engineering of Titanium Oxide. *Chem* **2020**, *6*, 3038–3053. [CrossRef]
23. Wang, S.; He, T.; Chen, P.; Du, A.; Ostrikov, K.; Huang, W.; Wang, L. In Situ Formation of Oxygen Vacancies Achieving Near-Complete Charge Separation in Planar $BiVO4$ Photoanodes. *Adv. Mater.* **2020**, *32*, 2001385. [CrossRef] [PubMed]

24. Li, P.; Zhou, Z.; Wang, Q.; Guo, M.; Chen, S.; Low, J.; Long, R.; Liu, W.; Ding, P.; Wu, Y.; et al. Visible-Light-Driven Nitrogen Fixation Catalyzed by Bi_5O_7Br Nanostructures: Enhanced Performance by Oxygen Vacancies. *J. Am. Chem. Soc.* **2020**, *142*, 12430–12439. [CrossRef] [PubMed]
25. Wei, Z.; Wang, W.; Li, W.; Bai, X.; Zhao, J.; Tse, E.C.M.; Phillips, D.L.; Zhu, Y. Steering Electron–Hole Migration Pathways Using Oxygen Vacancies in Tungsten Oxides to Enhance Their Photocatalytic Oxygen Evolution Performance. *Angew. Chem. Int. Ed.* **2021**, *60*, 8236–8242. [CrossRef] [PubMed]
26. Nam, Y.; Li, L.; Lee, J.Y.; Prezhdo, O.V. Strong Influence of Oxygen Vacancy Location on Charge Carrier Losses in Reduced TiO_2 Nanoparticles. *J. Phys. Chem. Lett.* **2019**, *10*, 2676–2683. [CrossRef] [PubMed]
27. Li, H.; Shi, J.; Zhao, K.; Zhang, L. Sustainable molecular oxygen activation with oxygen vacancies on the {001} facets of BiOCl nanosheets under solar light. *Nanoscale* **2014**, *6*, 14168–14173. [CrossRef]
28. Li, H.; Shang, J.; Zhu, H.; Yang, Z.; Ai, Z.; Zhang, L. Oxygen Vacancy Structure Associated Photocatalytic Water Oxidation of BiOCl. *ACS Catal.* **2016**, *6*, 8276–8285. [CrossRef]
29. Chen, F.; Liu, L.-L.; Zhang, Y.-J.; Wu, J.-H.; Huang, G.-X.; Yang, Q.; Chen, J.-J.; Yu, H.-Q. Enhanced full solar spectrum photocatalysis by nitrogen-doped graphene quantum dots decorated BiO_{2-x} nanosheets: Ultrafast charge transfer and molecular oxygen activation. *Appl. Catal. B Environ.* **2020**, *227*, 119218. [CrossRef]
30. Wang, S.; Ding, X.; Yang, N.; Zhan, G.; Zhang, X.; Dong, G.; Zhang, L.; Chen, H. Insight into the effect of bromine on facet-dependent surface oxygen vacancies construction and stabilization of Bi_2MoO_6 for efficient photocatalytic NO removal. *Appl. Catal. B Environ.* **2020**, *265*, 118585. [CrossRef]
31. Ma, Z.; Li, P.; Ye, L.; Zhou, Y.; Su, F.; Ding, C.; Xie, H.; Bai, Y.; Wong, P.K. Oxygen vacancies induced exciton dissociation of flexible BiOCl nanosheets for effective photocatalytic CO_2 conversion. *J. Mater. Chem. A* **2017**, *5*, 24995–25004. [CrossRef]
32. Wang, H.; Chen, S.; Yong, D.; Zhang, X.; Li, S.; Shao, W.; Sun, X.; Pan, B.; Xie, Y. Giant Electron-Hole Interactions in Confined Layered Structures for Molecular Oxygen Activation. *J. Am. Chem. Soc.* **2017**, *139*, 4737–4742. [CrossRef]
33. Rousseau, R.; Glezakou, V.-A.; Selloni, A. Theoretical insights into the surface physics and chemistry of redox-active oxides. *Nat. Rev. Mater.* **2020**, *5*, 460–475. [CrossRef]
34. Shi, Y.; Li, H.; Mao, C.; Zhan, G.; Yang, Z.; Ling, C.; Wei, K.; Liu, X.; Ai, Z.; Zhang, L. Manipulating Excitonic Effects in Layered Bismuth Oxyhalides for Photocatalysis. *ACS ES&T Engineer.* **2022**, *2*, 957–974.
35. Shi, Y.; Zhan, G.; Li, H.; Wang, X.; Liu, X.; Shi, L.; Wei, K.; Ling, C.; Li, Z.; Wang, H.; et al. Simultaneous Manipulation of Bulk Excitons and Surface Defects for Ultrastable and Highly Selective CO_2 Photoreduction. *Adv. Mater.* **2021**, *33*, 2100143. [CrossRef] [PubMed]
36. Wang, H.; Liu, W.; He, X.; Zhang, P.; Zhang, X.; Xie, Y. An Excitonic Perspective on Low-Dimensional Semiconductors for Photocatalysis. *J. Am. Chem. Soc.* **2020**, *142*, 14007–14022. [CrossRef] [PubMed]
37. Xu, X.; Yang, N.; Wang, P.; Wang, S.; Xiang, Y.; Zhang, X.; Ding, X.; Chen, H. Highly Intensified Molecular Oxygen Activation on $Bi@Bi_2MoO_6$ via a Metallic Bi-Coordinated Facet-Dependent Effect. *ACS Appl. Mater. Interfaces* **2020**, *12*, 1867–1876. [CrossRef]
38. Xu, X.; Ding, X.; Yang, X.; Wang, P.; Li, S.; Lu, Z.; Chen, H. Oxygen vacancy boosted photocatalytic decomposition of ciprofloxacin over Bi_2MoO_6: Oxygen vacancy engineering, biotoxicity evaluation and mechanism study. *J. Hazard. Mater.* **2019**, *364*, 691–699. [CrossRef] [PubMed]
39. Wang, S.; Xiong, Z.; Yang, N.; Ding, X.; Chen, H. Iodine-doping-assisted tunable introduction of oxygen vacancies on bismuth tungstate photocatalysts for highly efficient molecular oxygen activation and pentachlorophenol mineralization. *Chin. J. Catal.* **2020**, *41*, 1544–1553. [CrossRef]
40. Li, K.; Fang, X.; Fu, Z.; Yang, Y.; Nabi, I.; Feng, Y.; Bacha, A.-U.-R.; Zhang, L. Boosting photocatalytic chlorophenols reme-diation with addition of sulfite and mechanism investigation by in-situ DRIFTs. *J. Hazard. Mater.* **2020**, *398*, 123007.
41. Chang, Y.; Liu, Z.; Shen, X.; Zhu, B.; Macharia, D.K.; Chen, Z.; Zhang, L. Synthesis of Au nanoparticle-decorated carbon nitride nanorods with plasmon-enhanced photoabsorption and photocatalytic activity for removing various pollutants from water. *J. Hazard. Mater.* **2018**, *344*, 1188–1197.
42. Wang, Q.; Wang, W.; Zhong, L.; Liu, D.; Cao, X.; Cui, F. Oxygen vacancy-rich 2D/2D BiOCl-g-C_3N_4 ultrathin heterostructure nanosheets for enhanced visible-light-driven photocatalytic activity in environmental remediation. *Appl. Catal. B- Environ.* **2018**, *220*, 290–302.
43. Huang, S.; Tian, F.; Dai, J.; Tian, X.; Li, G.; Liu, Y.; Chen, Z.; Chen, R. Highly efficient degradation of chlorophenol over bismuth oxides upon near-infrared irradiation: Unraveling the effect of Bi-O-Bi-O defects cluster and $^1O_2$ involved process. *Appl. Catal. B-Environ.* **2021**, *298*, 120576.
44. Ma, H.-Y.; Zhao, L.; Guo, L.-H.; Zhang, H.; Chen, F.-J.; Yu, W.-C. Roles of reactive oxygen species (ROS) in the photocatalytic degradation of pentachlorophenol and its main toxic intermediates by TiO_2/UV. *J. Hazard. Mater.* **2019**, *369*, 719–726.
45. Li, H.; Zhang, L. Oxygen vacancy induced selective silver deposition on the 001 facets of BiOCl single-crystalline nanosheets for enhanced Cr(VI) and sodium pentachlorophenate removal under visible light. *Nanoscale* **2014**, *6*, 7805–7810.

Article

Kinetic Aspects of Benzene Degradation over TiO$_2$-N and Composite Fe/Bi$_2$WO$_6$/TiO$_2$-N Photocatalysts under Irradiation with Visible Light

Mikhail Lyulyukin [1,2], Nikita Kovalevskiy [1], Andrey Bukhtiyarov [1], Denis Kozlov [1] and Dmitry Selishchev [1,*]

[1] Boreskov Institute of Catalysis, Novosibirsk 630090, Russia; lyulyukin@catalysis.ru (M.L.); nikita@catalysis.ru (N.K.); avb@catalysis.ru (A.B.); kdv@catalysis.ru (D.K.)
[2] Ecology and Nature Management Department, Aircraft Engineering Faculty, Novosibirsk State Technical University, Novosibirsk 630073, Russia
* Correspondence: selishev@catalysis.ru; Tel.: +7-383-326-9429

Abstract: In this study, composite materials based on nanocrystalline anatase TiO$_2$ doped with nitrogen and bismuth tungstate are synthesized using a hydrothermal method. All samples are tested in the oxidation of volatile organic compounds under visible light to find the correlations between their physicochemical characteristics and photocatalytic activity. The kinetic aspects are studied both in batch and continuous-flow reactors, using ethanol and benzene as test compounds. The Bi$_2$WO$_6$/TiO$_2$-N heterostructure enhanced with Fe species efficiently utilizes visible light in the blue region and exhibits much higher activity in the degradation of ethanol vapor than pristine TiO$_2$-N. However, an increased activity of Fe/Bi$_2$WO$_6$/TiO$_2$-N can have an adverse effect in the degradation of benzene vapor. A temporary deactivation of the photocatalyst can occur at a high concentration of benzene due to the fast accumulation of non-volatile intermediates on its surface. The formed intermediates suppress the adsorption of the initial benzene and substantially increase the time required for its complete removal from the gas phase. An increase in temperature up to 140 °C makes it possible to increase the rate of the overall oxidation process, and the use of the Fe/Bi$_2$WO$_6$/TiO$_2$-N composite improves the selectivity of oxidation compared to pristine TiO$_2$-N.

Keywords: photocatalysis; visible light; N-doped TiO$_2$; Bi$_2$WO$_6$; composite photocatalyst; benzene degradation; thermoactivation

Citation: Lyulyukin, M.; Kovalevskiy, N.; Bukhtiyarov, A.; Kozlov, D.; Selishchev, D. Kinetic Aspects of Benzene Degradation over TiO$_2$-N and Composite Fe/Bi$_2$WO$_6$/TiO$_2$-N Photocatalysts under Irradiation with Visible Light. *Int. J. Mol. Sci.* **2023**, *24*, 5693. https://doi.org/10.3390/ijms24065693

Academic Editor: Junjiang Zhu

Received: 23 January 2023
Revised: 7 March 2023
Accepted: 14 March 2023
Published: 16 March 2023

Copyright: © 2023 by the authors. Licensee MDPI, Basel, Switzerland. This article is an open access article distributed under the terms and conditions of the Creative Commons Attribution (CC BY) license (https://creativecommons.org/licenses/by/4.0/).

1. Introduction

The development of advanced oxidation processes, such as photocatalytic oxidation (PCO), is permanently advanced to solve acute problems of environmental pollution and public health risk [1–4]. The PCO method has been shown to be effective in killing microorganisms and degrading hazardous chemical micropollutants in water and air environments [5–7]. Crystalline TiO$_2$ was the primary photocatalyst in this field for a long time because it could efficiently utilize optical radiation and generate charge carriers (i.e., electrons and holes) with high enough potentials to provide degradation of contaminants in an oxygen-containing medium. A key feature of TiO$_2$-mediated photocatalytic degradation is the possibility of complete oxidation of organic pollutants with the formation of harmless inorganic products such as carbon dioxide and water.

Despite the advantages, TiO$_2$ has a fundamental drawback due to the large width of its band gap. For instance, the band gap of anatase TiO$_2$ is 3.2 eV [8]. TiO$_2$ can absorb UV radiation (<390 nm), but it cannot utilize the majority (up to 95%) of solar radiation corresponding to the visible region. This drawback promotes the development of new visible-light active photocatalysts. In addition to other narrow-band semiconductors (e.g., g-C$_3$N$_4$, Ag$_3$PO$_4$, Bi$_2$MoO$_6$, BiVO$_4$, and ZnIn$_2$S$_4$ [9–13]), TiO$_2$ having an extended action spectrum due to modifications is a promising photocatalyst for the efficient degradation of pollutants under visible light [6,14].

Surface modification of TiO_2 with metal complexes or plasmonic metals provides a sensibilization effect and results in extending its absorption spectrum [15,16]. However, the efficiency of light absorption and generation of reactive species in this case is commonly not high enough due to the nature of electron excitation and the requirement for the transition of charge carriers to the energy bands of TiO_2 [8]. One of the efficient methods of TiO_2 modification is its doping with nitrogen, which results in the formation of additional energy levels in the band gap of TiO_2 and leads to a decrease in the minimum energy required for photoexcitation of electrons. TiO_2 doped with nitrogen (TiO_2-N) exhibits great visible-light activity in the degradation of organic pollutants, especially, under blue light [17–21].

Commonly, the quantum efficiency of N-doped TiO_2 in the visible region is not high enough due to rapid electron-hole recombination. An increase in the visible-light activity of TiO_2-N can be achieved by the surface modification of the photocatalyst with noble or other transition metals, which play the role of containers for electrons and improve charge separation due to the creation of an energy barrier [22]. For instance, a substantial increase in the visible-light activity was observed after surface modification of TiO_2-N with vanadium, iron, or copper species [23]. Noble metals also accelerate the transfer of electrons to adsorbed oxygen molecules, thus increasing the activity compared to that of pristine photocatalyst [24,25]. The combination of TiO_2-N with other narrow-band semiconductors (e.g., g-C_3N_4, CdS, MoS_2, Cu_2O) can improve the separation of charge carriers due to a heterojunction, thus increasing the activity of the photocatalyst under visible light. Many composite photocatalysts based on TiO_2-N have been described in the literature [26–30]. We have previously shown that Bi_2WO_6/TiO_2-N heterostructure prepared by hydrothermal method exhibits a high activity in the degradation of gaseous organic pollutants under visible light and has a high stability under power radiation for a long time [31]. Similar to TiO_2-N, the visible-light activity of the Bi_2WO_6/TiO_2-N composite can be substantially increased by depositing metal species (e.g., Fe) on its surface [32]. All three types of photocatalysts, namely TiO_2-N, Bi_2WO_6/TiO_2-N, and Fe/Bi_2WO_6/TiO_2-N, were selected as the main objects of this study to illustrate the effect of the photocatalyst's ability on the kinetic aspects of degradation of volatile organic compounds. The study is focused on benzene as a representative of aromatic compounds because this class of pollutants has a major harmful impact on human health. Analysis of pollutant adsorption and formation of intermediates and final degradation products is important for identification of the reaction pathways because they can change for heterostructures due to a change in the potentials of charge carriers photogenerated under visible light. Furthermore, the degradation of different classes of organic pollutants can occur through different pathways that would affect the overall reaction rate. To illustrate this statement in the paper, we discuss the kinetic aspects of the degradation of ethanol and benzene as nonaromatic and aromatic compounds, respectively. The kinetics of adsorption and degradation of both pollutants is studied in detail in a batch reactor using in situ IR spectroscopy. We show that under certain operating conditions the high photocatalytic ability of a material can have an adverse effect on the removal of aromatic pollutants from the air.

2. Results and Discussion
2.1. Characteristics of Synthesized Photocatalysts

Calcination of the precipitate formed after mixing aqueous solutions of titanium oxysulfate and ammonia results in the crystallization of anatase nanoparticles (Figure 1a). The main peaks at 2θ of 25.3°, 37.8°, 48.1°, 53.9°, 55.1°, and 62.8° correspond to the (101), (004), (200), (105), (211), and (204) planes of anatase TiO_2 (JCPDS card No.21-1272). The average size of anatase crystallites was estimated from the XRD data to be 20 nm. The formation of the rutile phase is not observed even under high-temperature treatment.

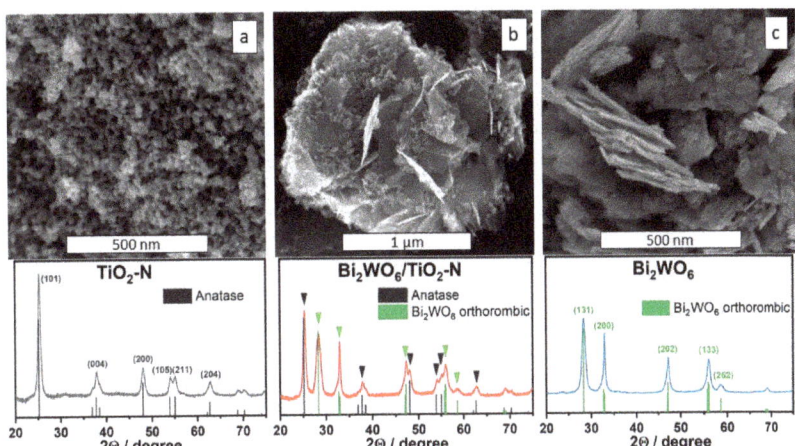

Figure 1. SEM micrographs and XRD patterns of synthesized TiO$_2$-N (**a**), Bi$_2$WO$_6$/TiO$_2$-N (**b**), and Bi$_2$WO$_6$ (**c**) photocatalysts.

Due to using ammonia as a precipitation agent, some nitrogen species are incorporated into the crystal structure of the formed TiO$_2$ [32–34]. According to the results of CHNS analysis, the total content of nitrogen in the solid titania sample after calcination in air at 450 °C is 0.36 ± 0.03 wt.%, which corresponds to a high value. The peak with a BE of 399.9 eV was observed in the photoelectron N1s spectral region of as-prepared TiO$_2$-N (Figure 2a). This peak can be attributed to a form of partially oxidized nitrogen located in the interstitial positions of the TiO$_2$ lattice [35]. The presence of nitrogen species was also confirmed by the EPR technique, which showed a signal of paramagnetic N centers (Figure S1 in the Supplementary Materials). According to the results of XPS analysis, titanium in the prepared TiO$_2$-N is present in the charge state of Ti^{4+} (Figure 2b).

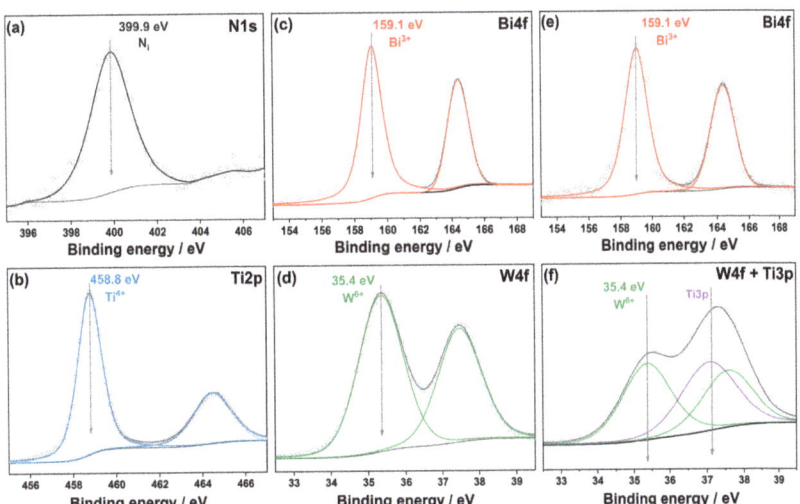

Figure 2. Photoelectron N1s (**a**) and Ti2p (**b**), Bi4f (**c**) and W4f (**d**), Bi4f (**e**) and W4f + Ti3p (**f**) spectral regions for TiO$_2$-N, Bi$_2$WO$_6$, and Bi$_2$WO$_6$/TiO$_2$-N samples, respectively. The experimental data points are marked with gray circles, the fitted data and Shirley-type backgrounds are shown with lines.

The nitrogen species create additional energy levels in the band gap of TiO$_2$, thus decreasing the minimum energy required for photoexcitation of electrons to the conduction band [36]. Figure 3a illustrates this statement and shows that TiO$_2$-N has an additional absorption shoulder in the range of 400–540 nm with a maximum at ca. 450 nm. According to the Tauc plot, shown in the inset in Figure 3b, the minimum excitation energy for TiO$_2$-N can be estimated to be 2.3 eV, which is substantially lower than the band gap of anatase (3.2 eV, Figure 3b). Therefore, TiO$_2$-N absorbs visible light with wavelengths up to 540 nm (mainly in the blue region). We have previously shown that the as-prepared TiO$_2$-N photocatalyst exhibits a high photocatalytic ability in the degradation of volatile organic compounds under both UV and visible light [32].

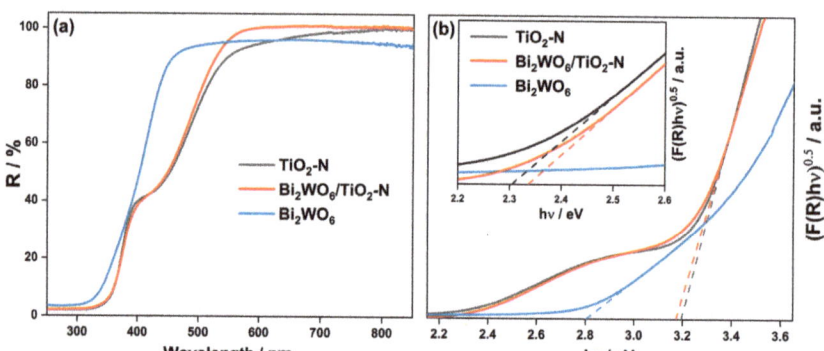

Figure 3. UV–vis diffuse reflectance spectra of TiO$_2$-N, Bi$_2$WO$_6$/TiO$_2$-N, and Bi$_2$WO$_6$ samples (**a**) and the corresponding Tauc plots (**b**). The inset in figure (**b**) shows an enlarged area of Tauc plots in the range of 2.2–2.6 eV.

Hydrothermal synthesis using aqueous solutions of Bi(NO$_3$)$_3$ and Na$_2$WO$_4$ results in the formation of orthorhombic Bi$_2$WO$_6$ with a lamellar structure presented in the form of nanoplates (Figure 1c). The main reflection peaks in the XRD pattern at 2Θ of 28.3°, 32.9°, 47.2°, 55.9°, and 58.6° correspond to the (131), (200), (202), (133), and (262) planes of orthorhombic Bi$_2$WO$_6$ (PDF 39-0256). XPS analysis confirms the chemical composition of the prepared Bi$_2$WO$_6$ sample because the corresponding peaks in the spectral regions of Bi4f and W4f (Figure 2c,d) are unambiguously attributed to the charge states of +3 for Bi and +6 for W [37]. The absence of other states for these elements confirms the formation of single-phase material without non-stochiometric species.

As a typical Aurivillius oxide, Bi$_2$WO$_6$ is a semiconducting material with a band gap narrower than that of anatase TiO$_2$. Figure 3a shows that Bi$_2$WO$_6$ can absorb a small portion of photons in the visible region (up to 440 nm) because its optical band gap corresponds to a value of 2.8 eV (Figure 3b). On the other hand, the absorption of visible light by Bi$_2$WO$_6$ is much weaker if compared with TiO$_2$-N material. Due to this reason, the optical properties of the Bi$_2$WO$_6$/TiO$_2$-N composite based on both materials (Figure 3a) are similar to the properties of single TiO$_2$-N. Figure 1b shows that titania nanoparticles in the Bi$_2$WO$_6$/TiO$_2$-N composite cover the external surface of agglomerated Bi$_2$WO$_6$ nanoplates and provide strong absorption of visible light. More illustrations of local composite structure can be found in Figure S2 (Supplementary Materials), which shows TEM micrographs of Bi$_2$WO$_6$/TiO$_2$-N obtained using HAADF imaging and EDX mapping techniques.

An approximation of the Tauc plot attributed to Bi$_2$WO$_6$/TiO$_2$-N sample (Figure 3b) gives the values of 3.17 eV as the energy of band-to-band excitation of electrons in anatase and 2.33 eV as the minimum energy required for excitation of electrons in this system due to the presence of nitrogen energy levels in the band gap of anatase.

Similarly to a single Bi_2WO_6, XPS analysis of Bi_2WO_6/TiO_2-N shows only the charge states of +3 and +6 for Bi and W elements, respectively (Figure 2e,f). No other states of these elements, as well as no change in the charge states of Ti and N (see Figure S3 in the Supplementary Materials), confirm no doping the TiO_2 lattice with Bi or W species under hydrothermal treatment.

The realization of heterojunction in Bi_2WO_6/TiO_2-N leads to a substantial increase in the visible-light activity of this composite compared to initial TiO_2-N [31]. For instance, the steady-state rate of CO_2 formation during the oxidation of acetone vapor under blue light (450 nm, 160 mW cm^{-2}) was 0.7 and 1.0 µmol min^{-1} for TiO_2-N and Bi_2WO_6/TiO_2-N, respectively. According to the literature data [38,39], the energy bands of Bi_2WO_6 are located lower than the bands of TiO_2 that makes possible a type II heterojunction between the semiconductors or a Z-scheme heterojunction. Both types of heterojunctions can improve the separation of photogenerated charge carriers and suppress their recombination, thus leading to an increase in the visible-light activity.

As mentioned in the Introduction, the photocatalytic ability of Bi_2WO_6/TiO_2-N can be improved by deposing a small amount of Fe species on its surface. In the Fe-modified photocatalysts, localization of photogenerated electrons occurs on the surface of the photocatalyst in Fe species, and the separation of charge carriers is enhanced due to an energy barrier. Bi_2WO_6/TiO_2-N was impregnated with an aqueous solution of $Fe(NO_3)_3$ to achieve a nominal Fe loading of 0.1 wt.%. This modification had no effect on the XRD pattern of Bi_2WO_6/TiO_2-N due to extremely low Fe content. The presence of Fe species on the surface of the photocatalyst was confirmed by XPS analysis when the Fe content was increased up to 0.3 wt.% (see Figure S3 in the Supplementary Materials).

The photocatalytic ability of all synthesized catalysts was preliminary checked in a continuous-flow setup as the steady-state rate of CO_2 formation during the oxidation of acetone vapor under blue light (450 nm, 160 mW cm^{-2}). The ability is increased as follows: TiO_2-N:Bi_2WO_6/TiO_2-N:Fe/Bi_2WO_6/TiO_2-N = 1:1.4:1.7. The next section describes the effect of photocatalyst ability on the kinetic aspects of the degradation of different classes of organic compounds.

2.2. Degradation of Volatile Organic Compounds

In this study, we investigate the photocatalytic degradation of ethanol and benzene vapor on the surface of TiO_2-N, Bi_2WO_6/TiO_2-N, and Fe/Bi_2WO_6/TiO_2-N photocatalysts, which substantially differ in their activities. These test organic compounds were selected due to the different types of intermediates formed during the degradation because they can affect the deactivation of the photocatalyst and, consequently, the kinetics of overall oxidation process [40,41]:

- The PCO of ethanol results in the formation of products that are easily desorbed from the surface of photocatalyst (especially acetaldehyde, which is the first in the sequence of oxidation products) [42,43];
- The PCO of benzene occurs without the formation of gas-phase intermediates but leads to strongly adsorbed species formed on the surface of the photocatalyst due to the polymerization of intermediate radicals [44–46]. Furthermore, oxidation pathways with the opening of the benzene ring lead to an increase in the total number of species, thus resulting in the gradual deactivation of the photocatalyst due to competition for adsorption sites with the initial compound [47,48].

A common pattern for the photocatalytic experiments in a batch reactor is that the injected liquid pollutant evaporates and partially adsorbs on the surface of the photocatalyst. Turning on the light leads to a decrease in the concentration of pollutant in the gas phase and a simultaneous increase in the amount of oxidation products. The intermediate oxidation products (if any) are completely removed from the gas phase during irradiation, while carbon oxides start to intensively release. The total amount of carbon oxides is commonly mass-balanced to the initial amount of compound injected into the reactor. It confirms the complete oxidation of the injected pollutant.

2.2.1. Degradation of Ethanol in Batch Reactor

Figure 4 shows the kinetic plots of the ethanol PCO over all samples: TiO_2-N, Bi_2WO_6/TiO_2-N, and $Fe/Bi_2WO_6/TiO_2$-N. The initial amount of ethanol injected into the reactor (1 µL) corresponds to its concentration of 57 µmol L^{-1}, but when the adsorption–desorption equilibrium is reached, the detected concentration of ethanol in the gas phase is 8–10 µmol L^{-1}. This fact is due to the high adsorption capacity of synthesized TiO_2-based photocatalyst toward ethanol. Complete removal of ethanol vapor from the gas phase during the irradiation of the photocatalyst occurs for 45–55 min both for TiO_2-N and Bi_2WO_6/TiO_2-N (Figure 4). In the case of $Fe/Bi_2WO_6/TiO_2$-N, the complete removal of ethanol occurs for 20 min.

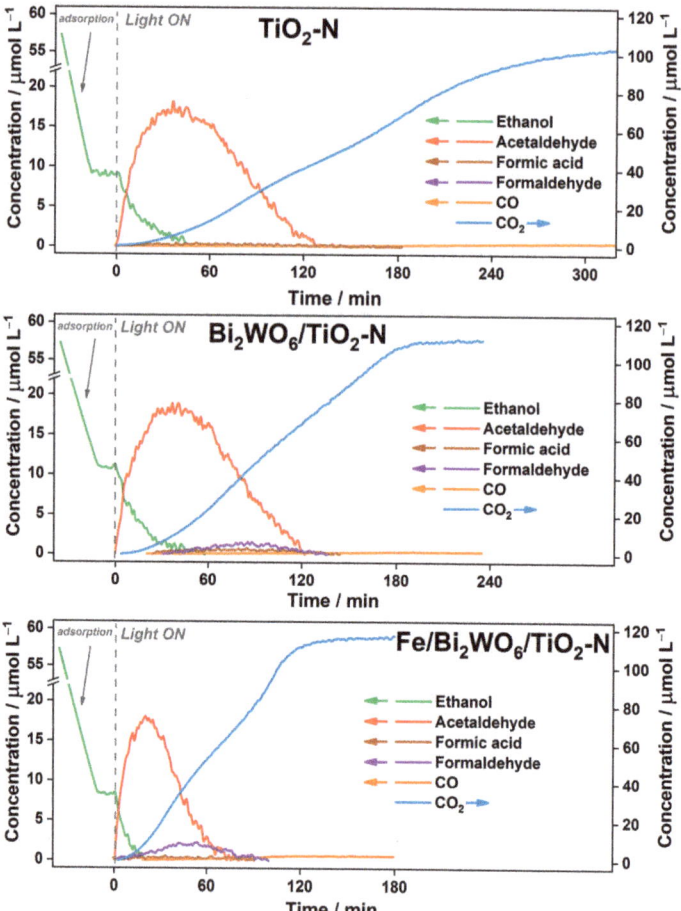

Figure 4. Kinetic plots of the reaction components during the PCO of ethanol over TiO_2-N, Bi_2WO_6/TiO_2-N, and $Fe/Bi_2WO_6/TiO_2$-N under blue light.

At the initial steps of ethanol oxidation, acetaldehyde is formed as a gas-phase intermediate. The rates of acetaldehyde formation are 0.95, 1.5, and 2.05 µmol L^{-1} min^{-1} for TiO_2-N, Bi_2WO_6/TiO_2-N, and $Fe/Bi_2WO_6/TiO_2$-N, respectively. Acetaldehyde is adsorbed less strongly than ethanol [42], thus it does not participate in the redox transformations on the surface of the photocatalyst while the adsorption sites are occupied by ethanol molecules. Oxidation of acetaldehyde boosts after the complete consumption of ethanol and an increase in vacant adsorption sites on the surface of the photocatalyst. Therefore, the time of maximum acetaldehyde concentration coincides with the time of complete removal of ethanol from the gas phase. For all studied photocatalysts, the maximum concentration of acetaldehyde detected in the gas phase is ca. 18 μmol L^{-1}. Figure 4 shows that the complete removal of acetaldehyde from the gas phase occurs for 100, 77, and 50 min for TiO$_2$-N, Bi$_2$WO$_6$/TiO$_2$-N, and Fe/Bi$_2$WO$_6$/TiO$_2$-N, respectively. In the case of TiO$_2$-N, formic acid (<0.45 μmol L^{-1}) is also detected in the gas phase. Furthermore, 1.6 μmol L^{-1} of formaldehyde and 0.5 μmol L^{-1} of formic acid are detected during the oxidation of ethanol over Bi$_2$WO$_6$/TiO$_2$-N, whereas Fe/Bi$_2$WO$_6$/TiO$_2$-N leads to the formation of formaldehyde in the concentration of 2 μmol L^{-1}.

CO$_2$ plots in Figure 4 show that the times required for complete conversion of ethanol to carbon oxides are 320, 180, and 120 min for TiO$_2$-N, Bi$_2$WO$_6$/TiO$_2$-N, and Fe/Bi$_2$WO$_6$/TiO$_2$-N, respectively, that corresponds to maximum CO$_2$ formation rate of 0.5, 0.68, and 1.4 μmol L^{-1} min^{-1}. This trend agrees with the results of acetone degradation under blue light, which are discussed in Section 2.1.

Additionally, the effect of the basic wavelength in the emission spectrum of LED on the kinetics of ethanol oxidation over TiO$_2$-N was studied to understand the reason for the formation of gaseous formaldehyde in the case of composite photocatalysts. Figure S4 in the Supplementary Materials shows that the rate of ethanol oxidation increases as the wavelength decreases, but the release of formaldehyde in the gas phase is not observed for all the experiments. This means that an increase in the activity of the photocatalyst increases the rates of all steps. No additional pathways with an intense formation of formaldehyde appear for the TiO$_2$-N photocatalyst. Therefore, the release of formaldehyde in the case of composite photocatalysts occurs due to the presence of the Bi$_2$WO$_6$ co-catalyst. In fact, gaseous formaldehyde is detected during the oxidation of ethanol vapor over Bi$_2$WO$_6$ (see Figure S5 in the Supplementary Materials). The low concentration of formaldehyde in this case is due to the very low activity of the Bi$_2$WO$_6$ photocatalyst. A high amount of formaldehyde in the case of and Bi$_2$WO$_6$/TiO$_2$-N composites may be due to a change in the potentials of photogenerated charge carriers and a difference in the oxidation rates over TiO$_2$-N and Bi$_2$WO$_6$ components.

2.2.2. Degradation of Benzene in Batch Reactor

The kinetic plots of the reaction components during the oxidation of benzene vapor over TiO$_2$-N, Bi$_2$WO$_6$/TiO$_2$-N, and Fe/Bi$_2$WO$_6$/TiO$_2$-N under irradiation with blue light are shown in Figure 5. As TiO$_2$-N poorly adsorbs benzene molecules on its surface, the concentration of benzene detected in the gas phase after evaporation is 37 μmol L^{-1} that corresponds to the theoretical value calculated from the amount of injected benzene. For the composite Bi$_2$WO$_6$/TiO$_2$-N and Fe/Bi$_2$WO$_6$/TiO$_2$-N photocatalysts, the starting concentration of benzene in the gas phase is 33–35 μmol L^{-1} due to the adsorption of benzene molecules on the surface of Bi$_2$WO$_6$. After turning the light source on, complete removal of benzene from the gas phase occurs after 380, 610, and 700 min for TiO$_2$-N, Bi$_2$WO$_6$/TiO$_2$-N, and Fe/Bi$_2$WO$_6$/TiO$_2$-N, respectively. We can see the opposite trend compared to the previous cases of acetone and ethanol degradation. The reasons for this trend will be discussed later.

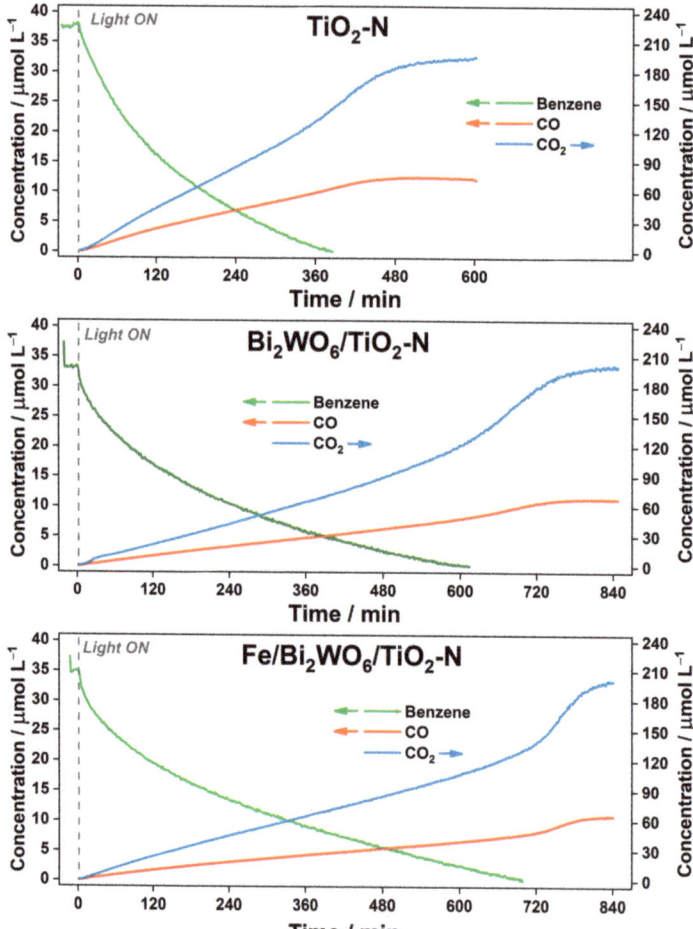

Figure 5. Kinetic plots of the reaction components during PCO of benzene (1.0 µL) over TiO$_2$-N, Bi$_2$WO$_6$/TiO$_2$-N, and Fe/Bi$_2$WO$_6$/TiO$_2$-N under blue light.

The accumulation of CO and CO$_2$ occurs simultaneously with the degradation of benzene (Figure 5). After the complete removal of benzene from the gas phase, the formation of final products is substantially enhanced. The rates of CO$_2$ formation for TiO$_2$-N are 0.35 µmol L^{-1} min^{-1} before the complete removal of benzene and 0.54 µmol L^{-1} min^{-1} after that. The corresponding values for Bi$_2$WO$_6$/TiO$_2$-N and Fe/Bi$_2$WO$_6$/TiO$_2$-N are 0.19 (and 0.55) µmol L^{-1} min^{-1} and 0.17 (and 0.82) µmol L^{-1} min^{-1}, respectively. These data show that the higher ability of photocatalyst leads to the lower rate of CO$_2$ formation at initial steps of reaction. After complete removal of benzene from the gas phase, the higher photocatalytic ability provides the higher rate of CO$_2$ accumulation. The reasons for the observed phenomenon may be related to the mechanism of benzene oxidation.

According to the literature data [44–46], the holes (h$^+$) photogenerated in the conduction band of the photocatalyst can directly react with adsorbed benzene molecules to form phenyl radical cations (C$_6$H$_5$$^{\bullet+}$), as well as oxidize adsorbed H$_2$O molecules to form OH$^{\bullet}$ radicals. C$_6$H$_5$$^{\bullet+}$ can further react with the adsorbed O$_2$ or O$_2$$^{\bullet-}$ to form a peroxide radical and, sequentially, phenol [49]. An alternative pathway involves the direct interaction of C$_6$H$_5$$^{\bullet+}$ with OH$^{\bullet}$ to form phenol and other hydroxylated intermediates (e.g., hydroquinone and benzoquinone) [50]. These compounds can be completely oxidized to CO_2 and H_2O through a series of steps that include opening the aromatic ring and oxidation of non-cyclic hydrocarbons. In parallel pathways, $C_6H_5^{\bullet+}$ can interact with other adsorbed benzene molecules to form carbon deposits due to polymerization. Accumulation of these deposits can substantially reduce the ability of photocatalyst to oxidize benzene molecules due to blocking its adsorption sites. Other transformation routes [51] include the reactions of photoinduced $OH^{\bullet}$ with adsorbed benzene molecules to form various types of alkyl radicals (e.g., $CH_3^{\bullet}$, $C_2H_5^{\bullet}$ and $C_6H_5^{\bullet}$), which also contribute to the overall decomposition of benzene [52].

Einaga et al. [45,46] previously showed using diffuse reflectance spectroscopy and other experimental techniques that a radical polymerization of aromatic compounds on the surface of TiO_2 during the degradation of benzene vapor leads to the formation of carbon deposits on the surface of photocatalyst and its strong deactivation. Indirect evidence for the formation of non-volatile intermediates in this study is that benzene is poorly adsorbed on the surface of the studied photocatalysts but ~30% of the formed CO_2 appears in the gas phase only after complete removal of benzene vapor. Therefore, it can be concluded that the non-linear form of the CO_2 kinetic plots in Figure 5 is due to the accumulation of non-volatile intermediates, which occupy available sites on the surface of photocatalyst and suppress further redox transformations.

To confirm the adverse effect of an excessive amount of benzene on the kinetics of its degradation, an additional experiment with 0.15 μL of injected benzene, instead of 1.0 μL, was carried out over TiO_2-N and $Fe/Bi_2WO_6/TiO_2$-N under the same conditions (Figure 6). When oxidizing 0.15 μL of benzene over TiO_2-N, the previously observed stage of CO_2 formation at a lower rate is absent, and the rate of the second ("fast") stage is the same as in the case of 1.0 μL (i.e., 0.55 μmol L^{-1} min^{-1}). When the same amount of benzene is oxidized over the $Fe/Bi_2WO_6/TiO_2$-N composite, the slope at the initial stage is 0.34 μmol L^{-1} min^{-1}, which is twice as high as the rate observed during the oxidation of 1 μL (Figure 5). This means that the coverage of photocatalyst surface has a drastic effect on the rate of benzene oxidation because non-volatile intermediates remain on the surface and can lead to a fast transfer and recombination of the photogenerated charge carriers [44]. As the adsorption of benzene on the composite $Fe/Bi_2WO_6/TiO_2$-N photocatalyst is slightly better and its photocatalytic ability is higher, a higher number of non-volatile species is formed on its surface at the initial steps, thus suppressing the degradation of benzene. After complete removal of benzene from the gas phase, the number of intermediates on the photocatalyst surface decreases, and the accumulation of final products increases. The rate of CO_2 accumulation during the "fast" oxidation stage for $Fe/Bi_2WO_6/TiO_2$-N (Figure 6) is ca. 1 μmol L^{-1} min^{-1}, which is substantially higher than the value observed during the oxidation of 1 μL of benzene.

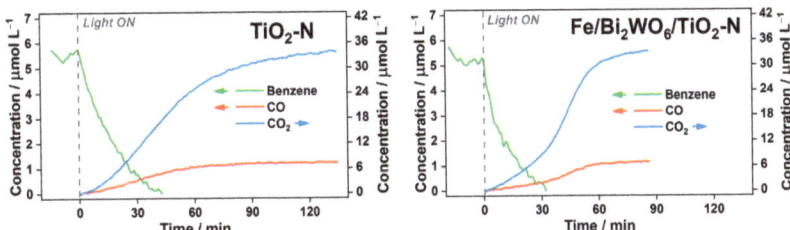

Figure 6. Kinetic plot of the reaction components during PCO of benzene (0.15 μL) over TiO_2-N and $Fe/Bi_2WO_6/TiO_2$-N under blue light.

Based on the results of all experiments performed in the batch reactor, it can be concluded that a multicomponent composite $Fe/Bi_2WO_6/TiO_2$-N system leads to an increase in the photocatalytic ability of the material compared to pristine TiO_2-N (Figure 6). However, the oxidation of aromatic compounds using this system may be associated with limitations because the high ability of photocatalyst promotes the occupation of available adsorption sites by non-volatile intermediates and its temporary deactivation. An illustration of the proposed deactivation mechanism by non-volatile intermediates for two types of photocatalysts (i.e., low-active and high-active photocatalysts) is shown in Figure 7.

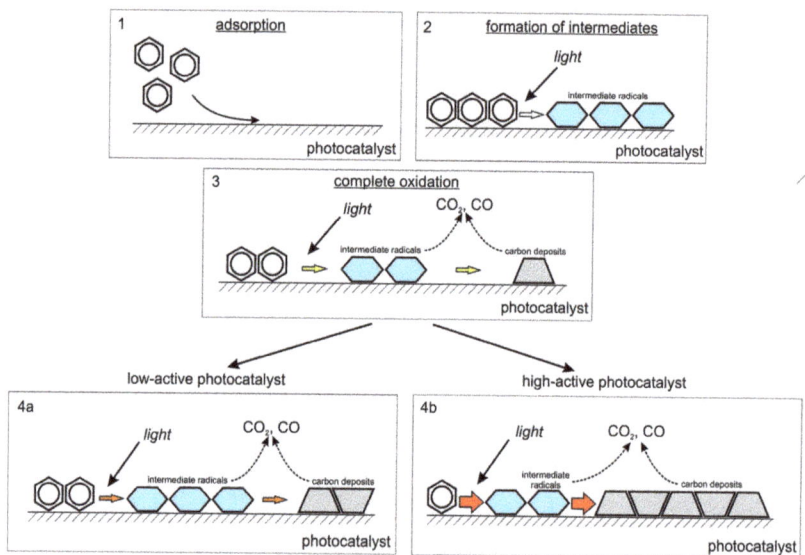

Figure 7. Proposed scheme of photocatalyst deactivation during the benzene degradation as a series of steps: 1—adsorption of benzene; 2—formation of intermediate oxidation products; 3—overall oxidation process, including the formation of final products; 4a and 4b—complete oxidation over low-active and high-active photocatalysts, respectively.

In contrast to the oxidation of ethanol vapor, in which the intermediates are desorbed from the surface of photocatalyst to the gas phase, the intermediates formed during the benzene degradation strongly suppress the adsorption of benzene molecules and, consequently, the rate of the overall oxidation process, estimated as the accumulation of final products. In real operational conditions, the concentration of benzene is rarely as high as in the experiments described above. The concentration would be at the level of its threshold limit value (TLV = 0.1 µmol L^{-1}), which is 400 times less than the amount used in the experiments. The results show that the visible-light oxidation of even 5.5 µmol L^{-1} of benzene, which exceeds the TLV by 55 times, occurs faster over Fe/Bi$_2$WO$_6$/TiO$_2$-N compared to pristine TiO$_2$-N. We believe that the problem of temporary photocatalyst deactivation can be solved by the design of a composite system with highly porous support (for example, activated carbon or zeolite), which is able to absorb a large amount of aromatic compounds and reduce their maximum concentration in the reaction mixture [53].

2.2.3. Degradation of Benzene in Continuous-Flow Reactor

The experiments in a batch reactor thoroughly simulate a situation of fast emission of a high amount of pollutant to the gas phase. However, a common case is the presence of an emission source that emits a small amount of pollutant at a constant rate (e.g., pieces of furniture can slowly emit organic compounds). The experiments under flow conditions are important to check the photocatalytic performance of materials for this case too. In contrast to a batch reactor, where concentrations of reaction components are changed for irradiation time, in a continuous-flow set-up, the concentrations of initial pollutant and formed products commonly reach steady-state values, which can be used for evaluation of the steady-state activity of photocatalyst. Additionally, the experiments under flow conditions allow us to check the long-term stability of photocatalysts under irradiation.

Figure 8 shows a change in the rate of CO_2 formation (μmol min^{-1}) during long-term benzene degradation in a continuous-flow setup over TiO_2-N and Fe/Bi_2WO_6/TiO_2-N irradiated with blue light. The results of these experiments confirm a high stability of both synthesized photocatalysts because their activities reach a steady-state level after irradiation for 10–12 h and further do not change substantially up to 26 h. It is important to note that both synthesized photocatalysts exhibit a high level of activity because the activity of commercially available TiO_2 P25 photocatalyst under the same conditions is 0.033 μmol min^{-1}, that is 10 times lower compared to the values for TiO_2-N and Fe/Bi_2WO_6/TiO_2-N (see Figure S6 in the Supplementary Materials).

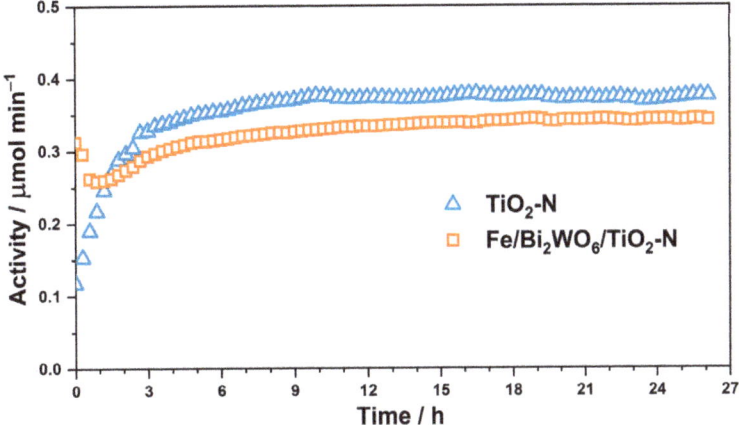

Figure 8. Stability of TiO_2-N and Fe/Bi_2WO_6/TiO_2-N during long-term benzene degradation under blue light.

The surface composition of the photocatalysts after long-term benzene degradation was checked using XPS technique. No change in the positions of peaks attributed to the main elements is detected in the XPS spectra of Fe/Bi_2WO_6/TiO_2-N before and after the long-term stability test (Figure 9). The same situation is observed for the single TiO_2-N sample (see Figure S7 and Table S1 in the Supplementary Materials). Additionally, Table 1 shows that the atomic ratios of the main elements on the surface of Fe/Bi_2WO_6/TiO_2-N are similar before and after the stability test that confirms a stable chemical composition of this photocatalyst.

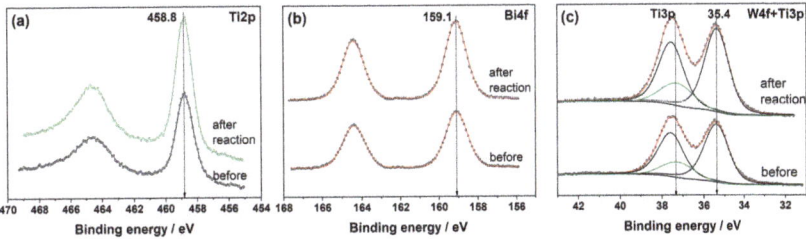

Figure 9. Photoelectron Ti2p (**a**), Bi4f (**b**), and W4f + Ti3p (**c**) spectral regions for the Fe/Bi_2WO_6/TiO_2-N photocatalyst before and after the stability test.

Table 1. Atomic ratios of the elements on the surface of Fe/Bi$_2$WO$_6$/TiO$_2$-N before and after the stability test.

	O/Ti	N/Ti	Bi/Ti	W/Ti
Before reaction	3.99	0.10	0.60	0.42
After reaction	3.81	0.10	0.61	0.41

Therefore, the kinetic data and physicochemical characteristics confirm that the synthesized photocatalysts are stable and can be successfully employed in the photocatalytic air purifiers for permanent removal and degradation of benzene micropollutants.

Another important parameter that affects the reaction rate is the temperature of the photocatalyst. The published information on the thermal activation of photocatalytic oxidation suggests some potential advantages for accelerating the process at high temperatures [54]. It was a reason to evaluate the steady-state activity of the synthesized photocatalysts at different temperatures from 40 °C to 140 °C. The results of these experiments for all studied photocatalysts are shown in Figure 10.

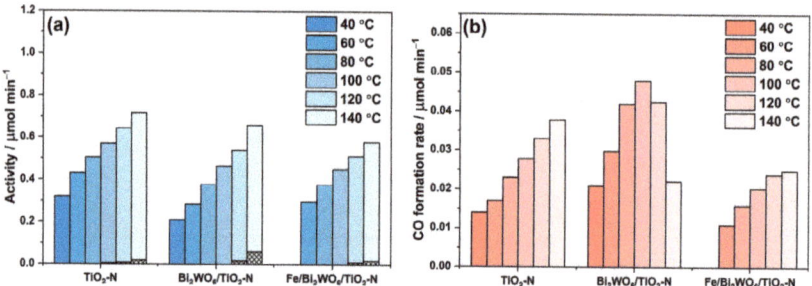

Figure 10. Effect of temperature on the steady-state photocatalytic activities of TiO$_2$-N, Bi$_2$WO$_6$/TiO$_2$-N, and Fe/Bi$_2$WO$_6$/TiO$_2$-N (**a**) and CO formation rates (**b**) during the benzene degradation under blue light. Cross-hatched area corresponds to the activity without irradiation of photocatalyst.

Several statements can be made considering the observed temperature dependencies:
- The activity of all samples in benzene degradation increases as the reaction temperature increases, even after overcoming a value of 80 °C, which is commonly regarded as the inflection point in the oxidation processes [55], because "the exothermic adsorption of reactant becomes disfavored and tends to become the rate limiting-step". This means that there is no actual limitation in the thermoactivation of benzene oxidation because benzene is poorly adsorbed on the photocatalyst surface.
- Under the conditions of this experiment, TiO$_2$-N is more active than the composite Bi$_2$WO$_6$/TiO$_2$-N and Fe/Bi$_2$WO$_6$/TiO$_2$-N photocatalysts over the whole temperature range (the reasons for that are discussed in the previous section). However, if we look at the formation rates of both final products: CO$_2$ as the product of complete oxidation (Figure 10a) and CO as the product of incomplete oxidation (Figure 10b), the Fe/Bi$_2$WO$_6$/TiO$_2$–N photocatalyst gives a lower amount of CO and provides more selective oxidation of benzene compared to other samples. This means that this photocatalyst would more effectively reduce the total hazard of air polluted with benzene vapor.

3. Materials and Methods

High purity grade titanium(IV) oxysulfate (TiOSO$_4$), bismuth(III) nitrate pentahydrate (Bi(NO$_3$)$_3$·5H$_2$O), and sodium tungstate dihydrate (Na$_2$WO$_4$·2H$_2$O) were purchased from Sigma-Aldrich (St. Louis, MO, USA). Reagent grade ammonium hydroxide solution (NH$_4$OH, 25%) and nitric acid (HNO$_3$, 65%) were purchased from AO Reachem Inc.

(Moscow, Russia). These reagents were used for the synthesis of photocatalysts as received without further purification. N-doped TiO$_2$ (TiO$_2$-N) was synthesized under neutral conditions by precipitation from TiOSO$_4$ aqueous solution using ammonia followed by washing of the precipitate with deionized water and its calcination in air at 450 °C. Parameters of synthesis were selected based on the results of our previous study [32] to prepare TiO$_2$-N with a high activity under visible light. Bi$_2$WO$_6$/TiO$_2$-N composite was synthesized by hydrothermal method using Bi(NO$_3$)$_3$ and Na$_2$WO$_4$ precursors according to our previously published technique [31]. A molar ratio between Bi$_2$WO$_6$ and TiO$_2$-N components (5:100) was adjusted by the initial amounts of precursors and TiO$_2$-N in suspension. The Bi$_2$WO$_6$/TiO$_2$-N composite was modified by depositing Fe species on its surface using an aqueous solution of iron nitrate. The estimated content of Fe was 0.1 wt.%. This sample is denoted in the paper as Fe/Bi$_2$WO$_6$/TiO$_2$-N.

The total content of nitrogen was measured by CHNS analysis using a Vario EL Cube elemental analyzer from Elementar Analysensysteme GmbH (Langenselbold, Germany). The crystal phases in the prepared photocatalysts were analyzed using powder X-ray diffraction (XRD). The data were collected in the 2θ range of 10–75° with a step of 0.05° and a collection time of 3 s using a D8 Advance diffractometer (Bruker, Billerica, MA, USA) equipped with a CuKα radiation source and a LynxEye position sensitive detector. SEM micrographs were received using an ultra-high-resolution Field-Emission SEM (FE-SEM) Regulus 8230 (Hitachi, Tokyo, Japan) at an accelerating voltage of 5 kV. X-ray photoelectron spectroscopy (XPS) analysis was performed using a SPECS photoelectron spectrometer (SPECS Surface Nano Analysis GmbH, Berlin, Germany) equipped with a PHOIBOS-150 hemi-spherical energy analyzer and an AlKα radiation source (hν = 1486.6 eV, 150 W). The binding energies (BE) were pre-calibrated using the lines of Au4f7/2 (84.0 eV) and Cu2p3/2 (932.67 eV) from metallic gold and copper foils. Peak fitting in the collected spectra was performed using XPSPeak 4.1 software (Informer Technologies Inc., Los Angeles, CA, USA). UV-Vis diffuse reflectance spectroscopy was used to examine the optical properties of the samples at room temperature. The spectra were taken on a Cary 300 UV-Vis spectrophotometer from Agilent Technologies Inc. (Santa Clara, CA, USA) equipped with a DRA-30I diffuse reflectance accessory and special pre-packed polytetrafluoroethylene as a reflectance standard in the range of 250–850 nm. The optical band gap was calculated using the Tauc method, assuming indirect allowed excitations.

The prepared photocatalysts were tested in the degradation of ethanol and benzene vapors under blue light. The kinetic aspects were studied at 25 °C in a 0.3 L batch reactor installed in the cell compartment of a Vector 22 FTIR spectrometer (Bruker, Billerica, MA, USA). The details of the experimental setup can be found elsewhere [56]. A sample of photocatalyst was uniformly deposited onto a 9 cm^2 glass support to obtain an even layer with an area density of 5 mg cm^{-2}. Then, this support was placed into the reactor with a quartz window. The reactor was purged for 45 min with purified air with a relative humidity of 20% and tightly closed. After that, the photocatalyst was irradiated for 30 min using a light-emitting diode (blue LED), which had a maximum in its emission spectrum at 445 nm (Figure 11a). The photon flux to the surface of photocatalyst provided by the blue LED was 120 μE min^{-1}. LED was turned off after this pretreatment required for the oxidation of all the organic species previously adsorbed on the surface of photocatalyst. Then, 1 μL of liquid ethanol or benzene was injected into the reactor, and IR spectra of the gas phase were collected periodically to analyze volatile compounds in the reactor. The blue LED was turned on again after 20–30 min when the adsorption–desorption equilibrium was reached. The irradiation continued until the total concentration of the C-containing products accumulated in the gas phase reached the theoretically estimated level. The starting point (i.e., 0 min) in all kinetic plots was placed to the moment of light on.

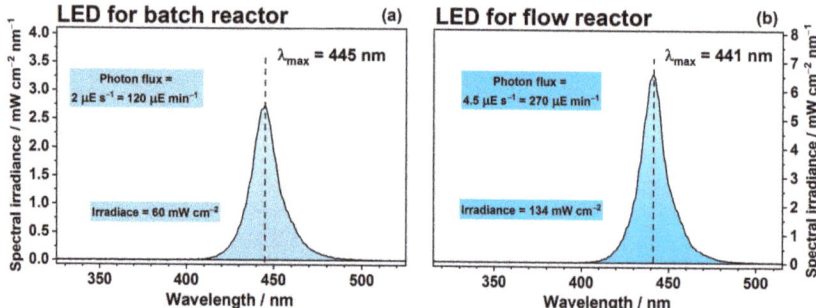

Figure 11. Emission spectra of LEDs used for the photocatalytic experiments in batch reactor (**a**) and in continuous-flow set-up (**b**). Colored rectangle areas show corresponding values of photon flux and specific irradiance of photocatalyst.

In addition to the experiments in the batch reactor, stability of photocatalysts and their steady-state activity were studied in a continuous-flow set-up operated under conditions as follows: the volume flow rate was 0.10 ± 0.02 L min^{-1}; the relative humidity of inlet air was 20%, the inlet concentration of benzene was 1–12 µmol L^{-1}, the temperature of photoreactor was 40–140 °C. The details of this continuous-flow set-up can be found elsewhere [54]. The photocatalyst was irradiated with a blue LED, which had a maximum in its emission spectrum at 441 nm (Figure 11b). The photon flux to the surface of the photocatalyst in this case was 270 µE min^{-1}. The steady-state rate of CO_2 formation (µmol min^{-1}) during benzene degradation was evaluated to compare the activity of photocatalysts. The results of preliminary experiments (see Figure S6 in the Supplementary Materials) showed that the rate of CO_2 formation for all studied photocatalysts slightly depends on the concentration of benzene in the region higher than 2 µmol L^{-1}. Therefore, the initial concentration of benzene was adjusted to 10 ± 0.5 µmol L^{-1} for the correct comparison of different photocatalysts. The temperature of photoreactor was varied from 40 °C to 140 °C to study its effect on the activity.

4. Conclusions

Photocatalysts based on N-doped TiO_2 (TiO_2-N) exhibit high activity in the degradation of volatile organic compounds under visible light. Ethanol and benzene are used as examples of non-aromatic and aromatic compounds, respectively, to study the kinetic aspects of the photocatalytic degradation. A combination of TiO_2-N with bismuth tungstate in a composite system (Bi_2WO_6/TiO_2-N) and subsequent modification of its surface with iron species (Fe/Bi_2WO_6/TiO_2-N) substantially increase the removal rate of the initial pollutant and formed intermediates from the gas phase compared to pristine TiO_2-N. However, the high photocatalytic ability of the composite system can have an adverse effect in the case of benzene because the rate of pollutant degradation is substantially decreased compared to a pristine TiO_2-N photocatalyst, especially at high concentrations. This effect occurs due to a fast accumulation of non-volatile intermediates, which occupy available adsorption sites on the surface of the photocatalyst, thus suppressing the adsorption of initial benzene and, consequently, reducing the rate of the overall oxidation process. The observed deactivation is reversible. When all benzene is removed from the gas phase, the surface of the photocatalyst is rapidly cleaned with the formation of a high amount of CO_2 in the gas phase. The rate of benzene degradation is monotonically increased as the reaction temperature is increased up to 140 °C. At the same time, this thermal activation does not increase the relative yield of CO as a by-product that allows a fast and efficient reduction in the total hazard of air polluted with benzene vapor.

Supplementary Materials: The following supporting information can be downloaded at: https://www.mdpi.com/article/10.3390/ijms24065693/s1.

Author Contributions: Conceptualization, D.S. and M.L.; methodology, N.K. and M.L.; software, M.L.; validation, N.K., M.L. and D.S.; formal analysis, N.K.; investigation, N.K. and A.B.; resources, D.K., M.L. and D.S.; data curation, M.L.; writing—original draft preparation, N.K. and M.L.; writing—review and editing, D.S.; visualization, M.L.; supervision, D.K.; project administration, D.S.; funding acquisition, M.L. and D.S. All authors have read and agreed to the published version of the manuscript.

Funding: Photocatalyst preparation and characterization was funded by the Russian Science Foundation, grant number 20-73-10135. Investigation of the kinetic aspects of pollutant degradation was funded by the Ministry of Science and High Education of the Russian Federation according to the project AAAA-A21-121011390006-0 at the Boreskov Institute of Catalysis.

Institutional Review Board Statement: Not applicable.

Informed Consent Statement: Not applicable.

Data Availability Statement: The data presented in this study are available on request from the corresponding author. The data are not publicly available due to privacy.

Acknowledgments: The research was performed using the equipment of the Center of Collective Use "National Center of Catalyst Research" at the Boreskov Institute of Catalysis. M.N. acknowledges RFBR for supporting the study on benzene degradation under different temperatures (project 19-33-60019). The authors thank Gaoke Zhang for his help and discussion of the results.

Conflicts of Interest: The authors declare no conflict of interest. The funders had no role in the design of the study; in the collection, analyses, or interpretation of data; in the writing of the manuscript, or in the decision to publish the results.

References

1. Mata, T.M.; Martins, A.A.; Calheiros, C.S.C.; Villanueva, F.; Alonso-Cuevilla, N.P.; Gabriel, M.F.; Silva, G.V. Indoor Air Quality: A Review of Cleaning Technologies. *Environments* **2022**, *9*, 118. [CrossRef]
2. Shayegan, Z.; Lee, C.-S.; Haghighat, F. TiO_2 Photocatalyst for Removal of Volatile Organic Compounds in Gas Phase—A Review. *Chem. Eng. J.* **2018**, *334*, 2408–2439. [CrossRef]
3. Salvadores, F.; Reli, M.; Alfano, O.M.; Kočí, K.; Ballari, M.D.L.M. Efficiencies Evaluation of Photocatalytic Paints Under Indoor and Outdoor Air Conditions. *Front. Chem.* **2020**, *8*, 551710. [CrossRef] [PubMed]
4. Yang, X.; Wang, D. Photocatalysis: From Fundamental Principles to Materials and Applications. *ACS Appl. Energy Mater.* **2018**, *1*, 6657–6693. [CrossRef]
5. Rashid, M.M.; Simončič, B.; Tomšič, B. Recent Advances in TiO_2-Functionalized Textile Surfaces. *Surf. Interfaces* **2021**, *22*, 100890. [CrossRef]
6. Chen, D.; Cheng, Y.; Zhou, N.; Chen, P.; Wang, Y.; Li, K.; Huo, S.; Cheng, P.; Peng, P.; Zhang, R.; et al. Photocatalytic Degradation of Organic Pollutants Using TiO_2-Based Photocatalysts: A Review. *J. Clean. Prod.* **2020**, *268*, 121725. [CrossRef]
7. Solovyeva, M.; Selishchev, D.; Cherepanova, S.; Stepanov, G.; Zhuravlev, E.; Richter, V.; Kozlov, D. Self-Cleaning Photoactive Cotton Fabric Modified with Nanocrystalline TiO_2 for Efficient Degradation of Volatile Organic Compounds and DNA Contaminants. *Chem. Eng. J.* **2020**, *388*, 124167. [CrossRef]
8. Daghrir, R.; Drogui, P.; Robert, D. Modified TiO_2 For Environmental Photocatalytic Applications: A Review. *Ind. Eng. Chem. Res.* **2013**, *52*, 3581–3599. [CrossRef]
9. Ong, W.-J.; Tan, L.-L.; Ng, Y.H.; Yong, S.-T.; Chai, S.-P. Graphitic Carbon Nitride ($g-C_3N_4$)-Based Photocatalysts for Artificial Photosynthesis and Environmental Remediation: Are We a Step Closer to Achieving Sustainability? *Chem. Rev.* **2016**, *116*, 7159–7329. [CrossRef]
10. Amirulsyafiee, A.; Khan, M.M.; Harunsani, M.H. Ag_3PO_4 and Ag_3PO_4–Based Visible Light Active Photocatalysts: Recent Progress, Synthesis, and Photocatalytic Applications. *Catal. Commun.* **2022**, *172*, 106556. [CrossRef]
11. Stelo, F.; Kublik, N.; Ullah, S.; Wender, H. Recent Advances in Bi_2MoO_6 Based Z-Scheme Heterojunctions for Photocatalytic Degradation of Pollutants. *J. Alloys Compd.* **2020**, *829*, 154591. [CrossRef]
12. Gu, M.; Yang, Y.; Zhang, L.; Zhu, B.; Liang, G.; Yu, J. Efficient Sacrificial-Agent-Free Solar H_2O_2 Production over All-Inorganic S-Scheme Composites. *Appl. Catal. B Environ.* **2023**, *324*, 122227. [CrossRef]
13. Qamar, M.A.; Shahid, S.; Javed, M.; Iqbal, S.; Sher, M.; Akbar, M.B. Highly Efficient $g-C_3N_4$/Cr-ZnO Nanocomposites with Superior Photocatalytic and Antibacterial Activity. *J. Photochem. Photobiol. Chem.* **2020**, *401*, 112776. [CrossRef]
14. Weon, S.; He, F.; Choi, W. Status and Challenges in Photocatalytic Nanotechnology for Cleaning Air Polluted with Volatile Organic Compounds: Visible Light Utilization and Catalyst Deactivation. *Environ. Sci. Nano* **2019**, *6*, 3185–3214. [CrossRef]
15. Wang, Z.; Liu, Y.; Huang, B.; Dai, Y.; Lou, Z.; Wang, G.; Zhang, X.; Qin, X. Progress on Extending the Light Absorption Spectra of Photocatalysts. *Phys. Chem. Chem. Phys.* **2014**, *16*, 2758. [CrossRef] [PubMed]

16. Park, H.; Park, Y.; Kim, W.; Choi, W. Surface Modification of TiO_2 Photocatalyst for Environmental Applications. *J. Photochem. Photobiol. C Photochem. Rev.* **2013**, *15*, 1–20. [CrossRef]
17. Asahi, R.; Morikawa, T.; Irie, H.; Ohwaki, T. Nitrogen-Doped Titanium Dioxide as Visible-Light-Sensitive Photocatalyst: Designs, Developments, and Prospects. *Chem. Rev.* **2014**, *114*, 9824–9852. [CrossRef] [PubMed]
18. Vaiano, V.; Sacco, O.; Sannino, D.; Ciambelli, P. Nanostructured N-Doped TiO_2 Coated on Glass Spheres for the Photocatalytic Removal of Organic Dyes under UV or Visible Light Irradiation. *Appl. Catal. B Environ.* **2015**, *170–171*, 153–161. [CrossRef]
19. Ansari, S.A.; Khan, M.M.; Ansari, M.O.; Cho, M.H. Nitrogen-Doped Titanium Dioxide (N-Doped TiO_2) for Visible Light Photocatalysis. *New J. Chem.* **2016**, *40*, 3000–3009. [CrossRef]
20. Bakar, S.A.; Ribeiro, C. Nitrogen-Doped Titanium Dioxide: An Overview of Material Design and Dimensionality Effect over Modern Applications. *J. Photochem. Photobiol. C Photochem. Rev.* **2016**, *27*, 1–29. [CrossRef]
21. Ariza-Tarazona, M.C.; Villarreal-Chiu, J.F.; Barbieri, V.; Siligardi, C.; Cedillo-González, E.I. New Strategy for Microplastic Degradation: Green Photocatalysis Using a Protein-Based Porous N-TiO_2 Semiconductor. *Ceram. Int.* **2019**, *45*, 9618–9624. [CrossRef]
22. Sun, Z.; Fang, Y. Electrical Tuning Effect for Schottky Barrier and Hot-Electron Harvest in a Plasmonic Au/TiO_2 Nanostructure. *Sci. Rep.* **2021**, *11*, 338. [CrossRef] [PubMed]
23. Higashimoto, S.; Tanihata, W.; Nakagawa, Y.; Azuma, M.; Ohue, H.; Sakata, Y. Effective Photocatalytic Decomposition of VOC under Visible-Light Irradiation on N-Doped TiO_2 Modified by Vanadium Species. *Appl. Catal. Gen.* **2008**, *340*, 98–104. [CrossRef]
24. Kamat, P.V. Photoinduced Transformations in Semiconductormetal Nanocomposite Assemblies. *Pure Appl. Chem.* **2002**, *74*, 1693–1706. [CrossRef]
25. Iwata, K.; Takaya, T.; Hamaguchi, H.; Yamakata, A.; Ishibashi, T.; Onishi, H.; Kuroda, H. Carrier Dynamics in TiO_2 and Pt/TiO_2 Powders Observed by Femtosecond Time-Resolved Near-Infrared Spectroscopy at a Spectral Region of 0.9–1.5 Mm with the Direct Absorption Method. *J. Phys. Chem. B* **2004**, *108*, 20233–20239. [CrossRef]
26. Fan, X.; Hua, N.; Jia, H.; Zhu, Y.; Wang, Z.; Xu, J.; Wang, C. Synthesis and Evaluation of Visible-Light Photocatalyst: Nitrogen-Doped TiO_2/Bi_2O_3 Heterojunction Structures. *Sci. Adv. Mater.* **2014**, *6*, 1892–1899. [CrossRef]
27. Yao, Y.; Qin, J.; Chen, H.; Wei, F.; Liu, X.; Wang, J.; Wang, S. One-Pot Approach for Synthesis of N-Doped TiO_2/$ZnFe_2O_4$ Hybrid as an Efficient Photocatalyst for Degradation of Aqueous Organic Pollutants. *J. Hazard. Mater.* **2015**, *291*, 28–37. [CrossRef]
28. Shi, X.; Fujitsuka, M.; Lou, Z.; Zhang, P.; Majima, T. In Situ Nitrogen-Doped Hollow-TiO_2/g-C_3N_4 Composite Photocatalysts with Efficient Charge Separation Boosting Water Reduction under Visible Light. *J. Mater. Chem. A* **2017**, *5*, 9671–9681. [CrossRef]
29. Zhao, Z.; Zhang, X.; Fan, J.; Xue, D.; Zhang, B.; Yin, S. N-TiO_2/g-C_3N_4/Up-Conversion Phosphor Composites for the Full-Spectrum Light-Responsive DeNOxphotocatalysis. *J. Mater. Sci.* **2018**, *53*, 7266–7278. [CrossRef]
30. Tang, X.; Wang, Z.; Huang, W.; Jing, Q.; Liu, N. Construction of N-Doped TiO_2/MoS_2 Heterojunction with Synergistic Effect for Enhanced Visible Photodegradation Activity. *Mater. Res. Bull.* **2018**, *105*, 126–132. [CrossRef]
31. Kovalevskiy, N.; Cherepanova, S.; Gerasimov, E.; Lyulyukin, M.; Solovyeva, M.; Prosvirin, I.; Kozlov, D.; Selishchev, D. Enhanced Photocatalytic Activity and Stability of Bi_2WO_6–TiO_2-N Nanocomposites in the Oxidation of Volatile Pollutants. *Nanomaterials* **2022**, *12*, 359. [CrossRef]
32. Kovalevskiy, N.; Svintsitskiy, D.; Cherepanova, S.; Yakushkin, S.; Martyanov, O.; Selishcheva, S.; Gribov, E.; Kozlov, D.; Selishchev, D. Visible-Light-Active N-Doped TiO_2 Photocatalysts: Synthesis from $TiOSO_4$, Characterization, and Enhancement of Stability Via Surface Modification. *Nanomaterials* **2022**, *12*, 4146. [CrossRef] [PubMed]
33. Bellardita, M.; Addamo, M.; Paola, A.D.; Palmisano, L.; Venezia, A.M. Preparation of N-Doped TiO_2: Characterization and Photocatalytic Performance under UV and Visible Light. *Phys. Chem. Chem. Phys.* **2009**, *11*, 4084–4093. [CrossRef]
34. Japa, M.; Tantraviwat, D.; Phasayavan, W.; Nattestad, A.; Chen, J.; Inceesungvorn, B. Simple Preparation of Nitrogen-Doped TiO_2 and Its Performance in Selective Oxidation of Benzyl Alcohol and Benzylamine under Visible Light. *Colloids Surf. Physicochem. Eng. Asp.* **2021**, *610*, 125743. [CrossRef]
35. Liu, S.J.; Ma, Q.; Gao, F.; Song, S.H.; Gao, S. Relationship between N-Doping Induced Point Defects by Annealing in Ammonia and Enhanced Thermal Stability for Anodized Titania Nanotube Arrays. *J. Alloys Compd.* **2012**, *543*, 71–78. [CrossRef]
36. Lee, S.; Cho, I.-S.; Lee, D.K.; Kim, D.W.; Noh, T.H.; Kwak, C.H.; Park, S.; Hong, K.S.; Lee, J.-K.; Jung, H.S. Influence of Nitrogen Chemical States on Photocatalytic Activities of Nitrogen-Doped TiO_2 Nanoparticles under Visible Light. *J. Photochem. Photobiol. Chem.* **2010**, *213*, 129–135. [CrossRef]
37. Zhu, Z.; Wan, S.; Zhao, Y.; Qin, Y.; Ge, X.; Zhong, Q.; Bu, Y. Recent Progress in Bi_2WO_6-Based Photocatalysts for Clean Energy and Environmental Remediation: Competitiveness, Challenges, and Future Perspectives. *Nano Sel.* **2021**, *2*, 187–215. [CrossRef]
38. Yang, C.; Huang, Y.; Li, F.; Li, T. One-Step Synthesis of Bi_2WO_6/TiO_2 Heterojunctions with Enhanced Photocatalytic and Superhydrophobic Property via Hydrothermal Method. *J. Mater. Sci.* **2016**, *51*, 1032–1042. [CrossRef]
39. Wang, R.; Xu, M.; Xie, J.; Ye, S.; Song, X. A Spherical TiO_2-Bi_2WO_6 Composite Photocatalyst for Visible-Light Photocatalytic Degradation of Ethylene. *Colloids Surf. Physicochem. Eng. Asp.* **2020**, *602*, 125048. [CrossRef]
40. He, F.; Muliane, U.; Weon, S.; Choi, W. Substrate-Specific Mineralization and Deactivation Behaviors of TiO_2 as an Air-Cleaning Photocatalyst. *Appl. Catal. B Environ.* **2020**, *275*, 119145. [CrossRef]
41. Salvadores, F.; Alfano, O.M.; Ballari, M.M. Kinetic Study of Air Treatment by Photocatalytic Paints under Indoor Radiation Source: Influence of Ambient Conditions and Photocatalyst Content. *Appl. Catal. B Environ.* **2020**, *268*, 118694. [CrossRef]

42. Sauer, M.L.; Ollis, D.F. Photocatalyzed Oxidation of Ethanol and Acetaldehyde in Humidified Air. *J. Catal.* **1996**, *158*, 570–582. [CrossRef]
43. Muggli, D.S.; McCue, J.T.; Falconer, J.L. Mechanism of the Photocatalytic Oxidation of Ethanol on TiO$_2$. *J. Catal.* **1998**, *173*, 470–483. [CrossRef]
44. d'Hennezel, O.; Pichat, P.; Ollis, D.F. Benzene and Toluene Gas-Phase Photocatalytic Degradation over H$_2$O and HCL Pretreated TiO$_2$: By-Products and Mechanisms. *J. Photochem. Photobiol. Chem.* **1998**, *118*, 197–204. [CrossRef]
45. Einaga, H.; Futamura, S.; Ibusuki, T. Photocatalytic Decomposition of Benzene over TiO$_2$ in a Humidified Airstream. *Phys. Chem. Chem. Phys.* **1999**, *1*, 4903–4908. [CrossRef]
46. Einaga, H. Heterogeneous Photocatalytic Oxidation of Benzene, Toluene, Cyclohexene and Cyclohexane in Humidified Air: Comparison of Decomposition Behavior on Photoirradiated TiO$_2$ Catalyst. *Appl. Catal. B Environ.* **2002**, *38*, 215–225. [CrossRef]
47. Kozlov, D.V. Titanium Dioxide in Gas-Phase Photocatalytic Oxidation of Aromatic and Heteroatom Organic Substances: Deactivation and Reactivation of Photocatalyst. *Theor. Exp. Chem.* **2014**, *50*, 133–154. [CrossRef]
48. Bui, T.D.; Kimura, A.; Higashida, S.; Ikeda, S.; Matsumura, M. Two Routes for Mineralizing Benzene by TiO$_2$-Photocatalyzed Reaction. *Appl. Catal. B Environ.* **2011**, *107*, 119–127. [CrossRef]
49. Zhuang, H.; Gu, Q.; Long, J.; Lin, H.; Lin, H.; Wang, X. Visible Light-Driven Decomposition of Gaseous Benzene on Robust Sn^{2+}-Doped Anatase TiO$_2$ Nanoparticles. *RSC Adv* **2014**, *4*, 34315–34324. [CrossRef]
50. Vikrant, K.; Park, C.M.; Kim, K.-H.; Kumar, S.; Jeon, E.-C. Recent Advancements in Photocatalyst-Based Platforms for the Destruction of Gaseous Benzene: Performance Evaluation of Different Modes of Photocatalytic Operations and against Adsorption Techniques. *J. Photochem. Photobiol. C Photochem. Rev.* **2019**, *41*, 100316. [CrossRef]
51. Bathla, A.; Vikrant, K.; Kukkar, D.; Kim, K.-H. Photocatalytic Degradation of Gaseous Benzene Using Metal Oxide Nanocomposites. *Adv. Colloid Interface Sci.* **2022**, *305*, 102696. [CrossRef] [PubMed]
52. He, F.; Ma, F.; Li, T.; Li, G. Solvothermal Synthesis of N-Doped TiO$_2$ Nanoparticles Using Different Nitrogen Sources, and Their Photocatalytic Activity for Degradation of Benzene. *Chin. J. Catal.* **2013**, *34*, 2263–2270. [CrossRef]
53. Selishchev, D.S.; Kolinko, P.A.P.; Kozlov, D.V. Adsorbent as an Essential Participant in Photocatalytic Processes of Water and Air Purification: Computer Simulation Study. *Appl. Catal. Gen.* **2010**, *377*, 140–149. [CrossRef]
54. Lyulyukin, M.; Kovalevskiy, N.; Prosvirin, I.; Selishchev, D.; Kozlov, D. Thermo-Photoactivity of Pristine and Modified Titania Photocatalysts under UV and Blue Light. *J. Photochem. Photobiol. Chem.* **2022**, *425*, 113675. [CrossRef]
55. Herrmann, J.-M. Heterogeneous Photocatalysis: Fundamentals and Applications to the Removal of Various Types of Aqueous Pollutants. *Catal. Today* **1999**, *53*, 115–129. [CrossRef]
56. Selishchev, D.; Svintsitskiy, D.; Kovtunova, L.; Gerasimov, E.; Gladky, A.; Kozlov, D. Surface Modification of TiO$_2$ with Pd Nanoparticles for Enhanced Photocatalytic Oxidation of Benzene Micropollutants. *Colloids Surf. Physicochem. Eng. Asp.* **2021**, *612*, 125959. [CrossRef]

Disclaimer/Publisher's Note: The statements, opinions and data contained in all publications are solely those of the individual author(s) and contributor(s) and not of MDPI and/or the editor(s). MDPI and/or the editor(s) disclaim responsibility for any injury to people or property resulting from any ideas, methods, instructions or products referred to in the content.

Article

Synthesis of Co₃O₄ Nanoparticles-Decorated Bi₁₂O₁₇Cl₂ Hierarchical Microspheres for Enhanced Photocatalytic Degradation of RhB and BPA

Syed Taj Ud Din [1], Wan-Feng Xie [1,2] and Woochul Yang [1,*]

1. Department of Physics, Dongguk University, Seoul 04620, Republic of Korea
2. School of Electronics and Information, University-Industry Joint Center for Ocean Observation and Broadband Communication, Qingdao University, Qingdao 266071, China
* Correspondence: wyang@dongguk.edu; Tel.: +82-02-2260-3444

Abstract: Three-dimensional (3D) hierarchical microspheres of $Bi_{12}O_{17}Cl_2$ (BOC) were prepared via a facile solvothermal method using a binary solvent for the photocatalytic degradation of Rhodamine-B (RhB) and Bisphenol-A (BPA). Co_3O_4 nanoparticles (NPs)-decorated BOC (Co_3O_4/BOC) heterostructures were synthesized to further enhance their photocatalytic performance. The microstructural, morphological, and compositional characterization showed that the BOC microspheres are composed of thin (~20 nm thick) nanosheets with a 3D hierarchical morphology and a high surface area. Compared to the pure BOC photocatalyst, the 20-Co_3O_4/BOC heterostructure showed enhanced degradation efficiency of RhB (97.4%) and BPA (88.4%). The radical trapping experiments confirmed that superoxide ($^{\bullet}O_2^-$) radicals played a primary role in the photocatalytic degradation of RhB and BPA. The enhanced photocatalytic performances of the hierarchical Co_3O_4/BOC heterostructure are attributable to the synergetic effects of the highly specific surface area, the extension of light absorption to the more visible light region, and the suppression of photoexcited electron-hole recombination. Our developed nanocomposites are beneficial for the construction of other bismuth-based compounds and their heterostructure for use in high-performance photocatalytic applications.

Keywords: $Co_3O_4/Bi_{12}O_{17}Cl_2$; heterojunction; photocatalysis; Rhodamine-B; Bisphenol-A

1. Introduction

Rapid industrialization and population growth has led to a tremendous increase in environmental pollutions. These pollutants mostly consist of hazardous Azo dyes and phenolic compounds. Rhodamine-B (RhB) cationic Azo dye is an anthraquinone derivative. It is highly stable and non-biodegradable in nature and is classified as a carcinogenic and neurotoxic substance [1]. Aside from dyes, the other frequently used compound is the colorless Bisphenol-A. It is a diphenylmethane derivative and a raw material that is widely used in the fabrication of numerous polymeric materials [2]. Long-term exposure of BPA causes endocrine, neurological, and reproductive developmental disorders [3]. Therefore, it is crucial to eradicate RhB and BPA before waste is discharged into water reservoirs and landfills. Pollution-free environmental remediation technologies to degrade these organic pollutants have attracted substantial attention [4]. Among them, visible-light-driven photocatalytic technology has emerged as the most promising approach for wastewater cleaning and pollutant removal [5].

Recently, bismuth-based nanomaterials, such as $BiPO_4$ [6], $Bi_2O_2CO_3$ [7–9], $Bi_4Ti_3O_{12}$ [10], Bi_2MoO_6 [11], Bi_2WO_6 [12], Bi_2O_3 [13], and BiOX (X = Cl, Br, I) [14,15], have attracted substantial attention for their use in photocatalytic applications, because O 2p and Bi 6s valence band hybridization not only narrows the bandgap but also enhances the mobility of photo-generated holes in the valence band. Similarly, a bismuth and oxygen-enriched bismuth-oxyhalide ($Bi_{12}O_{17}Cl_2$ (BOC)) is a typical tetragonal phase compound composed of a layered structure with an alternate stacking of $[Bi_2O_2]^{2+}$ sheets interleaved with $[Cl]^-$ groups, and it represents an important class of bismuth-based photocatalysts. The photocatalytic properties of nanobelts-like BOC were first reported by Xiao et al. in 2013 [16]. Since then, there have been many research efforts focused on the preparation of BOC with different morphologies, including nanobelts, nanosheets, and flower-like morphologies. For instance, Wang and colleagues [17] prepared BOC nanobelts through a solvothermal treatment using $Bi(NO_3)_3 \cdot 5H_2O$, NH_4Cl, and NaOH as precursors in a solvent consisting of ethylene glycol (EG) and water for the photocatalytic degradation of BPA. Liu et al. [18] presented two-dimensional (2D)-BOC nanosheets oriented along the [002] direction which showed enhanced photocatalytic RhB degradation. Similarly, Fang et al. [19] prepared 3D BOC hierarchical nanostructures using a coprecipitation method followed by calcination, and these nanostructures demonstrated high photocatalytic efficiency for RhB degradation. Among the morphologies detailed above, the flower-shaped BOC has excellent characteristics, including high surface area, good adsorption capability, and maximum light absorption. Therefore, it is essential to develop a facile method to fabricate a 3D flower-like BOC. Despite these advantages, the 3D BOC still need to resolve the issues of a rapid electron-hole recombination rate and inappropriate redox potentials. Therefore, various research strategies have been developed to overcome these issues, including the fabrication of heterojunctions [20], element doping [21], noble metal deposition [22], and graphene decoration [23]. Among these methods, heterojunction preparation is the most effective approach because of the fast transfer rate of photo-generated electron and holes (e^--h^+), which facilitates the separation of the photo-generated e^--h^+ pairs in the photocatalysts, which play quite an important role in enhancing the photocatalytic activity of photocatalysts. For instance, He et al. [24] obtained a BOC/β-Bi_2O_3 composite with flower-like micro/nano architectures that demonstrated good photocatalytic activity for the degradation of 4-tert-butyphenol under visible light. In another study, Huang et al. [25] prepared BiOI@BOC heterojunction photocatalysts with high exposure of the active BiOI (001) facet, which exhibited excellent photocatalytic performance for RhB and BPA degradation. Moreover, BOC heterostructures with non-bismuth-based compounds, such as CoAl-LDH/BOC [26] and Ag_2O/BOC p-n junction catalysts [27], have also been reported; both exhibit significant photo-degradation efficiency under visible light irradiation.

Co_3O_4 is a traditional p-type semiconductor (band gap, E_g = 1.2~2.6 eV) with interesting electronic, magnetic, sensing, and catalytic properties [28]. In particular, Co_3O_4-based heterojunctions have yielded high photocatalytic activity; for example, the 0D/2D Co_3O_4/TiO_2 heterojunction photocatalyst has exhibited enhanced photocatalytic activity under visible light irradiations [29]. In another study, Dai et al. [30] synthesized a Co_3O_4/BOC photocatalyst that showed effective visible-light-driven RhB photodegradation due to the more positive value of the valence band potential of Co_3O_4 relative to BOC. However, BOC is an n-type semiconductor, as indicated by its positive slope in the Mott-Schottky plot [31], and the more positive valence band (VB) potential than that of the Co_3O_4 counterpart. Therefore, the combination of Co_3O_4 with BOC is favorable for the formation of the p-n heterojunction. As a result, the photo-generated holes on the VB of BOC could be easily transferred to the VB of Co_3O_4 under light illumination, thus resulting in practical separation of photo-generated e^--h^+ pairs of BOC and Co_3O_4, which would be beneficial for a photocatalyst in terms of photo-degradation efficiency. However, to our knowledge, there is a lack of research into using a Co_3O_4/BOC hierarchical microsphere photocatalyst for the degradation of RhB and BPA.

Herein, the synthesis of a BOC hierarchical microsphere and its decoration with Co_3O_4 nanoparticles (NPs) via a solvothermal method have been reported. The heterojunction formation of the Co_3O_4 NPs-decorated BOC (Co_3O_4/BOC) was evaluated through structural, morphological, spectroscopic, and electrochemical investigations. The photocatalytic degradation efficiency of the hierarchical microsphere Co_3O_4/BOC heterojunction was evaluated against RhB and BPA aqueous pollutants. The results showed that the 20-Co_3O_4/BOC heterostructure had an outstanding degradation efficiency of RhB (97.4%) and BPA (88.4%)

after 140 min and 175 min of visible light irradiation, respectively, compared to pure BOC and other composite samples. The improved photocatalytic degradation performance could be ascribed to the synergetic effects of the larger active area of hierarchical microsphere, the extended light absorption to visible light range with Co_3O_4 NPs, and the suppression of e^--h^+ recombination caused by the *p-n* junction formation of Co_3O_4/BOC.

2. Results and Discussion

2.1. Structural, Morphological, and Elemental Analyses

Figure 1 shows the morphology of the BOC fabricated with different volume ratios of ethylene glycol (EG) and ethyl alcohol (EtOH). BOC-1 (Figure 1a) synthesized in EG only and BOC-3 (Figure 1c) prepared in mixtures of EG and EtOH both have uniform 3D architectures, whereas BOC-2 (Figure 1b) prepared in pure EtOH solution shows a nanoparticle-like morphology with almost no agglomeration. Compared to BOC-1, the morphology of BOC-3 was more regular with a flower-like shape, and the size of the microflower was about 4 μm (Figure 1c). In addition, BOC-3 was also prepared with different solvothermal reaction times to understand the microspherical morphology growth and optimize the reaction time (discussed in Appendix A). BOC-3 synthesized following 6 h of solvothermal treatment was composed of many ultrathin nanosheets (inset of Figure 1), which is beneficial for photocatalytic degradation due to the increased specific surface area. Figure 1d shows the N_2 adsorption-desorption isotherms of BOC-1 (dark), BOC-2 (red), and BOC-3 (blue). The isotherm plots showed type-IV isotherm and hysteresis loop curves [32]. The S_{BET} values of BOC-1, BOC-2, and BOC-3 were 8.914, 8.160, and 15.720 m^2/g, respectively, indicating that BOC-3 had the largest specific surface area.

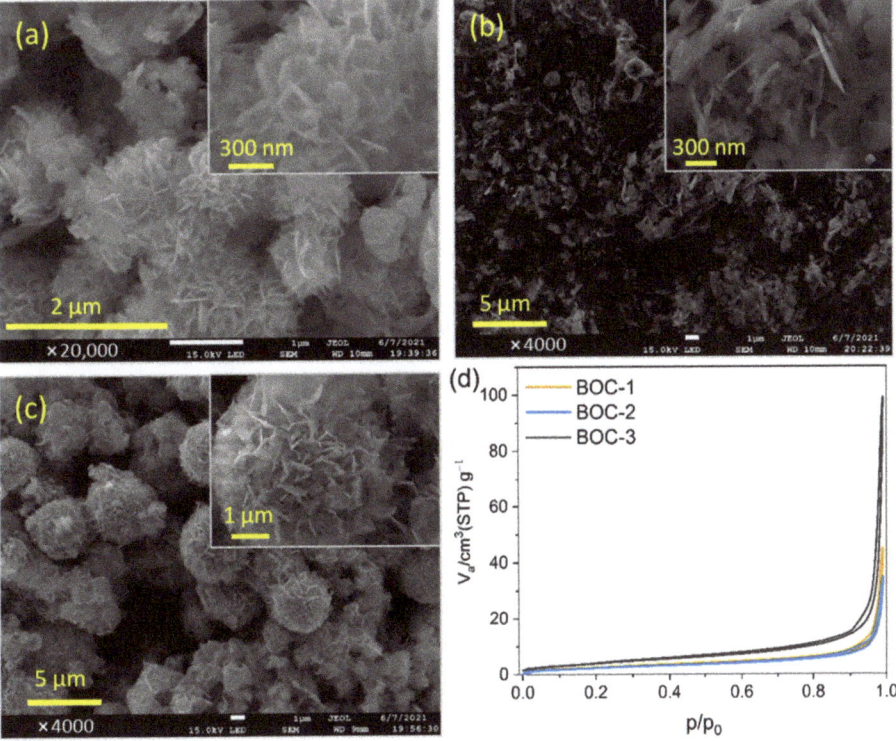

Figure 1. FESEM images of (**a**) BOC-1, (**b**) BOC-2, and (**c**) BOC-3; insets show the respective SEM image with high resolution. (**d**) N_2 adsorption–desorption isotherms of all samples.

Figure 2a shows a FESEM image of the 20-Co_3O_4/BOC synthesized with 20 mg of Co_3O_4 NPs and BOC-3. Its hierarchical morphology was almost identical to that of BOC-3, and it was not affected by the incorporation of Co_3O_4 NPs during the synthesis process. To investigate the existence of the nanosized Co_3O_4 in the BOC hierarchical morphology, HRTEM measurements of 20-Co_3O_4/BOC were performed, as shown in Figure 2b. The results confirmed that the Co_3O_4 NPs, which had an approximate diameter of 10 nm, were decorated on the surface of BOC. In the HRTEM, the fringe spacing of 0.23 nm belonged to the (222) crystal plane of Co_3O_4. By contrast, the fringe spacings of 0.272 nm and 0.31 nm, which respectively refer to the (200) and (117) crystal planes of BOC, have also been observed. Further, a high-angle annular dark-field (HAADF) image of 20-Co_3O_4/BOC was obtained (Figure 2c), in which the tiny black spots identified across the BOC surface indicated Co_3O_4 NPs; this finding was further confirmed by EDX analysis in Figure 2d,e. These results confirmed the 0D/3D morphology of the prepared photocatalyst might be conducive to the photocatalytic performance of the Co_3O_4/BOC heterostructure. Moreover, the obtained EDS mapping spectrum (Figure A2 in Appendix B) confirmed the presence of bismuth (Bi), oxygen (O), chlorine (Cl), and cobalt (Co) elements in the 20-Co_3O_4/BOC sample. The at.% and wt.% of the elements are also shown (inset table in Figure A2). Figure 2f presents the N_2 adsorption–desorption isotherm plots for 20-Co_3O_4/BOC. The estimated S_{BET} using N_2 isotherms was found to be 14.873 m^2/g, which was close to the S_{BET} value of BOC-3. This result shows that the surface area of 20-Co_3O_4/BOC was slightly affected by the incorporation of Co_3O_4 NPs.

The XRD patterns of the pristine BOC-3, Co_3O_4, and 20-Co_3O_4/BOC are shown in Figure 2g. The observed peaks in the XRD patterns of BOC-3 and Co_3O_4 matched the BOC and Co_3O_4 crystallites (JCPDS cards #37-0702 and #42-1467), respectively. The 20-Co_3O_4/BOC samples showed all characteristic peaks of BOC-3. However, the characteristic peaks of Co_3O_4 were not observed in the prepared composite samples. The absence of Co_3O_4 characteristic peaks was attributed to the low amount of Co_3O_4 compared to BOC in 20-Co_3O_4/BOC composite samples. The structural properties of 20-Co_3O_4/BOC were further investigated using Raman spectroscopy, as shown in Figure 2h. All characteristic peaks of BOC have been observed in the Raman spectrum of pure BOC-3 [33]. The observed peak at 165.80 cm^{-1} belonged to the A_{1g} internal stretching of the Bi-Cl bond [33,34]. Because of the oxygen-rich nature of BOC-3, the observed peaks in the range from 200 cm^{-1} to 500 cm^{-1} belong to the vibrational modes of Bi and O bonding. Among them, The peak at 470.51 cm^{-1} is the characteristic vibrational mode of BOC-3, which belongs to O-Bi-O bending modes. Further, the peak at 598.80 cm^{-1} represents Cl-Cl stretching modes [35]. Regarding Co_3O_4 NPs, all the characteristic peaks of Co_3O_4 appeared in the Raman spectra, indicating the successful formation of Co_3O_4 NPs, along with an extra peak at 481.15 cm^{-1} from the glass substrate. These observed peaks belonged to the F_{2g} and E_g modes of the combined vibrations of the tetrahedral site and octahedral oxygen vibrations [36,37]. The Raman spectra of Co_3O_4 NPs-decorated BOC were also obtained. The Raman spectra of 20-Co_3O_4/BOC showed all the characteristic peaks of both Co_3O_4 and BOC-3 samples, along with an extra peak at 307.50 cm^{-1}. Since both Co_3O_4 and BOC-3 are oxygen-rich compounds, the interconnection of Co_3O_4 and BOC through oxygen bonding led to a new peak formation in the 20-Co_3O_4/BOC sample. The observed intense peak belongs to the Bi-O(1) rocking and weak O(2) breathing modes in the 20-Co_3O_4/BOC heterostructures, thus confirming the successful heterojunction formation [38].

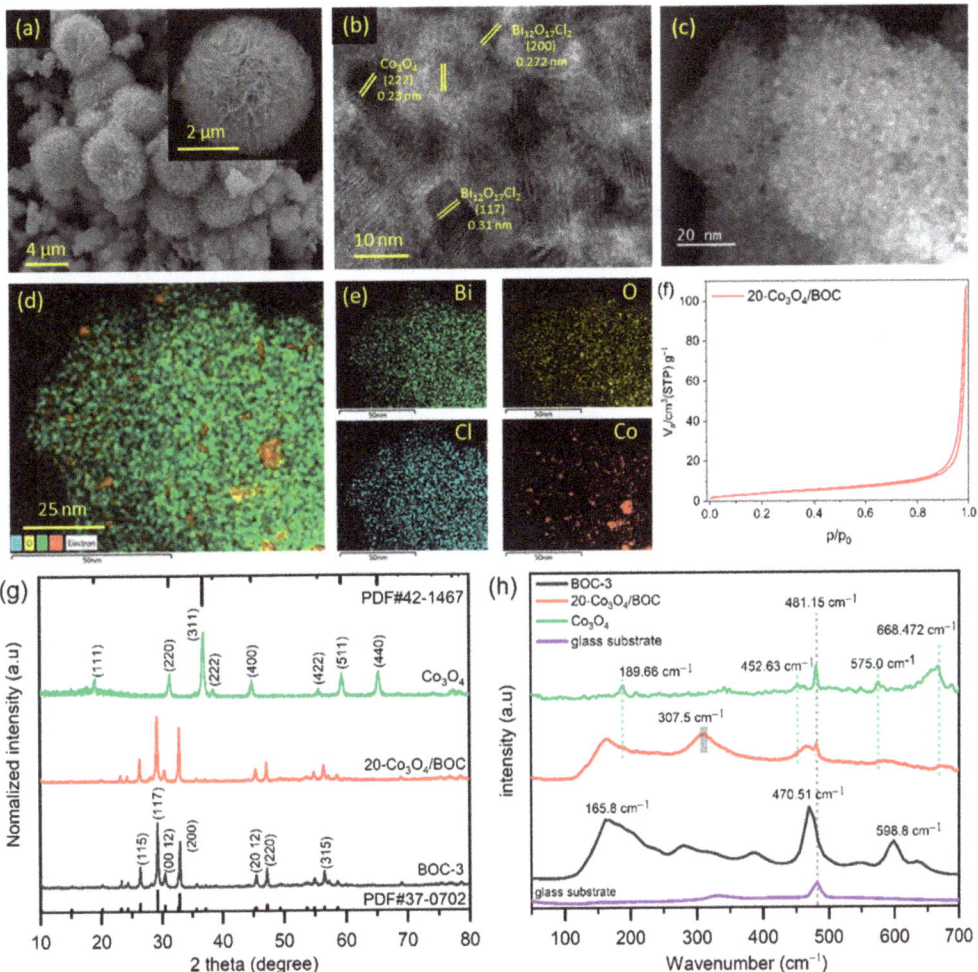

Figure 2. (**a**) FESEM image, (**b**) HRTEM image, (**c**) HAADF image, (**d**) EDX mapping, and (**e**) elemental EDX mapping of 20-Co$_3$O$_4$/BOC sample, (**f**) N$_2$ adsorption-desorption isotherm of 20-Co$_3$O$_4$/BOC, (**g**) XRD patterns, and (**h**) Raman spectra of Co$_3$O$_4$, BOC-3, and 20-Co$_3$O$_4$/BOC micro flowers.

To analyze the chemical composition and chemical state of the elements, we performed X-ray photoelectron spectroscopy (XPS) of the 20-Co$_3$O$_4$/BOC heterostructure photocatalyst (Appendix C). The XPS survey spectrum clearly demonstrated that all peaks were attributable to Bi, O, Cl, and Co elements, revealing that the heterostructure consisted of Bi, O, Cl, and Co elements, as shown in Figure A3 (See Appendix C). The high-resolution XPS spectra of Bi 4f, C 1s, O 1s, and Co 2p for the heterostructure are respectively shown in Figure A3b–e. The two strong peaks at 159.63 and 164.93 eV were assigned to Bi 4f$_{7/2}$ and Bi 4f$_{5/2}$, respectively, which are the features of Bi^{3+} in BOC (Figure A3b). As depicted in Figure A3c, the O 1s profile could be deconvoluted into three peaks, thus indicating the existence of three different kinds of O species in the sample. The peaks observed at 529.674 and 530.568 eV were assigned to the lattice oxygen metal bonds and hydroxyl ($^\bullet$OH) functional groups in 20-Co$_3$O$_4$/BOC, respectively [39]; the peak at 531.592 eV corresponded to oxygen vacancies in Co$_3$O$_4$ in the 20-Co$_3$O$_4$/BOC composite sample [40]. Figure A3d shows the spectrum of Cl 2p, which contained diverse peaks at 198.58 and 200.139 eV, respectively. These can be attributed to Cl $2p_{3/2}$ and Cl $2p_{1/2}$ of the Cl$^-$ ions in the corresponding 20-Co_3O_4/BOC sample [41]. In Figure A3e, the Co 2p peak of 20-Co_3O_4/BOC showed Co $2p_{3/2}$ and Co $2p_{1/2}$ spin-orbit doublets. The peaks at 782.25 and 793.50 eV in 20-Co_3O_4/BOC corresponded to Co^{2+} ions, whereas the peaks observed at 779.50 and 792.50 eV were assigned to Co^{3+} ions, therefore indicating the coexistence of Co^{2+} and Co^{3+} in both samples [42].

2.2. Photocatalytic Performance

The photocatalytic performance of BOC-1, BOC-3, 10-Co_3O_4/BOC, 20-Co_3O_4/BOC, and 40-Co_3O_4/BOC was explored by degrading RhB dye in aqueous solution under visible light, as shown in Figure 3a,b. Figure 3a indicates that 20-Co_3O_4/BOC outperformed BOC-3, 10-Co_3O_4/BOC, and 40-Co_3O_4/BOC by decomposing RhB dye solution in 140 min. Further, the degradation rate of RhB in the presence of each photocatalyst could be determined by the pseudo-first-order kinetic model, as expressed in Equation (1).

$$\ln C/C_0 = kt, \quad (1)$$

where k, C_0, and C represent the reaction rate constant, initial concentration, and remaining concentration at time t, respectively. Figure 3b shows the reaction rate and degradation efficiency of all photocatalysts used for RhB degradation after 140 min. 20-Co_3O_4/BOC had the highest degradation rate of 2.21×10^{-2}/min and an efficiency of 97.4%. Moreover, we compared the RhB photocatalytic degradation performance of the single BOC and 20-Co_3O_4/BOC photocatalyst with previously reported, similarly structured semiconducting photocatalysts, as shown in Table A1 (In Appendix E). Both synthesized BOC and Co_3O_4/BOC hierarchical microspheres possessed relatively higher degradation performance under similar test conditions. This enhanced efficiency might be related to the high surface area of hierarchically structured BOC and the formation of a Co_3O_4/BOC heterojunction with Co_3O_4 NPs-decoration.

Moreover, the photocatalytic degradation of BPA was performed to evaluate the photocatalytic activity of BOC and Co_3O_4/BOC, as shown in Figure 3c,d. The BPA degradation results showed that 20-Co_3O_4/BOC efficiently decomposed BPA aqueous pollutant solution in 170 min. Moreover, the degradation of BPA in the presence of each photocatalyst followed the 1st-order reaction kinetic model. Figure 3d shows the reaction rate and degradation efficiency after 110 min for all the photocatalysts used for BPA degradation. 20-Co_3O_4/BOC had the highest degradation rate of 1.66×10^{-2}/min and an efficiency of 88.4%, followed by BOC-3, BOC-1, and BOC-2.

We further conducted a reusability test for the 20-Co_3O_4/BOC in the presence of RhB and BPA pollutants, as shown in Figure A4 (in Appendix D). During RhB and BPA degradation, a consistent decrease in the degradation of RhB and BPA occurred after the 3rd cycle of degradation, and its degradation performance was slightly reduced. This slight reduction in photocatalytic degradation may be attributable to the adsorption of RhB and BPA molecules on the surface of the 20-Co_3O_4/BOC sample.

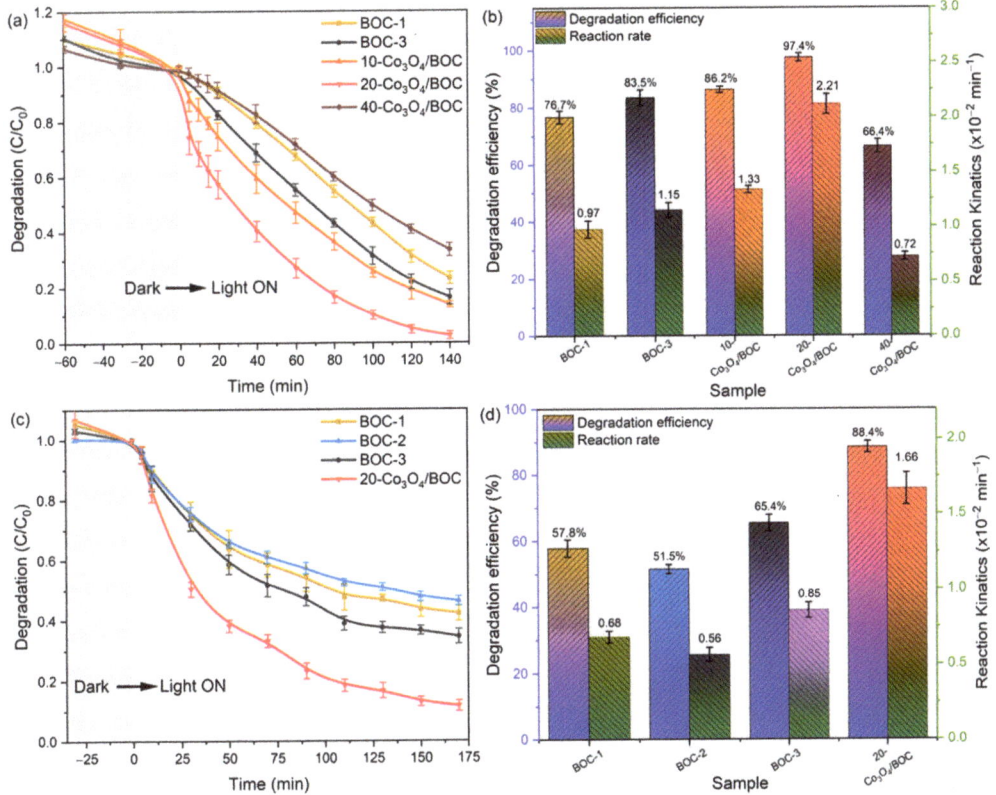

Figure 3. (**a**) Photocatalytic degradation curves of RhB, (**b**) its degradation efficiency with reaction rate constant, in the presence of BOC-1, BOC-3, 10-Co$_3$O$_4$/BOC, 20-Co$_3$O$_4$/BOC, and 40-Co$_3$O$_4$/BOC, (**c**) photocatalytic degradation of BPA, and (**d**) its degradation efficiency with reaction rate constant in the presence of BOC-1, BOC-2, BOC-3, and 20-Co$_3$O$_4$/BOC.

2.3. Analysis of Enhanced Photocatalytic Activity of Co$_3$O$_4$/BOC

Photocatalytic activity is mainly attributed to light absorption capacity and the separation and transfer efficiency of photoinduced charge carriers. Firstly, photoluminescence (PL) measurements were conducted to investigate the recombination rate of photo-induced e$^-$-h$^+$ pairs in BOC-3 and 20-Co$_3$O$_4$/BOC photocatalysts, as shown in Figure 4a. The PL emission intensity of 20-Co$_3$O$_4$/BOC was lower than that of BOC-3, thus indicating reduced recombination of e$^-$-h$^+$ pairs in the 20-Co$_3$O$_4$/BOC photocatalyst. The transient photocurrent response was also measured to provide further support to the efficient separation of photo-generated charges, as shown in Figure 4b. The 20-Co$_3$O$_4$/BOC had a higher photocurrent response than BOC-3. The higher photocurrent response of 20-Co$_3$O$_4$/BOC was attributed to the higher separation efficiency of the excitons and its longer lifetime. Moreover, the EIS Nyquist plot was obtained to examine the electrode/electrolyte interfacial charge transfer resistance, as shown in Figure 4c. 20-Co$_3$O$_4$/BOC showed a smaller arc radius than BOC-3. The inset in Figure 4c shows the circuit of the sample-solution in the EIS measurements. According to the model of the circuit, R$_s$ is related to uncompensated solution resistance, R$_p$ is related to the porosity of the electrode, and R$_{ct}$ represents the charge transfer resistance at the interface [43]. The R$_{ct}$ values for BOC and 20-Co$_3$O$_4$/BOC are 1.58×10^{-4} Ω and 82.93×10^{-4} Ω, respectively. The smaller R$_{ct}$ value of 20-Co$_3$O$_4$/BOC represents its lower charge transfer resistance, which is beneficial for high redox reactions during photocatalysis.

Further, the UV-vis DRS of Co_3O_4, BOC-3, and 20-Co_3O_4/BOC were measured to investigate the optical absorption ability, as shown in Figure 4d. Co_3O_4 showed absorption throughout the whole UV and visible range. The absorption edges for BOC-3 were located around 520 nm. In comparison, 20-Co_3O_4/BOC showed enhanced absorption in the visible range after loading Co_3O_4 NPs on BOC. The high absorption of 20-Co_3O_4/BOC was attributed to the strong contribution of Co_3O_4 in 20-Co_3O_4/BOC to the absorption of visible light. Ultimately, these results suggest that the formation of the heterojunction in 20-Co_3O_4/BOC heterostructure could effectively suppress the recombination of the photoexcited charge carriers and enhance the visible light absorption ability. Therefore, the photocatalytic performance could effectively be improved by the 20-Co_3O_4/BOC heterostructure.

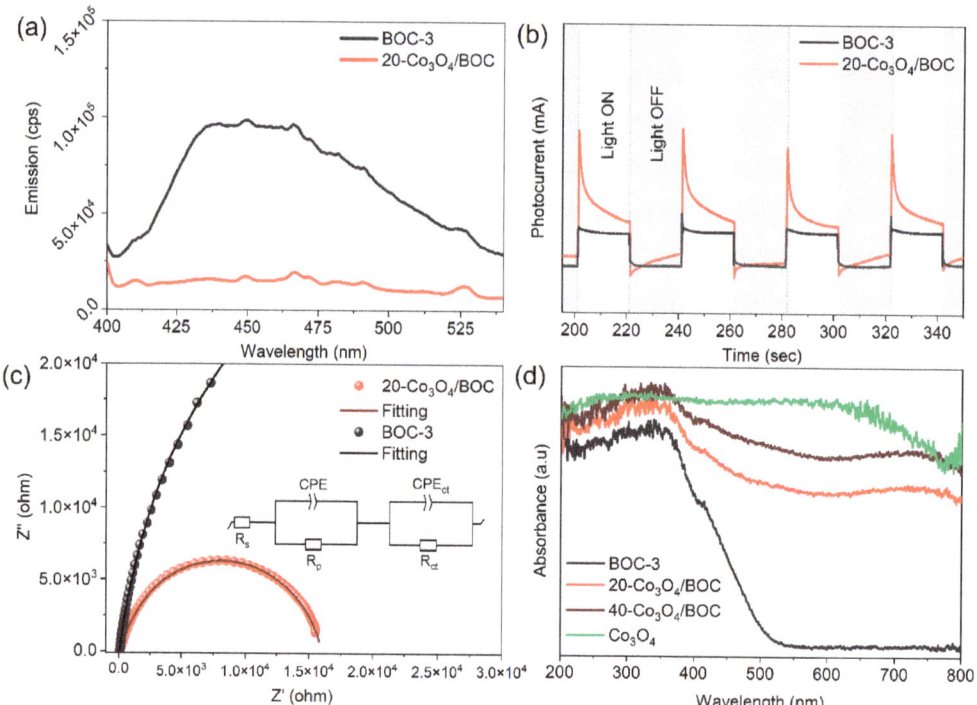

Figure 4. (**a**) Photoluminescence spectra, (**b**) Transient photocurrent response, (**c**) EIS Nyquist plots of BOC-3 and 20-Co_3O_4/BOC samples, and (**d**) UV–Vis diffuse reflectance spectra of Co_3O_4, BOC-3, 20-Co_3O_4/BOC, and 40-Co_3O_4/BOC.

2.4. Interfacial Charge Transfer Behavior and Photocatalytic Reaction Mechanism

The electronic structures of Co_3O_4, BOC, and Co_3O_4/BOC were analyzed by UV-Vis diffuse reflectance spectra (DRS), the Mott–Schottky (MS) plot, and valence band (VB) XPS measurements to elucidate the photocatalytic mechanism of the 20-Co_3O_4/BOC heterostructure during the photodegradation of RhB and BPA, as shown in Figure 5. First, the optical bandgap energies of BOC-3 and Co_3O_4 could be obtained through curve fitting of the Tauc plot of $(\alpha h\nu)^{n/2}$ versus $h\nu$ (Figure 5a), where n = 4 for the Co_3O_4 direct band gap semiconductor and n = 1 for the indirect band gap semiconductor [44,45]. The obtained bandgap energies were 2.34 and 2.25 eV for BOC-3 and Co_3O_4, respectively.

Secondly, the Fermi energy (E_f) levels of the prepared Co_3O_4 and BOC-3 were obtained using MS analysis, as shown in Figure 5b,c, respectively. The MS plots of Co_3O_4 and BOC-3 showed negative and positive slopes, thus indicating p-type and n-type semiconducting behaviors, respectively [46,47]. Further, by extrapolating MS plots, the flatband potentials of Co_3O_4 and BOC-3 were found to be +0.037 and −0.51 V, respectively, vs. the standard calomel electrode (SCE). Note that the flatband potential (E_{fb}) of the n-type and p-type semiconductors represents the E_f level. The E_f level vs. the normal hydrogen electrode (NHE) scale could be calculated using Equation (2). The resultant E_f of Co_3O_4 and BOC-3 were +0.277 and −0.266 eV vs. NHE, respectively

$$E_{fb} \text{ (vs. NHE)} = E_{fb} \text{ (vs. SCE)} + 0.244 \text{ (eV)}, \qquad (2)$$

Valence band (VB) XPS measurement was conducted to determine the VB potentials of Co_3O_4 and BOC-3, as shown in Figure 5d. The VB XPS spectra revealed the VB maxima of 0.86 and 1.69 eV for Co_3O_4 and BOC-3, respectively. Thus, the VB potentials with respect to the E_f level were calculated to be 1.137 and 1.424 eV vs. NHE, respectively (Figure 6a). The CB minima of Co_3O_4 and BOC-3 were determined using the following Equation (3).

$$E_{CB} \text{ (vs. NHE)} = E_{VB} \text{ (vs. NHE)} - E_g, \qquad (3)$$

where E_{CB}, E_{VB}, and E_g denote the sample's CB potential, VB potential, and bandgap, respectively. As a result, the calculated CB potentials for Co_3O_4 and BOC-3 were −1.113 and −0.916 eV vs. NHE, respectively.

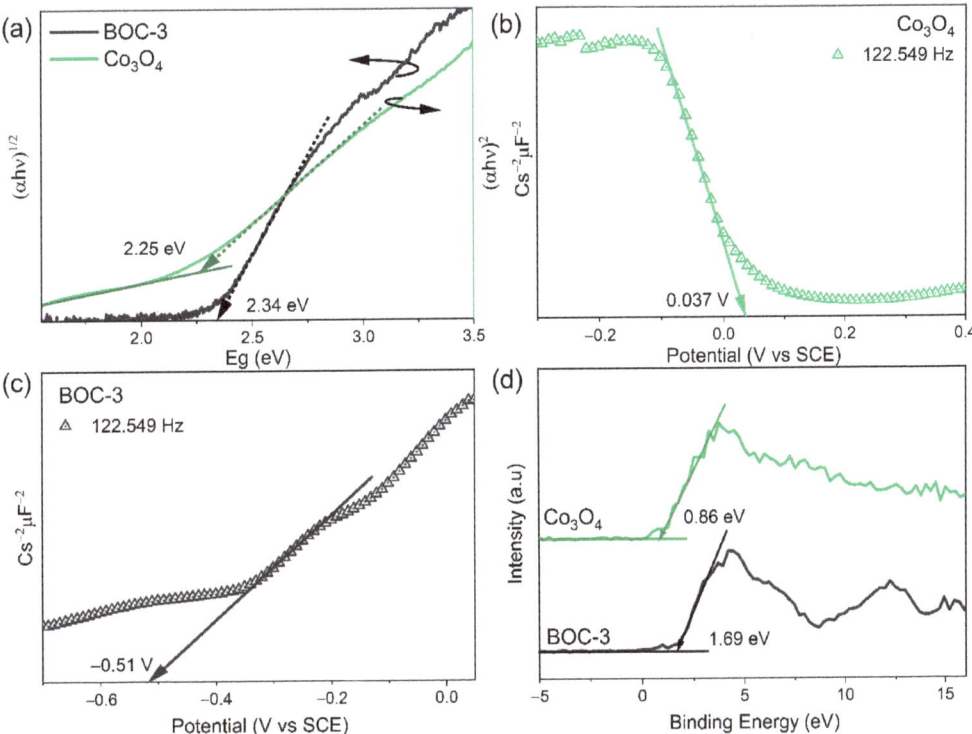

Figure 5. (**a**) Tauc plots, (**b**,**c**) Mott–Schottky plots, and (**d**) valence band XPS of Co_3O_4 and BOC-3.

Therefore, based on the above analysis, we can obtain the band structures of Co_3O_4 and BOC in NHE scale before contact, as shown in Figure 6. When Co_3O_4 and BOC form the Co_3O_4/BOC heterojunction after they come into contact, the electrons will spontaneously migrate from BOC to Co_3O_4 through the Co_3O_4/BOC interface to align with the Fermi level because BOC has a higher E_f level than Co_3O_4. The migration of these electrons results in the band bending upward for BOC and downward for Co_3O_4, near the interface of BOC and Co_3O_4, respectively, as shown in Figure 6b. These band bendings lead to a depletion region at the Co_3O_4 and BOC-3 interface, thus resulting in the generation of an internal electric field (IEF) from BOC toward Co_3O_4 at the interface. After light is irradiated on the photocatalyst, the charge carriers simultaneously excite from VB to the CB in Co_3O_4 and BOC-3 to produce photo-generated e^- and h^+ pairs. Then, the photoexcited electrons in the CB of Co_3O_4 will quickly migrate to the CB of BOC-3, whereas the remaining holes in the VB of BOC-3 will migrate to the VB of Co_3O_4 because of the IEF directed from BOC to Co_3O_4 in the 20-Co_3O_4/BOC, which is a typical charge transport of a type-II heterostructure. As a result, the e^--h^+ recombination is suppressed in the heterostructure system. Moreover, the unique 0D/3D morphology of 20-Co_3O_4/BOC will provide more active catalytic reaction centers and increase the active sites. Thus, the developed heterostructure could be suggested to be beneficial in photocatalytic degradation.

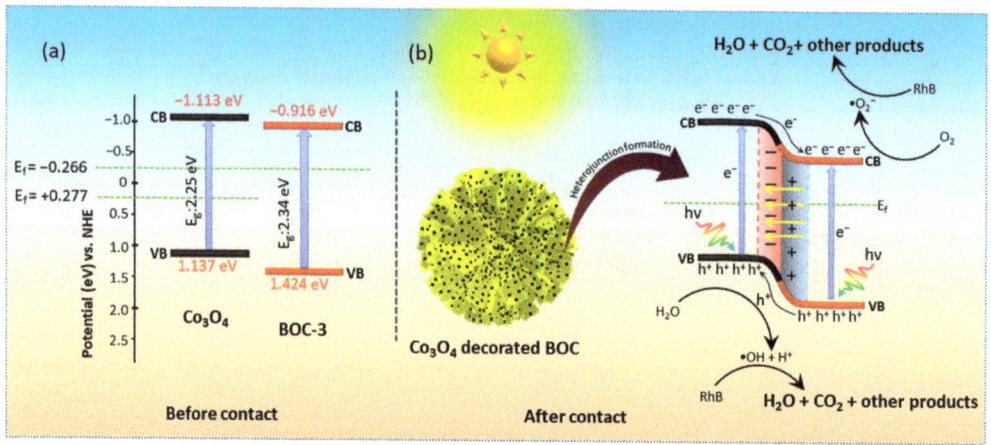

Figure 6. (a) Energy-level diagrams of Co_3O_4 and BOC and (b) interfacial charge transfer behavior and redox reaction process in 20-Co_3O_4/BOC heterostructure under visible-light irradiation.

Finally, photocatalytic active radical detection experiments were conducted to investigate the photocatalytic reaction mechanism and validate our proposed heterojunction formation, as depicted in Figure 7. Here, BQ, IPA, and KI were used as scavengers of $^\bullet O_2^-$ radicals, $^\bullet OH$ radicals, and hole (h^+), respectively, which are produced during the photocatalytic degradation of RhB [48]. Since 20-Co_3O_4/BOC composite decomposed the RhB dye efficiently compared to the other photocatalysts (shown in Figure 3a), a 20-Co_3O_4/BOC photocatalyst sample was chosen for active radical detection. As shown in Figure 7, in the presence of BQ and IPA, the degradation efficiency of RhB was significantly reduced from 98% (no scavenger) to 68.5% and 86.2%, respectively. However, no effect on the degradation efficiency was observed when using KI as an h^+ scavenger. Therefore, $^\bullet O_2^-$ and $^\bullet OH$ radicals were found to be the reactive species with increased generation of $^\bullet O_2^-$ during the photodegradation of RhB.

Based on the radical scavenger experiments, the possible reaction mechanism was proposed as shown in reaction (4)–(10); the excited electrons on BOC reduced the oxygen molecule (O_2) into $^\bullet O_2^-$ radicals, and then the $^\bullet O_2^-$ radicals reacted with e- and hydrogen ions (H^+), ultimately resulting in the generation of hydrogen peroxide (H_2O_2) radical, which was further reduced to •OH and hydroxyl ion (OH⁻). Then, the •O₂⁻, •OH, and OH⁻ finally decomposed RhB and BPA into small chain molecules.

$$Co_3O_4 + h\nu \rightarrow Co_3O_4^* \ (e^- + h^+) \quad (4)$$

$$BOC + h\nu \rightarrow BOC^* \ (h^+ + e^-) \quad (5)$$

$$Co_3O_4^* \ (e^- + h^+) + BOC^* \ (h^- + e^+) \rightarrow Co_3O_4^* \ (h^+) + BOC^* \ (e^-) \quad (6)$$

$$O_2^- + e^- \rightarrow {}^\bullet O_2^- \quad (7)$$

$$O_2^- + e^- + 2H^+ \rightarrow H_2O_2^- \quad (8)$$

$$H_2O_2 + e^- + H^+ \rightarrow {}^\bullet OH + OH^- \quad (9)$$

$$RhB/BPA + {}^\bullet OH/OH^-/O_2^- \rightarrow decomposed\ products \quad (10)$$

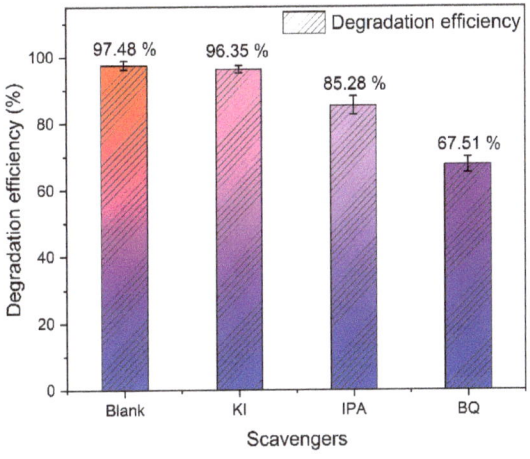

Figure 7. Active radicals detection in the presence of various scavengers during RhB degradation.

3. Materials and Methods

3.1. Chemicals

Bismuth nitrate pentahydrate (Bi(NO₃)₃·5H₂O, 99%), Cobalt acetate tetrahydrate (Co(CH₃COO)₂·4H₂O), Potassium chloride (KCl, 99%), Ethylene glycol (EG) ((CH₂OH)₂, 99%), Dimethylformamide (DMF) (C₃H₇NO), Rhodamine-B (RhB) (C₂₂H₂₄N₂O₈, 99%), Bisphenol-A (BPA) (C₁₅H₁₆O₂, 99%), 1,4-benzoquinone (BQ) (C₆H₄O₂, 99%), and Fluorine doped tin oxide (FTO) glass were purchased from Sigma Aldrich Inc. (St. Louis, MO, USA). Ethanol (EtOH) (C₂H₅OH, 99%) and Isopropyl alcohol (IPA) (C₃H₈O, 99%) were purchased from DAEJUN Co., Ltd (Daejun, Korea). All reagents were used without any further purification.

3.2. Preparation of Co₃O₄ Nanoparticles

The Co₃O₄ NPs were prepared by the solvothermal method [29]. In a typical synthesis, 80 mg of Co(CH₃COO)₂·4H₂O was dissolved in 60 mL of EtOH using a magnetic stirrer. The prepared solution was then transferred to a 100 mL Teflon-lined autoclave and heated at 150 °C for 4 h in a thermal oven to initiate the solvothermal reaction. After completion of the reaction, the autoclave was allowed to cool down to room temperature, at which point the raw Co₃O₄ was washed with EtOH and centrifuged at 15,000 rpm for 30 min. Following centrifugation, the collected sample was dried at 60 °C overnight to obtain the Co₃O₄ NPs.

3.3. Preparation of BOC and Co$_3$O$_4$/BOC

BOC was prepared by a solvothermal method, as shown in Scheme 1a. In a typical synthesis method, 2.186 g (4.5 mmol) of Bi(NO$_3$)$_3$·5H$_2$O was dissolved in 17.5 mL of ethanol by ultrasonication, followed by stirring. The resulting solution was referred to as solution-A. At the same time, 1.5 mmol of KCl was dissolved in 17.5 mL of EG by ultrasonication, followed by stirring for 30 min. The resulting solution was referred to as solution-B. Next, solution-B was added dropwise to solution-A and then constantly stirred for 30 min. Then, the resulting combined solution was transferred to a 50 mL Teflon autoclave and heated for 6 h at 160 °C. The product obtained in this way was washed and dried overnight at 60 °C. Afterward, the gray powder was collected and calcinated in a muffle furnace at 450 °C for 1 h to obtain the targeted BOC hierarchical microspheres. The Co$_3$O$_4$/BOC samples were prepared using the identical synthesis procedure with the addition of Co$_3$O$_4$ NPs (10, 20, and 40 mg) into the combined solution (see Scheme 1b). Moreover, the BOC sample was prepared in pure EG and EtOH solutions to investigate the roles of EG, EtOH, and EG-EtOH mixed solvents during BOC synthesis. The samples prepared in EG, EtOH, and EG-ETOH mixed solvents were labeled as BOC-1, BOC-2, and BOC-3, and the sample prepared with the addition of 10, 20, and 40 mg of Co$_3$O$_4$ NPs were labeled as 10-Co$_3$O$_4$/BOC, 20-Co$_3$O$_4$/BOC, and 40-Co$_3$O$_4$/BOC, respectively.

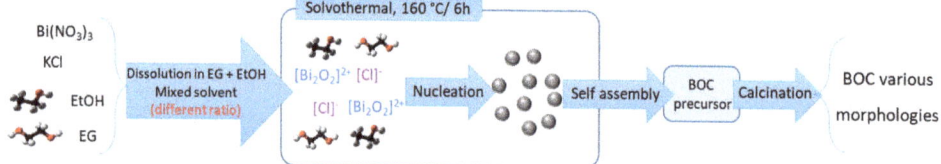

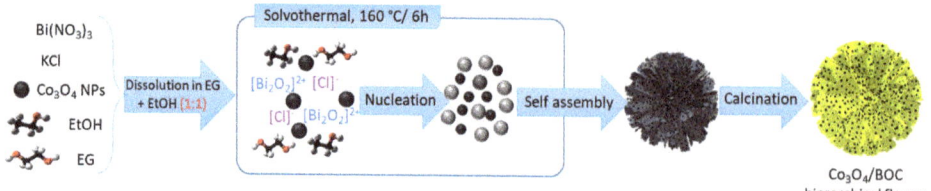

Scheme 1. Synthesis of (**a**) nanoflower-like BOC, and (**b**) 0D/3D Co$_3$O$_4$/BOC hierarchical micro flower.

3.4. Characterization of Samples

The crystal structure and phase purity were characterized by powder X-ray diffraction (XRD) in the 2θ range from 10~80° (2° min^{-1}) using an X-ray diffractometer (Rigaku D/MAX-2500) with a Cu Kα irradiation source (λ = 1.54178 Å) and X-ray power of 40 kV/30 mA. Micro-Raman spectroscopy (XperRAM100, Nanobase Inc., Seoul, Korea) equipped with a monochromatic laser source (wavelength of 532 nm and power of 6 mW) was used to characterize the crystalline phase. The morphologies were examined using a field emission scanning electron microscope (SEM) (JSM-6700F, Jeol Ltd., Tokyo, Japan) and a transmission electron microscope (TEM) ("NEOARM "/JEM-ARM200F, Jeol Ltd.) equipped with an energy dispersive spectroscope (EDX). X-ray photoemission spectroscopy (XPS) (Veresprobe II, ULVAC-PHI Inc., Kanagawa, Japan) with a Monochromatic Al Kα X-ray source was used to examine the chemical compositions of the samples. The specific surface area was volumetrically assessed by measuring the nitrogen adsorption/desorption isotherms at 77 K using Microtrac, BELsorp-mini II. UV–vis diffuse reflectance spectra

(DRS) were obtained using a spectrometer (V-750, Jasco Inc., Tokyo, Japan) equipped with a 60 mm integrating sphere while using BaSO$_4$ as a reference. Photoluminescence spectra (PL) were collected using a spectrophotometer (FS5 fluorescence, Edinburg, United Kingdom) with an excitation wavelength of 375 nm. Electrochemical impedance spectroscopy (EIS) was performed using a three-electrode workstation (VSP Potentiostat, Biologic, Seyssinet-Pariset, France). A pt square plate (1 × 2 cm^2) and a standard calomel were used as counter and reference electrodes, respectively. A clean fluorine-doped tin oxide (FTO) glass with an active surface area of 0.8 cm^2 was used as a substrate for the working electrode, whereas an aqueous solution of 0.5 M Na$_2$SO$_4$ (80 mL) was used as the electrolyte. In the preparation of the working electrode, 0.1 mg of photocatalyst was added to 1 mL of DMF solution and sonicated for 1 h. Then, 50 µL of the dispersed solution was drop-casted on FTO and annealed at 160 °C for 1 h, which was further used in the electrochemical investigation.

3.5. Photocatalytic Activity Measurements

The photocatalytic characteristics of the samples were examined using a visible light source with a 300 W Xenon lamp (1000 W/m^2) (CEL-HXF300, CEAULIGHT Co., Beijing, China) equipped with a UV-IR cutoff filter (420 nm > λ > 780 nm). A double wall jacket beaker with a surface area of 80 cm^2 connected to a water chiller was used to perform the photocatalytic degradation measurement of the photocatalysts. The height from the surface of the pollutant solution to the light source was kept at 30 cm. In this study, RhB dye and BPA colorless pollutants were used to evaluate the degradation efficiency of the synthesized photocatalysts. Briefly, 20 and 40 mg of photocatalyst was used to degrade 40 mL of RhB dye (20 ppm) and BPA colorless pollutant (10 ppm), respectively. Before initiating the photocatalytic experiment, the RhB and BPA aqueous solutions were stirred for 60 min and 30 min, respectively, in dark conditions to attain an adsorption–desorption equilibrium of the photocatalysts. To investigate the photocatalytic degradation rate, 2 mL solution was taken from the RhB or BPA solution after a specific interval, and this solution was then centrifuged for 3 min at 5000 rpm to separate the photocatalyst, after which the absorbance spectrum of the supernatant using UV-visible spectrophotometer was measured. Further, radical trapping experiments were conducted to determine the dominant radical species involved in the photocatalytic decomposition of RhB and BPA. IPA, BQ, and KI of 2 mmol were used as trapping reagents to explore the active species, such as, $^\bullet O_2^-$, $^\bullet OH$ radicals, and h$^+$, respectively.

4. Conclusions

In this work, we successfully developed a 3D hierarchical BOC microsphere and a 0D/3D-Co$_3$O$_4$/BOC heterojunction photocatalyst composed of BOC decorated with Co$_3$O$_4$ NPs using a simple solvothermal synthesis method. The developed heterostructure showed a conventional type II charge transport phenomenon across Co$_3$O$_4$ and BOC-3 by forming a *p-n* heterojunction. The 0D/3D hierarchical morphology of the Co$_3$O$_4$/BOC could increase active sites because of its high surface area, suppressing e$^-$-h$^+$ recombination, and improving visible light absorption. The results of mechanistic studies have proven that the generation of an IEF from BOC-3 to Co$_3$O$_4$ led to a *p-n* junction and the formation of a type II heterojunction. Thus, benefiting from the above properties, the Co$_3$O$_4$/BOC sample demonstrated a higher reaction rate and a higher degradation efficiency than bare Co$_3$O$_4$ and BOC during RhB and BPA degradation. Conclusively, our developed photocatalyst should be considered a good candidate for pollutant degradation.

Author Contributions: S.T.U.D. contributed to the main idea and the writing of the original draft. W.-F.X. contributed to the writing—review and editing. W.Y. contributed to the writing—review and editing, supervision, project management, and funding acquisition. All authors have read and agreed to the published version of the manuscript.

Funding: This research was supported by the National Research Foundation of Korea (NRF) grant funded by the Korean government (MSIT) (No. 2022R1F1A1074441 and 2021H1D3A2A01100019).

Institutional Review Board Statement: Not applicable.

Informed Consent Statement: Not applicable.

Data Availability Statement: Not applicable.

Conflicts of Interest: The authors declare no conflict of interest.

Appendix A

We also prepared BOC-3 in different solvothermal reaction times to explore the formation mechanism of the BOC microspherical morphology. Figure A1 represents the SEM images of the BOC samples prepared with solvothermal reaction times of 3, 6, 9, 12, and 18 h. The formation of a sphere-like morphology composed of ultrathin nanosheets could be observed after 3 h of solvothermal reaction (Figure A1a). When the solvothermal reaction time increased to 6 h, the nanosheets grew further, thus increasing the diameter of the microspheres with clear visibility of ultrathin nanosheets (Figure A1b). When the reaction time was increased to 9 h, the microspheres were covered with external BOC nanosheets, and the nanosheets did not grow further (Figure A1c). The BOC nanosheets were covered with BOC nanosheets after increasing the hydrothermal reaction time to 12 h and 18 h, respectively (Figure A1d,e). These results indicate that the formation of BOC microspheres quickly underwent nucleation, growth, and self-assembly.

Based on the above results, the formation of the BOC samples in the different solvent ratios proceeded as illustrated in Scheme 1. It could be concluded that $Bi(NO_3)_3 \cdot 5H_2O$ will be hydrolyzed into $[Bi_2O_2]^{2+}$ after dissolution in EtOH. Secondly, Cl^- ions generated from KCl in EG solution could react with $[Bi_2O_2]^{2+}$ to form Cl–Bi–O–Bi–Cl nuclei through the coulomb force during the solvothermal reaction [49], thus resulting in homogenous nucleation. As the reaction proceeded in the first 3 h of the solvothermal reaction time at 160 °C, the BOC nanosheets formed. These nanosheets formations could potentially have occurred due to the bonding of the OH functional group to Bi^{3+} ions in Cl–Bi–O–Bi–Cl complex, which grew vertically along the c axis due to its intrinsic crystal structure. Since BOC has a known silane-type structure with a space group of P4/nmm and lattice constants of a = 5.4 Å and c = 35.20 Å, which are related to $\sqrt{2}a_s \times \sqrt{2}b_s \times c_s$ supercell, its layered structure was constructed through the combination of the 6-fold metal-oxygen (Bi-O) layer separated by the Cl layer [50]. Therefore, the formation of the nanosheets and their regulation with the OH functional group were considered to be favorable [51]. Thirdly, as the reaction progressed, the obtained nanosheets were assembled into microspheres, which could decrease the surface energy and achieve a stable structure. Finally, the hierarchical microspheres grew further and the size of the samples gradually increased.

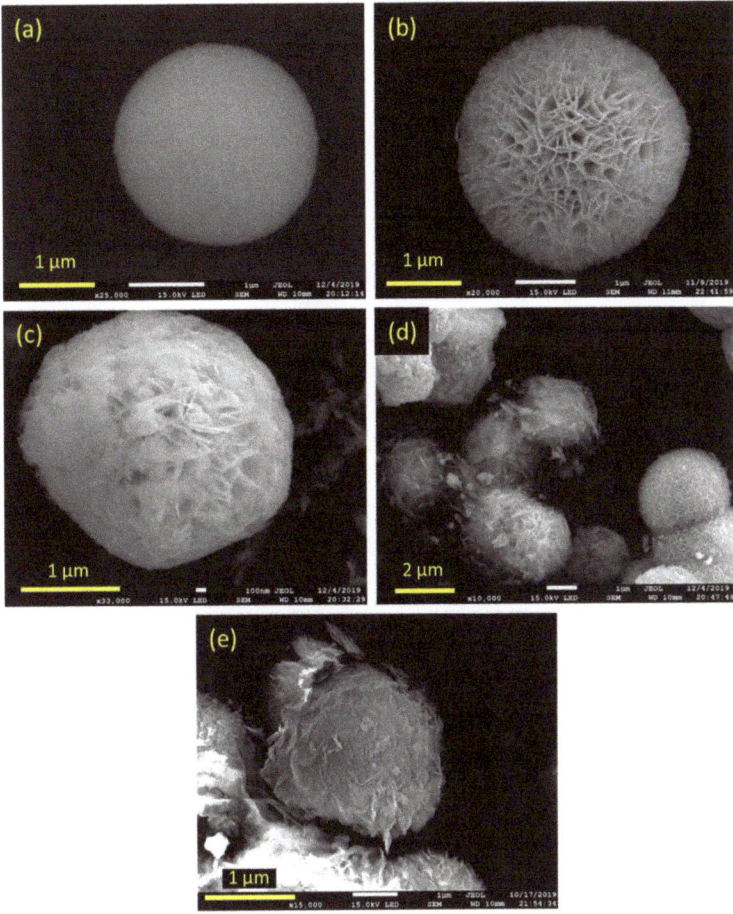

Figure A1. SEM images of BOC samples prepared after (**a**) 3, (**b**) 6, (**c**) 9, (**d**) 12, and (**e**) 18 h of solvothermal reaction time.

Appendix B

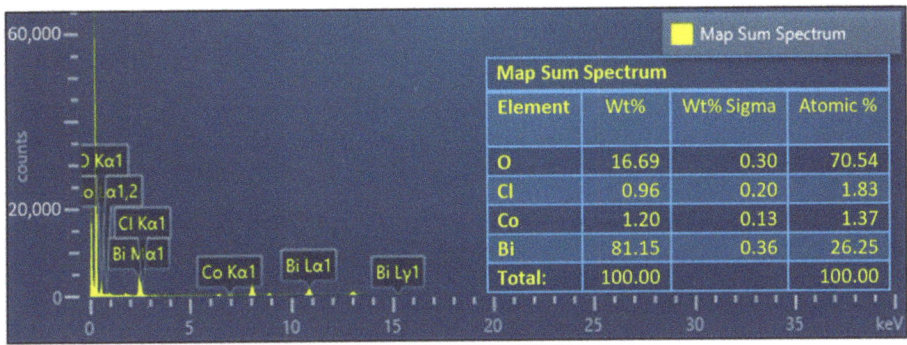

Figure A2. EDX spectrum of 20-Co_3O_4/BOC, and its elemental and atomic wt.% in the inset table.

Appendix C

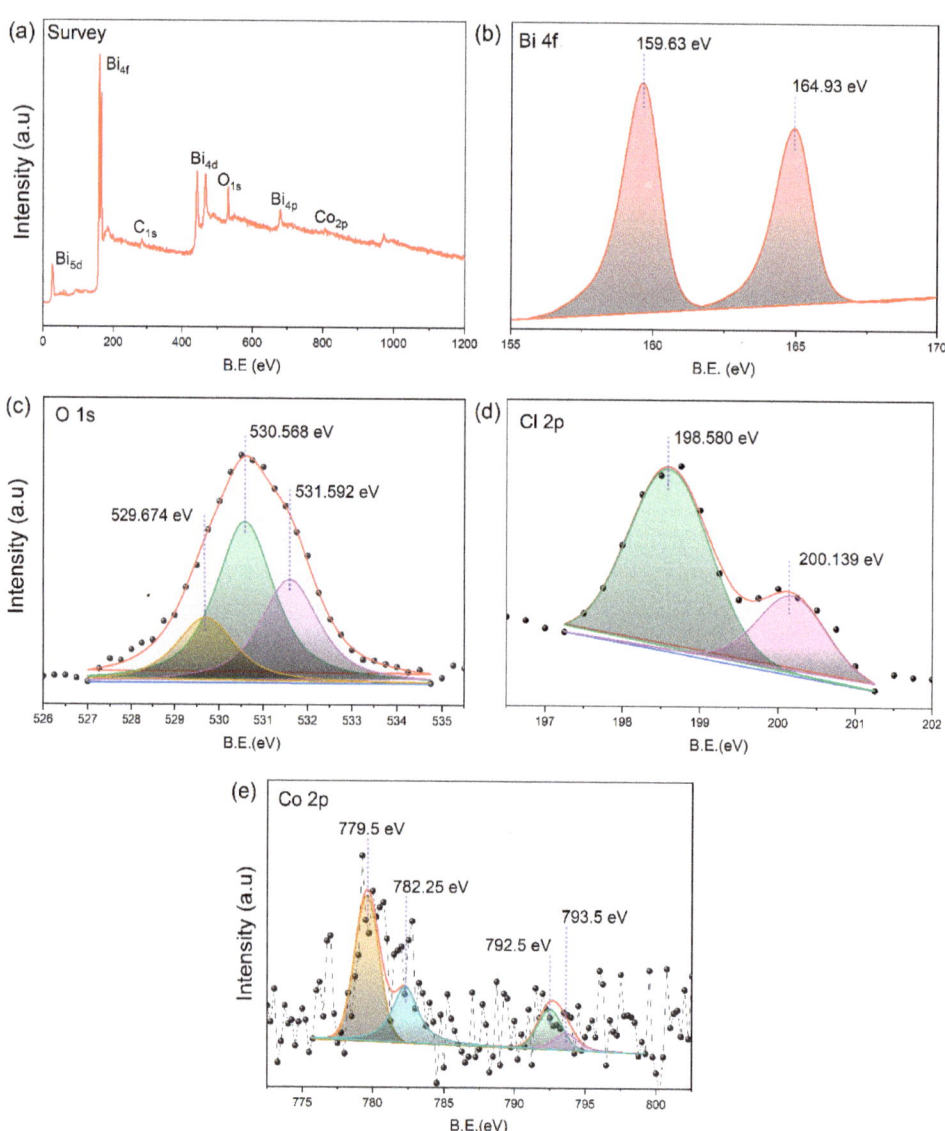

Figure A3. XPS survey spectra (**a**), core level spectra of (**b**) Bi 4f, (**c**) O 1s, (**d**) Cl 2p, and (**e**) Co 2p spectra for the prepared 20-Co_3O_4/BOC samples.

Appendix D

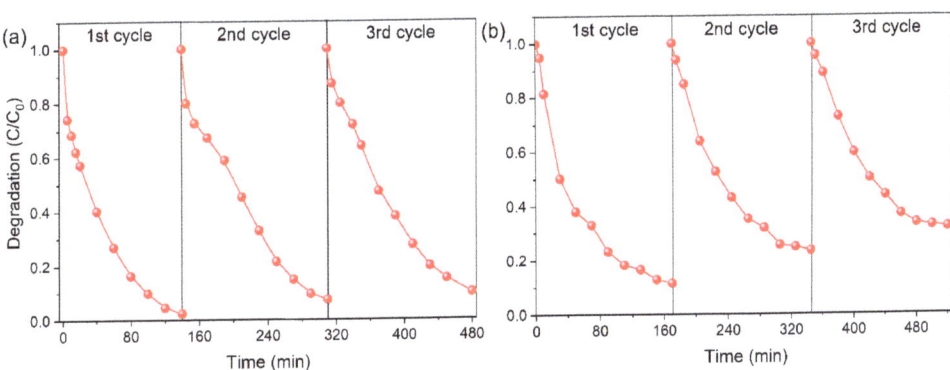

Figure A4. Cyclic photocatalytic degradation of (**a**) RhB and (**b**) BPA in the presence of 20-Co_3O_4/BOC sample.

Appendix E

Table A1. Comparison of photocatalytic degradation of RhB in the presence of pure BOC and BOC composite sample with recently reported results.

Photocatalyst	Rate Constant (min^{-1})	Degradation Efficiency (%)	Degradation Time (min)	Lamp (C)	Amount of Catalyst (mg)	Solution Volume (mL)	Solution Concentration (mg/L)	Ref
$Bi_{12}O_{17}Cl_2$ $Co_3O_4/Bi_{12}O_{17}Cl_2$	0.0115 min^{-1} 0.021 min^{-1}	83.5% 97.4%	140	300 W Xenon lamp, (420–720 nm)	20	40	20	This work
$Bi_{12}O_{17}Cl_2$ $Ag/Bi_{12}O_{17}Cl_2$	1.93 × 10^{-3} min^{-1} 10.3 × 10^{-3} min^{-1}	40% 93%	240	300 W Xenon lamp, (420–720 nm)	40	80	15	[52]
$Bi_{12}O_{17}Cl_2$ Bi-$Bi_{12}O_{17}Cl_2$	1.16 × 10^{-2} min^{-1} 2.19 × 10^{-2} min^{-1}	83% 99%	150	300 W Xenon lamp, (400–780 nm)	80	80	20	[53]
$Bi_{12}O_{17}Cl_2$ Fe(III)-modified $Bi_{12}O_{17}Cl_2$	0.074 min^{-1} 0.157 min^{-1}	– –	20	300 W Xenon lamp, (400 nm)	10	50	10	[54]
$Bi_{12}O_{17}Cl_2$ Graphene/$Bi_{12}O_{17}Cl_2$	0.0613 min^{-1} 0.160 min^{-1}	78% –	20	300 W Xenon lamp, (400 nm)	30	50	10	[23]
$Bi_{12}O_{17}Cl_2$ 2D/2D g-C_3N_4/$Bi_{12}O_{17}Cl_2$	0.102 min^{-1} 0.353 min^{-1}	83% 99%	20	300 W Xenon lamp, (400 nm)	30	50	5	[55]
ZnO/NiO	0.019 min^{-1}	–	120	320 mW/cm^2. LED light	50	50	10	[56]
g-C_3N_4	1.11 min^{-1}	99%	90	150 watt halogen lamp	50	100	5	[57]
Carbon dot implanted g-C_3N_4	0.48 min^{-1}	–	80	300 W Xenon lamp, (320–780 nm)	10	30	15	[58]
$BiVO_4$-Ni/$AgVO_3$	0.133 min^{-1}	–	30	300 W Xenon lamp, (400 nm)	30	50	10	[59]
Fe-BiOBr + H_2O_2	0.0646 min^{-1}	98.23%	60	350 W xenon lamp (420 nm)	30	50	20	[60]
$SrSnO_3$/g-C_3N_4	k1: 0.0083^{-1} k2: 0.0348^{-1}	97.3	250	Sun Light	100	100	5	[61]

References

1. Al-Buriahi, A.K.; Al-Gheethi, A.A.; Senthil Kumar, P.; Radin Mohamed, R.M.S.; Yusof, H.; Alshalif, A.F.; Khalifa, N.A. Elimination of Rhodamine B from Textile Wastewater Using Nanoparticle Photocatalysts: A Review for Sustainable Approaches. *Chemosphere* **2022**, *287*, 132162. [CrossRef]
2. Corrales, J.; Kristofco, L.A.; Steele, W.B.; Yates, B.S.; Breed, C.S.; Williams, E.S.; Brooks, B.W. Global Assessment of Bisphenol A in the Environment: Review and Analysis of Its Occurrence and Bioaccumulation. *Dose-Response* **2015**, *13*, 1559325815598308. [CrossRef]
3. Wright-Walters, M.; Volz, C.; Talbott, E.; Davis, D. An Updated Weight of Evidence Approach to the Aquatic Hazard Assessment of Bisphenol A and the Derivation a New Predicted No Effect Concentration (Pnec) Using a Non-Parametric Methodology. *Sci. Total Environ.* **2011**, *409*, 676–685. [CrossRef]
4. Sadeghfar, F.; Zalipour, Z.; Taghizadeh, M.; Taghizadeh, A.; Ghaedi, M. Chapter 2—Photodegradation Processes. In *Interface Science and Technology*; Ghaedi, M., Ed.; Photocatalysis: Fundamental Processes and Applications; Elsevier: Amsterdam, The Netherlands, 2021; Volume 32, pp. 55–124.
5. Taghizadeh, A.; Taghizadeh, M.; Sabzehmeidani, M.M.; Sadeghfar, F.; Ghaedi, M. Chapter 1—Electronic Structure: From Basic Principles to Photocatalysis. In *Interface Science and Technology*; Ghaedi, M., Ed.; Photocatalysis: Fundamental Processes and Applications; Elsevier: Amsterdam, The Netherlands, 2021; Volume 32, pp. 1–53.
6. Kumar, R.; Raizada, P.; Khan, A.A.P.; Nguyen, V.-H.; Van Le, Q.; Ghotekar, S.; Selvasembian, R.; Gandhi, V.; Singh, A.; Singh, P. Recent Progress in Emerging $BiPO_4$-Based Photocatalysts: Synthesis, Properties, Modification Strategies, and Photocatalytic Applications. *J. Mater. Sci. Technol.* **2022**, *108*, 208–225. [CrossRef]
7. Li, L.; Gao, H.; Liu, G.; Wang, S.; Yi, Z.; Wu, X.; Yang, H. Synthesis of Carnation Flower-like $Bi_2O_2CO_3$ Photocatalyst and Its Promising Application for Photoreduction of Cr(VI). *Adv. Powder Technol.* **2022**, *33*, 103481. [CrossRef]
8. Li, L.; Gao, H.; Yi, Z.; Wang, S.; Wu, X.; Li, R.; Yang, H. Comparative Investigation on Synthesis, Morphological Tailoring and Photocatalytic Activities of $Bi_2O_2CO_3$ Nanostructures. *Colloids Surf. A: Physicochem. Eng. Asp.* **2022**, *644*, 128758. [CrossRef]
9. Li, L.; Sun, X.; Xian, T.; Gao, H.; Wang, S.; Yi, Z.; Wu, X.; Yang, H. Template-Free Synthesis of $Bi_2O_2CO_3$ Hierarchical Nanotubes Self-Assembled from Ordered Nanoplates for Promising Photocatalytic Applications. *Phys. Chem. Chem. Phys.* **2022**, *24*, 8279–8295. [CrossRef]
10. Cheng, T.; Gao, H.; Liu, G.; Pu, Z.; Wang, S.; Yi, Z.; Wu, X.; Yang, H. Preparation of Core-Shell Heterojunction Photocatalysts by Coating CdS Nanoparticles onto $Bi_4Ti_3O_{12}$ Hierarchical Microspheres and Their Photocatalytic Removal of Organic Pollutants and Cr(VI) Ions. *Colloids Surf. A Physicochem. Eng. Asp.* **2022**, *633*, 127918. [CrossRef]
11. Guo, J.; Shi, L.; Zhao, J.; Wang, Y.; Tang, K.; Zhang, W.; Xie, C.; Yuan, X. Enhanced Visible-Light Photocatalytic Activity of Bi_2MoO_6 Nanoplates with Heterogeneous Bi_2MoO_6-X@Bi_2MoO_6 Core-Shell Structure. *Appl. Catal. B Environ.* **2018**, *224*, 692–704. [CrossRef]
12. Huang, C.; Chen, L.; Li, H.; Mu, Y.; Yang, Z. Synthesis and Application of Bi_2WO_6 for the Photocatalytic Degradation of Two Typical Fluoroquinolones under Visible Light Irradiation. *RSC Adv.* **2019**, *9*, 27768–27779. [CrossRef] [PubMed]
13. Shi, Q.; Zhang, Y.; Sun, D.; Zhang, S.; Tang, T.; Zhang, X.; Cao, S. Bi_2O_3-Sensitized TiO_2 Hollow Photocatalyst Drives the Efficient Removal of Tetracyclines under Visible Light. *Inorg. Chem.* **2020**, *59*, 18131–18140. [CrossRef]
14. Guo, J.; Li, X.; Liang, J.; Yuan, X.; Jiang, L.; Yu, H.; Sun, H.; Zhu, Z.; Ye, S.; Tang, N.; et al. Fabrication and Regulation of Vacancy-Mediated Bismuth Oxyhalide towards Photocatalytic Application: Development Status and Tendency. *Coord. Chem. Rev.* **2021**, *443*, 214033. [CrossRef]
15. Zhu, G.; Hojamberdiev, M.; Zhang, S.; Din, S.T.U.; Yang, W. Enhancing Visible-Light-Induced Photocatalytic Activity of BiOI Microspheres for NO Removal by Synchronous Coupling with Bi Metal and Graphene. *Appl. Surf. Sci.* **2019**, *467–468*, 968–978. [CrossRef]
16. Xiao, X.; Jiang, J.; Zhang, L. Selective Oxidation of Benzyl Alcohol into Benzaldehyde over Semiconductors under Visible Light: The Case of $Bi_{12}O_{17}Cl_2$ Nanobelts. *Appl. Catal. B Environ.* **2013**, *142–143*, 487–493. [CrossRef]
17. Wang, C.-Y.; Zhang, X.; Qiu, H.-B.; Wang, W.-K.; Huang, G.-X.; Jiang, J.; Yu, H.-Q. Photocatalytic Degradation of Bisphenol A by Oxygen-Rich and Highly Visible-Light Responsive $Bi_{12}O_{17}Cl_2$ Nanobelts. *Appl. Catal. B Environ.* **2017**, *200*, 659–665. [CrossRef]
18. Liu, X.; Xing, Y.; Liu, Z.; Du, C. Enhanced Photocatalytic Activity of $Bi_{12}O_{17}Cl_2$ Preferentially Oriented Growth along [200] with Various Surfactants. *J. Mater. Sci.* **2018**, *53*, 14217–14230. [CrossRef]
19. Fang, K.; Shi, L.; Wang, F.; Yao, L. The Synthesis of 3D $Bi_{12}O_{17}Cl_2$ Hierarchical Structure with Visible-Light Photocatalytic Activity. *Mater. Lett.* **2020**, *277*, 128352. [CrossRef]
20. Zhang, Y.; Di, J.; Zhu, X.; Ji, M.; Chen, C.; Liu, Y.; Li, L.; Wei, T.; Li, H.; Xia, J. Chemical Bonding Interface in $Bi_2Sn_2O_7$/BiOBr S-Scheme Heterojunction Triggering Efficient N_2 Photofixation. *Appl. Catal. B Environ.* **2023**, *323*, 122148. [CrossRef]
21. Guo, M.; He, H.; Cao, J.; Lin, H.; Chen, S. Novel I-Doped $Bi_{12}O_{17}Cl_2$ Photocatalysts with Enhanced Photocatalytic Activity for Contaminants Removal. *Mater. Res. Bull.* **2019**, *112*, 205–212. [CrossRef]
22. Zhang, M.; Bi, C.; Lin, H.; Cao, J.; Chen, S. Construction of Novel Au/$Bi_{12}O_{17}Cl_2$ Composite with Intensive Visible Light Activity Enhancement for Contaminants Removal. *Mater. Lett.* **2017**, *191*, 132–135. [CrossRef]
23. Ma, J.; Shi, L.; Hou, L.; Yao, L.; Lu, C.; Geng, Z. Fabrication of Graphene/$Bi_{12}O_{17}Cl_2$ as an Effective Visible-Light Photocatalyst. *Mater. Res. Bull.* **2020**, *122*, 110690. [CrossRef]

24. He, G.; Xing, C.; Xiao, X.; Hu, R.; Zuo, X.; Nan, J. Facile Synthesis of Flower-like $Bi_{12}O_{17}Cl_2/\beta$-Bi_2O_3 Composites with Enhanced Visible Light Photocatalytic Performance for the Degradation of 4-Tert-Butylphenol. *Appl. Catal. B Environ.* **2015**, *170–171*, 1–9. [CrossRef]
25. In Situ Assembly of $BiOI@Bi_{12}O_{17}Cl_2$ P-n Junction: Charge Induced Unique Front-Lateral Surfaces Coupling Heterostructure with High Exposure of BiOI {001} Active Facets for Robust and Nonselective Photocatalysis. *Appl. Catal. B Environ.* **2016**, *199*, 75–86. [CrossRef]
26. Guo, J.; Sun, H.; Yuan, X.; Jiang, L.; Wu, Z.; Yu, H.; Tang, N.; Yu, M.; Yan, M.; Liang, J. Photocatalytic Degradation of Persistent Organic Pollutants by Co-Cl Bond Reinforced $CoAl$-$LDH/Bi_{12}O_{17}Cl_2$ Photocatalyst: Mechanism and Application Prospect Evaluation. *Water Res.* **2022**, *219*, 118558. [CrossRef] [PubMed]
27. Liu, T.; Shi, L.; Wang, Z.; Liu, D. Preparation of $Ag_2O/Bi_{12}O_{17}Cl_2$ p-n Junction Photocatalyst and Its Photocatalytic Performance under Visible and Infrared Light. *Colloids Surf. A Physicochem. Eng. Asp.* **2022**, *632*, 127811. [CrossRef]
28. Kim, K.-H.; Choi, Y.-H. Surface Oxidation of Cobalt Carbonate and Oxide Nanowires by Electrocatalytic Oxygen Evolution Reaction in Alkaline Solution. *Mater. Res. Express* **2022**, *9*, 034001. [CrossRef]
29. Wang, Y.; Zhu, C.; Zuo, G.; Guo, Y.; Xiao, W.; Dai, Y.; Kong, J.; Xu, X.; Zhou, Y.; Xie, A.; et al. 0D/2D Co_3O_4/TiO_2 Z-Scheme Heterojunction for Boosted Photocatalytic Degradation and Mechanism Investigation. *Appl. Catal. B Environ.* **2020**, *278*, 119298. [CrossRef]
30. Dai, X.; Cui, L.; Yao, L.; Shi, L. Facile Construction of Novel $Co_3O_4/Bi_{12}O_{17}Cl_2$ Heterojunction Composites with Enhanced Photocatalytic Performance. *J. Solid State Chem.* **2021**, *297*, 122066. [CrossRef]
31. Quan, Y.; Wang, B.; Liu, G.; Li, H.; Xia, J. Carbonized Polymer Dots Modified Ultrathin $Bi_{12}O_{17}Cl_2$ Nanosheets Z-Scheme Heterojunction for Robust CO_2 Photoreduction. *Chem. Eng. Sci.* **2021**, *232*, 116338. [CrossRef]
32. Wang, L.; Min, X.; Sui, X.; Chen, J.; Wang, Y. Facile Construction of Novel $BiOBr/Bi_{12}O_{17}Cl_2$ Heterojunction Composites with Enhanced Photocatalytic Performance. *J. Colloid Interface Sci.* **2020**, *560*, 21–33. [CrossRef]
33. Zhu, J.; Fan, J.; Cheng, T.; Cao, M.; Sun, Z.; Zhou, R.; Huang, L.; Wang, D.; Li, Y.; Wu, Y. Bilayer Nanosheets of Unusual Stoichiometric Bismuth Oxychloride for Potassium Ion Storage and CO_2 Reduction. *Nano Energy* **2020**, *75*, 104939. [CrossRef]
34. Xu, Y.; Ma, Y.; Ji, X.; Huang, S.; Xia, J.; Xie, M.; Yan, J.; Xu, H.; Li, H. Conjugated Conducting Polymers PANI Decorated $Bi_{12}O_{17}Cl_2$ Photocatalyst with Extended Light Response Range and Enhanced Photoactivity. *Appl. Surf. Sci.* **2019**, *464*, 552–561. [CrossRef]
35. Superior Visible Light Hydrogen Evolution of Janus Bilayer Junctions via Atomic-Level Charge Flow Steering | Nature Communications. Available online: https://www.nature.com/articles/ncomms11480 (accessed on 15 October 2022).
36. Diallo, A.; Beye, A.C.; Doyle, T.B.; Park, E.; Maaza, M. Green Synthesis of Co_3O_4 Nanoparticles via Aspalathus Linearis: Physical Properties. *Green Chem. Lett. Rev.* **2015**, *8*, 30–36. [CrossRef]
37. Wang, Y.; Wei, X.; Hu, X.; Zhou, W.; Zhao, Y. Effect of Formic Acid Treatment on the Structure and Catalytic Activity of Co_3O_4 for N_2O Decomposition. *Catal. Lett.* **2019**, *149*, 1026–1036. [CrossRef]
38. Lopes Matias, J.A.; Sabino da Silva, E.B.; Raimundo, R.A.; Ribeiro da Silva, D.; Oliveira, J.B.L.; Morales, M.A. $(Bi_{13}Co_{11})Co_2O_{40}$–$Co_3O_4$ Composites: Synthesis, Structural and Magnetic Properties. *J. Alloy. Compd.* **2021**, *852*, 156991. [CrossRef]
39. Gu, Y.; Guo, B.; Yi, Z.; Wu, X.; Zhang, J.; Yang, H. Synthesis of a Self-Assembled Dual Morphologies Ag-$NPs/SrMoO_4$ Photocatalyst with LSPR Effect for the Degradation of Methylene Blue Dye. *ChemistrySelect* **2022**, *7*, e202201274. [CrossRef]
40. Urgunde, A.B.; Kamboj, V.; Kannattil, H.P.; Gupta, R. Layer-by-Layer Coating of Cobalt-Based Ink for Large-Scale Fabrication of OER Electrocatalyst. *Energy Technol.* **2019**, *7*, 1900603. [CrossRef]
41. Din, S.T.U.; Lee, H.; Yang, W. Z-Scheme Heterojunction of 3-Dimensional Hierarchical Bi_3O_4Cl/Bi_5O_7I for a Significant Enhancement in the Photocatalytic Degradation of Organic Pollutants (RhB and BPA). *Nanomaterials* **2022**, *12*, 767. [CrossRef]
42. Li, R.; Hu, B.; Yu, T.; Chen, H.; Wang, Y.; Song, S. Insights into Correlation among Surface-Structure-Activity of Cobalt-Derived Pre-Catalyst for Oxygen Evolution Reaction. *Adv. Sci.* **2020**, *7*, 1902830. [CrossRef]
43. Yu, L.; Lei, T.; Nan, B.; Kang, J.; Jiang, Y.; He, Y.; Liu, C.T. Mo Doped Porous Ni–Cu Alloy as Cathode for Hydrogen Evolution Reaction in Alkaline Solution. *RSC Adv.* **2015**, *5*, 82078–82086. [CrossRef]
44. Yang, Y.; Zhao, S.; Bi, F.; Chen, J.; Wang, Y.; Cui, L.; Xu, J.; Zhang, X. Highly Efficient Photothermal Catalysis of Toluene over Co_3O_4/TiO_2 p-n Heterojunction: The Crucial Roles of Interface Defects and Band Structure. *Appl. Catal. B Environ.* **2022**, *315*, 121550. [CrossRef]
45. Rational Design of Carbon-Doped Carbon Nitride/$Bi12O17Cl2$ Composites: A Promising Candidate Photocatalyst for Boosting Visible-Light-Driven Photocatalytic Degradation of Tetracycline | ACS Sustainable Chemistry & Engineering. Available online: https://pubs.acs.org/doi/full/10.1021/acssuschemeng.8b00782 (accessed on 2 November 2022).
46. Guerrero-Araque, D.; Acevedo-Peña, P.; Ramírez-Ortega, D.; Lartundo-Rojas, L.; Gómez, R. SnO_2–TiO_2 Structures and the Effect of CuO, CoO Metal Oxide on Photocatalytic Hydrogen Production. *J. Chem. Technol. Biotechnol.* **2017**, *92*, 1531–1539. [CrossRef]
47. Patel, M.; Park, W.-H.; Ray, A.; Kim, J.; Lee, J.-H. Photoelectrocatalytic Sea Water Splitting Using Kirkendall Diffusion Grown Functional Co_3O_4 Film. *Sol. Energy Mater. Sol. Cells* **2017**, *171*, 267–274. [CrossRef]
48. Zhu, G.; Hojamberdiev, M.; Zhang, W.; Taj Ud Din, S.; Joong Kim, Y.; Lee, J.; Yang, W. Enhanced Photocatalytic Activity of Fe-Doped $Bi_4O_5Br_2$ Nanosheets Decorated with Au Nanoparticles for Pollutants Removal. *Appl. Surf. Sci.* **2020**, *526*, 146760. [CrossRef]
49. Xiong, J.; Cheng, G.; Li, G.; Qin, F.; Chen, R. Well-Crystallized Square-like 2D BiOCl Nanoplates: Mannitol-Assisted Hydrothermal Synthesis and Improved Visible-Light-Driven Photocatalytic Performance. *RSC Adv.* **2011**, *1*, 1542–1553. [CrossRef]

50. Kato, D.; Tomita, O.; Nelson, R.; Kirsanova, M.A.; Dronskowski, R.; Suzuki, H.; Zhong, C.; Tassel, C.; Ishida, K.; Matsuzaki, Y.; et al. Bi12O17Cl2 with a Sextuple Bi—O Layer Composed of Rock-Salt and Fluorite Units and Its Structural Conversion through Fluorination to Enhance Photocatalytic Activity. Available online: https://onlinelibrary.wiley.com/doi/10.1002/adfm.202204112 (accessed on 6 October 2022).
51. Chen, Y.; Liu, G.; Dong, L.; Liu, X.; Liu, M.; Wang, X.; Gao, C.; Wang, G.; Teng, Z.; Yang, W.; et al. A Microwave-Assisted Solvothermal Method to Synthesize BiOCl Microflowers with Oxygen Vacancies and Their Enhanced Photocatalytic Performance. *J. Alloy. Compd.* **2023**, *930*, 167331. [CrossRef]
52. Chang, F.; Wang, X.; Luo, J.; Wang, J.; Xie, Y.; Deng, B.; Hu, X. Ag/$Bi_{12}O_{17}Cl_2$ Composite: A Case Study of Visible-Light-Driven Plasmonic Photocatalyst. *Mol. Catal.* **2017**, *427*, 45–53. [CrossRef]
53. Chang, F.; Lei, B.; Zhang, X.; Xu, Q.; Chen, H.; Deng, B.; Hu, X. The Reinforced Photocatalytic Performance of Binary-Phased Composites Bi-$Bi_{12}O_{17}Cl_2$ Fabricated by a Facile Chemical Reduction Protocol. *Colloids Surf. A Physicochem. Eng. Asp.* **2019**, *572*, 290–298. [CrossRef]
54. Meng, X.; Shi, L.; Yao, L.; Zhang, Y.; Cui, L. Fe (III) Clusters Modified $Bi_{12}O_{17}Cl_2$ Nanosheets Photocatalyst for Boosting Photocatalytic Performance through Interfacial Charge Transfer Effect. *Colloids Surf. A Physicochem. Eng. Asp.* **2020**, *594*, 124658. [CrossRef]
55. Shi, L.; Si, W.; Wang, F.; Qi, W. Construction of 2D/2D Layered g-C_3N_4/$Bi_{12}O_{17}Cl_2$ Hybrid Material with Matched Energy Band Structure and Its Improved Photocatalytic Performance. *RSC Adv.* **2018**, *8*, 24500–24508. [CrossRef]
56. Ma, L.; Ai, X.; Chen, Y.; Liu, P.; Lin, C.; Lu, K.; Jiang, W.; Wu, J.; Song, X. Improved Photocatalytic Activity via N-Type ZnO/p-Type NiO Heterojunctions. *Nanomaterials* **2022**, *12*, 3665. [CrossRef] [PubMed]
57. Baranowska, D.; Kędzierski, T.; Aleksandrzak, M.; Mijowska, E.; Zielińska, B. Influence of Hydrogenation on Morphology, Chemical Structure and Photocatalytic Efficiency of Graphitic Carbon Nitride. *Int. J. Mol. Sci.* **2021**, *22*, 13096. [CrossRef] [PubMed]
58. Zhang, L.; Zhang, J.; Xia, Y.; Xun, M.; Chen, H.; Liu, X.; Yin, X. Metal-Free Carbon Quantum Dots Implant Graphitic Carbon Nitride: Enhanced Photocatalytic Dye Wastewater Purification with Simultaneous Hydrogen Production. *Int. J. Mol. Sci.* **2020**, *21*, 1052. [CrossRef] [PubMed]
59. Gao, Y.; Liu, F.; Chi, X.; Tian, Y.; Zhu, Z.; Guan, R.; Song, J. A Mesoporous Nanofibrous $BiVO_4$-Ni/$AgVO_3$ Z-Scheme Heterojunction Photocatalyst with Enhanced Photocatalytic Reduction of Cr6+ and Degradation of RhB under Visible Light. *Appl. Surf. Sci.* **2022**, *603*, 154416. [CrossRef]
60. An, W.; Wang, H.; Yang, T.; Xu, J.; Wang, Y.; Liu, D.; Hu, J.; Cui, W.; Liang, Y. Enriched Photocatalysis-Fenton Synergistic Degradation of Organic Pollutants and Coking Wastewater via Surface Oxygen Vacancies over Fe-BiOBr Composites. *Chem. Eng. J.* **2023**, *451*, 138653. [CrossRef]
61. de Sousa Filho, I.A.; Arana, L.R.; Doungmo, G.; Grisolia, C.K.; Terrashke, H.; Weber, I.T. $SrSnO_3$/g-C_3N_4 and Sunlight: Photocatalytic Activity and Toxicity of Degradation Byproducts. *J. Environ. Chem. Eng.* **2020**, *8*, 103633. [CrossRef]

Article

Rhombohedral/Cubic In$_2$O$_3$ Phase Junction Hybridized with Polymeric Carbon Nitride for Photodegradation of Organic Pollutants

Xiaorong Cai [1,†], Yaning Wang [1], Shuting Tang [1], Liuye Mo [1], Zhe Leng [1,*], Yixian Zang [1], Fei Jing [1,†] and Shaohong Zang [1,2,*]

[1] Institute of Innovation & Application, National Engineering Research Center For Marine Aquaculture, Zhejiang Ocean University, Zhoushan 316022, China
[2] Donghai Laboratory, Zhoushan 316021, China
* Correspondence: lengzhe@zjou.edu.cn (Z.L.); shzang@zjou.edu.cn (S.Z.)
† These authors contributed equally to this work.

Abstract: In recent studies, phase junctions constructed as photocatalysts have been found to possess great prospects for organic degradation with visible light. In this study, we designed an elaborate rhombohedral corundum/cubic In$_2$O$_3$ phase junction (named MIO) combined with polymeric carbon nitride (PCN) via an in situ calcination method. The performance of the MIO/PCN composites was measured by photodegradation of Rhodamine B under LED light (λ = 420 nm) irradiation. The excellent performance of MIO/PCN could be attributed to the intimate interface contact between MIO and PCN, which provides a reliable charge transmission channel, thereby improving the separation efficiency of charge carriers. Photocatalytic degradation experiments with different quenchers were also executed. The results suggest that the superoxide anion radicals (O$_2^-$) and hydroxyl radicals (·OH) played the main roles in the reaction, as opposed to the other scavengers. Moreover, the stability of the MIO/PCN composites was particularly good in the four cycling photocatalytic reactions. This work illustrates that MOF-modified materials have great potential for solving environmental pollution without creating secondary pollution.

Keywords: photocatalysis; photodegradation; carbon nitride; phase junction; MOF

1. Introduction

Organic dye pollutants in wastewater have constantly been a concern with the development of society [1–3]. Their toxicity and carcinogenicity always threaten ecological balance and biological health. The processing methods of organic dyes generally include adsorption, physical/chemical precipitation, biological methods, and photodegradation. Among them, solar-driven degradation by semiconductor photocatalysts has great potential in resolving organic dye pollution due to its convenience, eco-friendliness, and low cost [4,5]. It is well known that the most important part of photocatalysis is the catalysts because they accelerate the reaction process and improve the degradation efficiency in organic dye degradation reactions [6,7]. Therefore, photocatalysts working under visible light irradiation with outstanding photodegradation efficiency still need further exploration.

As an organic representative, polymeric carbon nitride (PCN) is a star material in photocatalysis owing to its suitable energy band position, excellent stability, and simple synthetic applications [8–11]. The extended π-conjugated systems consisting by sp^2-hybridized C and N atoms have been widely used for studies on energy and the environment [12]. The suitable bandgap of PCN (2.7 eV) enables it to harvest visible light and surmount the endothermic character of water-splitting reactions (theoretically, 1.23 eV) [13]. However, as a non-metal photocatalyst, the insufficient capacity of charge carrier transfer results in the unsatisfactory photocatalytic ability of PCN [12,14]. Many strategies have been used to remedy this issue, such as morphological control, element doping, cocatalyst loading, band structure engineering, and heterojunction construction [15–18]. Among various methods, heterojunction construction has been proven to be an easy yet effective method to accelerate the migration of charge carriers [19–21]. Therefore, coupling with proper semiconductors can also improve the photoactivity of PCN.

Previous reports revealed that the phase junction of polymorph semiconductors plays a vital role in charge separation [22,23]. Photo-induced charge transfer between two phases is driven by the built-in electric field in the phase junction, resulting in enhanced photocatalysis. Taking TiO_2 as an example, Li's group established that the photocatalytic activity is directly influenced by the surface phase structure [24,25]. And the phase junction consisting of anatase and rutile particles performed better in photocatalytic H_2 evolution. Hao et al. fabricated a novel CdS phase junction with bonding region-width-control and resolved photocorrosion and phase exclusion of CdS [26]. The best performance reached as much as 60-fold that of the single cubic or hexagonal phase. Liu et al. reported a black/red phase junction phosphorus with faster charge transport properties benefiting from the appropriate band structures [27]. The theoretical and experimental data have indicated that different kinds of phase junction materials stand out in photo-to-electron conversion efficiency.

Recently, a metal organic framework (MOF)-derived rhombohedral/cubic In_2O_3 phase junction (named MIO) was reported for solar-driven water splitting [28]. Because it has no cytotoxicity or cellular ROS generation, and is easy to obtain, MIO has exhibited a growth potential in photocatalysis, such as H_2 production, CO_2 reduction, and pollutant degradation [29]. Theoretical calculations have illustrated that the photo-generated electrons transfer from $c-In_2O_3$ to $rh-In_2O_3$ was efficient in preventing the recombination of charge carriers. Previous research has mainly focused on cubic In_2O_3-based semiconductor ($c-In_2O_3$), but rarely on its phase junction [30,31]. For example, Wang et al. designed $ZnIn_2S_4-In_2O_3$ nanotubes with good stability for CO_2 reduction [32]. Li's group reported core-shell In_2O_3@Carbon nanoparticles for photocatalytic hydrogen evolution [33]. The improved accessibility between $c-In_2O_3$ and carbon nanoparticles not only favored the efficient separation of charge carriers, but also enhanced the optical absorption. Sun et al. synthesized $G-C_3N_4/In_2O_3$ composites for effective formaldehyde detection [34]. Xu et al. reported a carbon-doped $In_2O_3/g-C_3N_4$ heterojunction for photoreduction of CO_2 [35]. Jin and Uddin et al. reported $c-In_2O_3$ hybridization with boron-doped and oxygen-doped carbon nitride for photodegradation, respectively [36,37]. Both of them showed a superior kinetic degradation rate rather than either In_2O_3 or PCN alone. Although the investigation of cubic In_2O_3-heterojunction in photocatalysis has made some progress, the In_2O_3 phase junction has not been looked at in detail.

In this study, we designed a MOF-derived phase junction In_2O_3/PCN (named MIO/PCN, Scheme 1) heterojunction prepared by an in situ method in which the two precursors of MIO and PCN were mixed and then calcined at 500 °C. The MIO/PCN heterojunctions exhibited better visible light absorption, more active sites, and faster charge transfer. The optimal photodegradation activity of MB by MIO/PCN was about 95- and 19-fold that of the MIO and PCN, respectively. In addition, the MIO/PCN composites exhibited excellent stability after four-cycle photodegradation. The main active species were determined to be superoxide anion radicals (O_2^-) and hydroxyl radicals ($\cdot OH$) by adding scavengers to the degradation reaction, and were further confirmed by EPR analysis. Based on UV-Vis DRS, PL/TRPL, and photoelectrochemical tests, the possible photodegradation reaction mechanism of MIO/PCN is discussed. This present strategy may promote new ideas for exploring the application field of the phase junction oxides and other related materials.

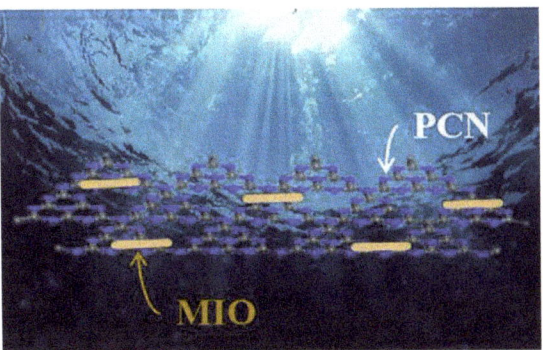

Scheme 1. Diagram of the MIO/PCN sample.

2. Results and Discussion

The crystal structure of x wt% MIO/PCN composites was confirmed by X-ray diffraction (XRD) analysis. As observed from the XRD pattern in Figure 1a, the crystal structure of the MIL-68(In)-NH$_2$ precursor and MIO were in accordance with the previous report and no other characteristic peak was detected, respectively [31]. MIO calcined at 500 °C is composed of the mixed phase of rh-In$_2$O$_3$ (PDF No. 22-0336) and c-In$_2$O$_3$ (PDF No. 06-0416) [28]. And no other peaks appeared, indicating good purity of the In$_2$O$_3$ phase-junction. In Figure 1b, there were two characteristic diffraction peaks of the pure PCN sample located at 2θ = 12.8° and 27.3°, which corresponded to the (100) and (002) planes, respectively (JCPDS card No. 87-1526). The former was assigned to the packing motif of heptazine units in plane and the latter originated from the stacking of the conjugated aromatic system in the interlayer [18]. In the XRD patterns of MIO/PCN composites, the peak intensity of MIO became more muscular while the content of MIO increased. It was identified that MIO and PCN existed in the MIO/PCN composites. In addition, the peaks of the pure PCN sample located at 2θ = 27.3° moved to high degree, indicating that the spacing d was reduced according to the Bragg equation. That is because of the interaction between PCN and MIO and is beneficial to charge transfer [8].

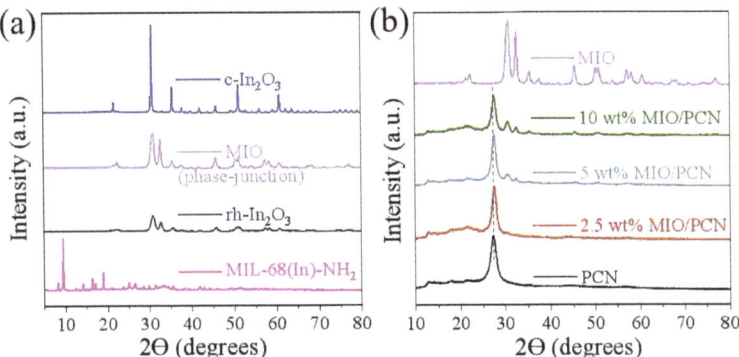

Figure 1. XRD spectra of (**a**) MIL-68(In)-NH$_2$ and different phases of In$_2$O$_3$, and (**b**) pure PCN, MIO, and x wt% MIO/PCN composites.

The FT-IR spectra of PCN, MIO, and MIO/PCN are shown in Figure 2 to illustrate the information pertaining to functional groups. In the FT-IR spectra of PCN, the broad peaks located at 3000–3600 cm^{-1} indicate the stretching vibration of O-H, N-H, and hydrogen-bonding interactions [38,39]. The characteristic peaks at 1200–1600 cm^{-1} correspond to the C-N skeleton of PCN. The breathing mode of the triazine units appeared at 800 cm^{-1}, meaning the presence of -NH and -NH$_2$ groups [39]. In that of MIO, the characteristic peaks at 570 cm^{-1} were determined to be In-O asymmetric stretching, and the peaks at 3000–3400 cm^{-1} correspond to the stretching vibration of the -OH hydrogen bond [11,34]. As shown in the FT-IR spectra of MIO/PCN, all the characteristic peaks of PCN appeared, indicating the main structure of PCN was not changed. Although the In-O peak was not observed, possibly because of the small amount of MIO in composites and the weak peak intensity, the characteristic peaks of PCN and MIO appeared simultaneously at 3000–3600 cm^{-1}. The FT-IR spectra further confirmed the formation of MIO/PCN.

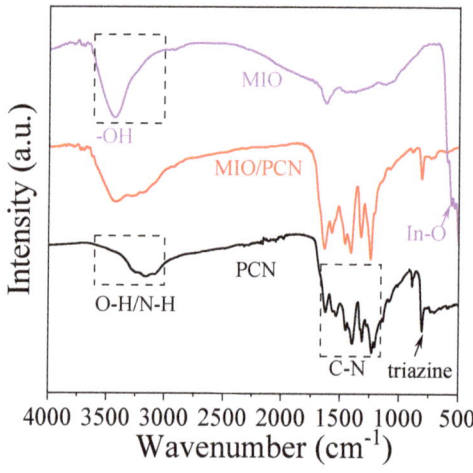

Figure 2. FT-IR spectra of pure PCN, MIO, and 2.5 wt% MIO/PCN composites.

SEM and TEM images shown in Figures 3 and S1 are to illustrate the morphology of the composites. In Figure S1, the shape of MIL-68(In)-NH$_2$ is a rod with a smooth surface and that of PCN is a nanosheet, which was in accordance with the previous reports. After in situ calcination loading, the morphology of MIO/PCN had no noticeable change compared with pure PCN, whereas MIO could not be observed. In the TEM images of MIO/PCN (Figure 3a–c), the rod-like MIO was surrounded by nanosheets of carbon nitride. This was mainly due to the in situ method, which mixed the two precursors of carbon nitride and indium oxide first, which were then calcined together. Moreover, that is why only PCN could be observed in the SEM images of MIO/PCN. In Figure 3d, the HRTEM image of MIO/PCN, two lattice fringes could be observed in MIO/PCN with layer distances of 0.274 nm and 0.292 nm, corresponding to (110) rh-In$_2$O$_3$ crystal planes and (222) c-In$_2$O$_3$ crystal planes, respectively. In addition, amorphous PCN nanosheets were observed. The EDX mapping in Figure 3e showed that C, N, In, and O were mainly distributed across the whole nanocomposites. This phenomenon may be caused by the in situ method. The analysis of XRD, FT-IR spectra, SEM, and TEM images confirmed the formation of MIO/PCN composites.

To analyze the chemical components and chemical states in 2.5 wt% MIO/PCN, XPS research was then conducted. Consistent with the EDX results, In, O, C, and N elements were all detected in the XPS survey spectra (Figure S2), which further proves the co-existence of MIO and PCN in the composites. High-resolution XPS spectrum of the elements in MIO/PCN was also carried out. In Figure 4a, there were three peaks of C 1s at the binding energy of 288.2, 286.4, and 284.8 eV. The first peak was vested in N-C=N of the triazine ring, the second peak was attributed to C-O=C bond, while the last ones were vested in the C=C group of PCN [17,38]. In Figure 4b, the peaks of N 1s were divided into three characteristic peaks. The characteristic peak at 401.0 eV was caused by uncondensed

C-N-H groups on the surface of PCN [9,10]. The last two forms of N 1s, located at 400.1 and 398.6 eV, belonged to the tertiary N-(C)$_3$ groups and sp^2-hybridized nitrogen (C=N-C) in aromatic triazine rings, respectively [39]. Both of them and sp^2-C constituted the heptazine C$_6$N$_7$ units of PCN [40]. In Figure 4c, the two XPS peaks located at 444.8 and 452.5 eV corresponded to the spin-orbit coupling of In 3d$_{5/2}$ and In 3d$_{3/2}$ of MIO [41]. The peak in Figure 4d at 532.0 eV belongs to O$_C$ orbitals fitted to chemisorbed oxygen species. The XPS analysis indicated the co-existence of MIO and PCN in the composites.

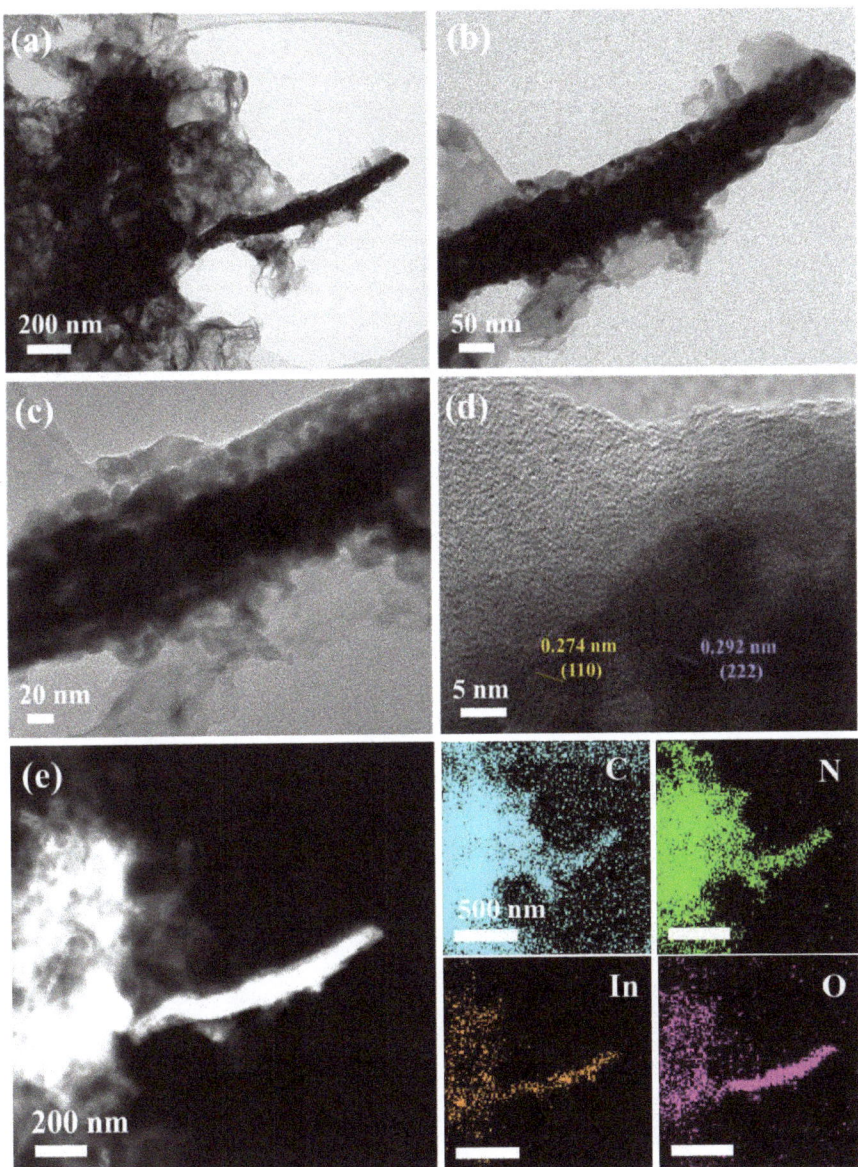

Figure 3. (**a**–**d**) TEM images, (**e**) EDX mapping of 2.5 wt% MIO/PCN composites.

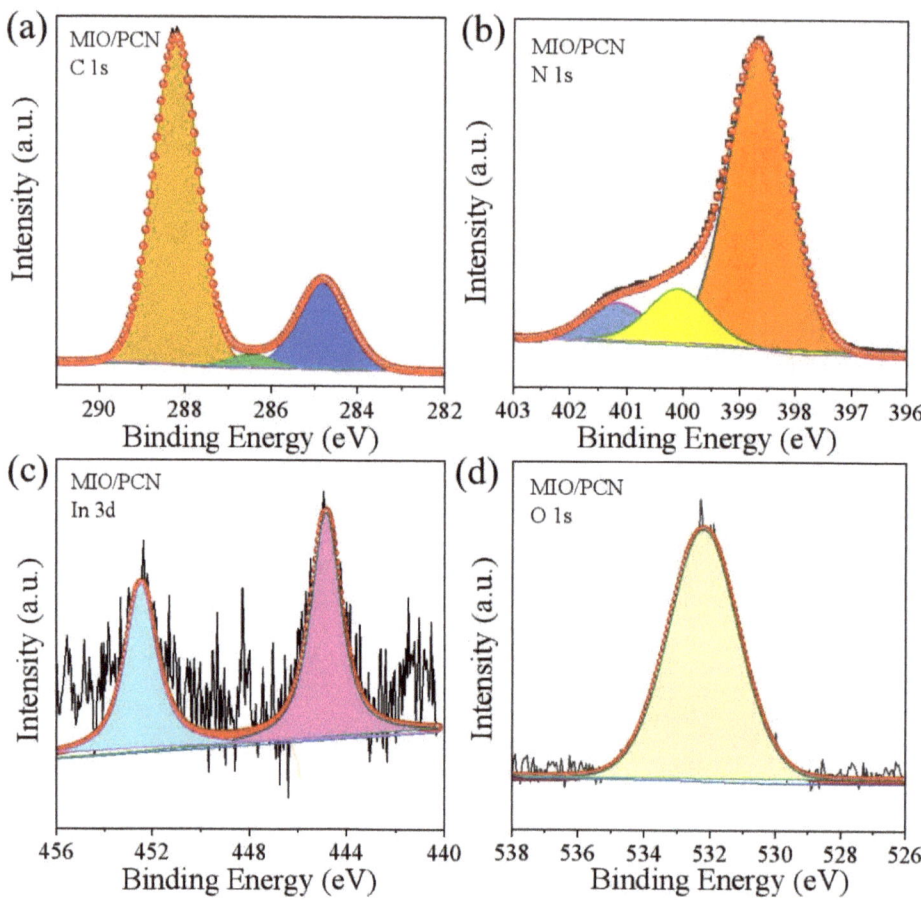

Figure 4. XPS spectra of 2.5 wt% MIO/PCN (**a**) C 1s, (**b**) N 1s, (**c**) In 3d, (**d**) O 1s.

The performance of the MIO/PCN composites was measured by photodegradation RhB under LED light irradiation (λ = 420 nm). It can be seen in Figure 5a that the adsorption of RhB on pure PCN, MIO, or x wt% MIO/PCN was faint. In the absence of a catalyst, the photodegradation of RhB hardly occurred. This further illustrated the crucial importance of a catalyst in photodegradation of RhB. However, the photocatalytic activity of the pure MIO sample showed poor degradation capacity (17%) for RhB while that of PCN was 90% in 60 min. Experimental results indicated that controlling the ratio of MIO in the MIO/PCN sample was of great importance in achieving optimal photocatalytic degradation [42]. The 2.5 wt% MIO/PCN sample exhibited the best activity, reaching almost 100% degradation within 50 min under LED light irradiation (λ = 420 nm). The photodegradation efficiency of MIO/PCN samples was obviously enhanced compared to that of pure MIO. The photocatalytic stability of the 2.5 wt% MIO/PCN sample was tested by cycle degradation of RhB (Figure 5b). After four runs of continuous reaction, the 2.5 wt% MIO/PCN sample still exhibited stable photodegradation efficiency. The experimental results indicate that the MIO/PCN sample had good stability and the enhanced photodegradation activity of RhB was due to the introduction of MIO in the composites.

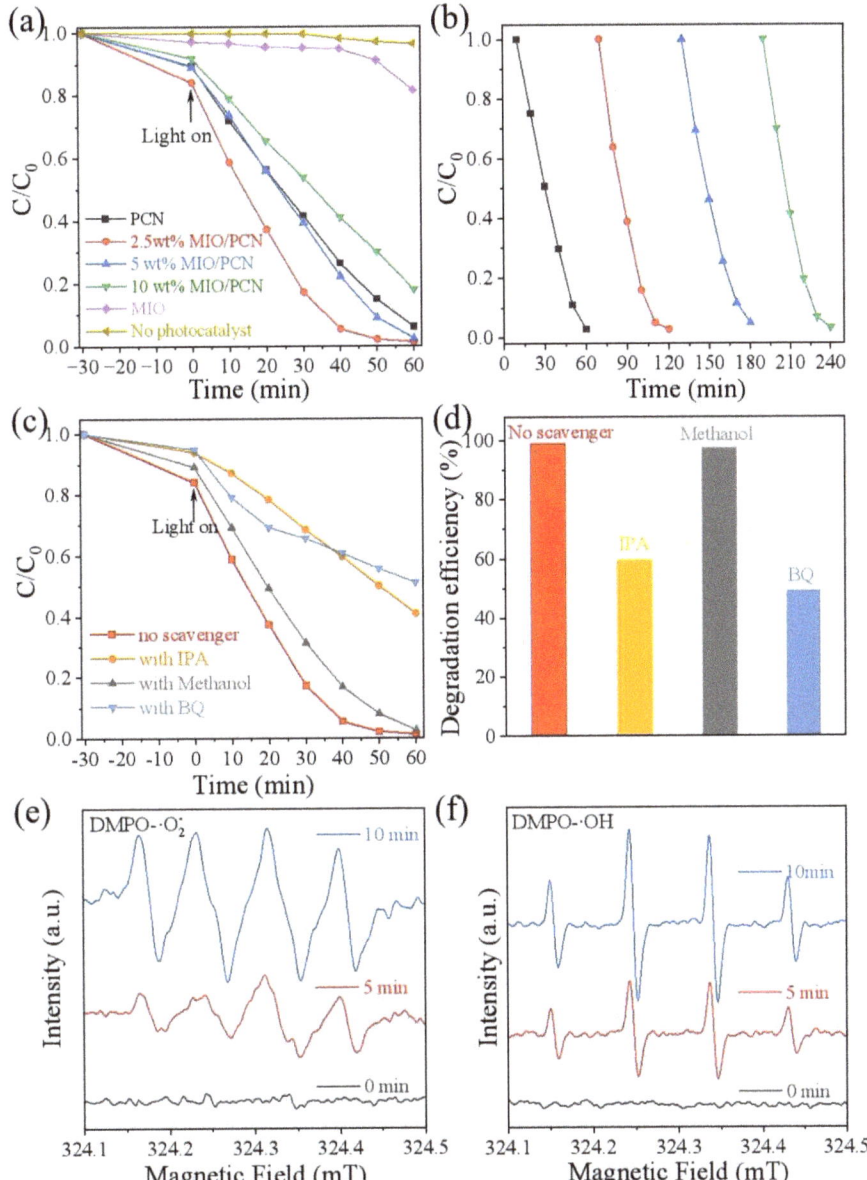

Figure 5. Photodegradation of RhB under LED light irradiation (λ = 420 nm) (**a**) by x wt% MIO/PCN samples, (**b**) four cycle tests by 2.5 wt% MIO/PCN, (**c**,**d**) different scavengers on the degradation of RhB of 2.5 wt% MIO/PCN, EPR spectra of 2.5 wt% MIO/PCN for the detection of (**e**) DMPO-·O_2^-, (**f**) DMPO-·OH.

Photodegradation of RhB occurred because of the active species produced in MIO/PCN during the reaction. Therefore, the main active species were tested under the same photocatalytic degradation condition except for adding different scavengers. The scavengers included 1, 4-benzoquinone (BQ), methyl alcohol (MeOH), and isopropyl alcohol (IPA), which capture superoxide radicals ($\cdot O_2^-$), holes (h^+), and hydroxyl radicals ($\cdot OH$), respectively [43]. Figure 5c,d shows that significantly decreased efficiency of the photodegradation occurred when BQ and IPA were added to the system. Additionally, there was no obvious change after the addition of MeOH in photocatalytic degradation. The photocatalytic experiment with scavengers suggested that $\cdot O_2^-$ and OH were the major active species in the photodegradation reaction of RhB [44].

The generation of the main active species on the MIO/PCN composite under visible-light irradiation was also probed by a 5,5-dimethyl-1-pyrroline N-oxide (DMPO) spin-trapping electron paramagnetic resonance (EPR) technique. In Figure 5e, when the light was off, no EPR signal appeared. When the light was on, a strong EPR signal of DMPO-·O_2^- appeared. Additionally, when the light was kept on for another five minutes, the signals were measurably enhanced. As shown in Figure 5f, the characteristic signals of the DMPO-·OH radical emerged after visible light irradiation. It can be concluded that both ·O_2^- and ·OH are the major reactive species in photodegradation of RhB by 2.5 wt% MIO/PCN photocatalyst. The EPR spectra is in accordance with the results of the capture experiments of active species [45]. By the above experimental studies, the possible mechanism was speculated as follows [46,47]:

$$MIO/PCN + h\nu \rightarrow e^- + h^+$$

$$e^- + O_2 \rightarrow \cdot O_2^-$$

$$h^+ + H_2O/OH^- \rightarrow \cdot OH + H^+$$

$$O_2^-, OH + RhB \rightarrow products$$

The optical and photoelectrical properties of MIO/PCN were investigated to explore the mechanism of enhanced photoactivity. In Figure 6a, there was a red shift of the DRS curve with increasing content of MIO in MIO/PCN. Broadened visible light absorption is vitally essential for increasing photocatalytic activity [48,49]. PL and TRPL spectra were used to explore the charge-carrier separation and migration behavior. It can be observed from Figure 6b that the fluorescence intensity of 2.5 wt% MIO/PCN was apparently quenched after being modified by MIO, indicating that the charge recombination was efficiently suppressed [17]. The average lifetime of PCN and 2.5 wt% MIO/PCN composite in Figure 6c was 2.54 ns and 1.32 ns. The rapid charge-carrier transfer from MIO to PCN resulted in a shorter PL lifetime [14,46]. Furthermore, the charge separation was further explored by the electrochemical impedance spectra (EIS). The evidently decreased Nyquist radius in Figure 6d illustrates that the resistance of 2.5 wt% MIO/PCN was much smaller than that of PCN [50,51]. The reduced resistance was conducive to the rapid migration of charge carriers. These results convincingly prove that the construction of MIO/PCN could promote charge separation and transfer.

In order to exclude the influence of structural change of the photocatalyst, XRD analysis was conducted. As shown in Figure 7a, the structure of the MIO/PCN did not change before or after photodegradation of RhB. This indicates that the impact of structural changes of photocatalysts on their enhanced photoactivity can be excluded. In Figure 7b, the BET surface areas of PCN and MIO were determined as 72 and 45 m^2g^{-1}, respectively, while that of the MIO/PCN sample was 65 m^2g^{-1}. Compared with pure PCN, the slightly decreased specific surface areas of MIO/PCN was due to the addition of MIL-68(In)-NH$_2$ by the in situ calcination [7,32]. The combination of the two block substances resulted in the decrease of BET-specific surface area. Therefore, the MIO/PCN composites performed well in photodegradation of RhB because of boosting solar absorption, fasting charge transfer, and being suppressed charge recombination. Based on the above discussion, a possible mechanism of charge transfer route in the MIO/PCN composites was proposed, as shown in Scheme 2. Irradiated by visible light, the electrons (e^-) excited by light in the valence band (VB) of PCN transferred to the conduction band (CB) of PCN, and then flowed to the CB of MIO. Certainly, part of the electrons and holes recombined before the reaction. In addition, the electrons that migrated to the surface of photocatalysts reacted with O_2 to form $\cdot O_2^-$. The holes in the CB reacted with H_2O/OH^- to generate $\cdot OH$. The two active species, $\cdot O_2^-/OH$, attacked the RhB pollutants and generated harmless products, even including water and carbon dioxide.

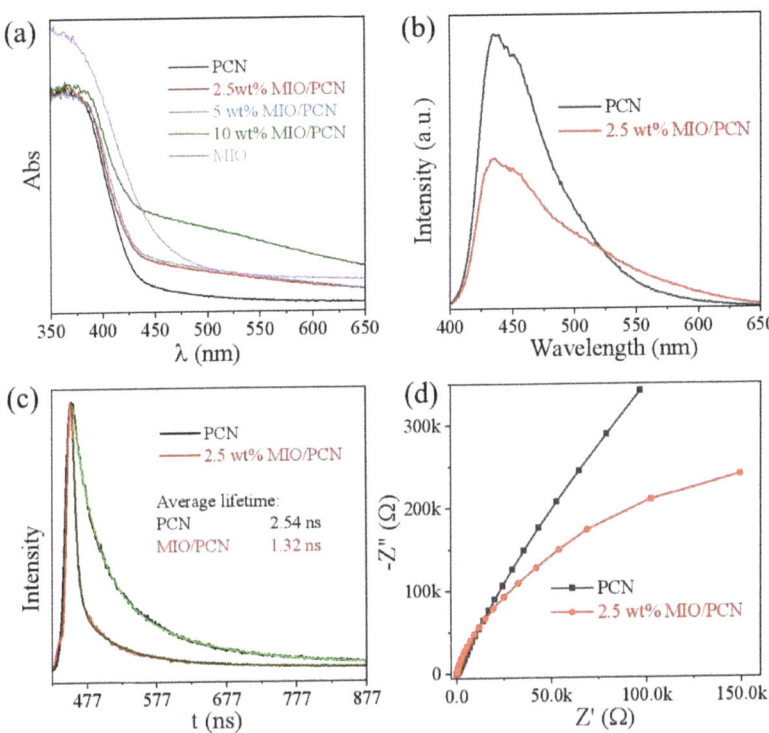

Figure 6. (**a**) UV/Vis DRS spectra, (**b**) room temperature PL spectra, (**c**) time-resolved PL spectra recorded at 298 K, (**d**) EIS Nyquist plots of pure PCN and MIO/PCN composites.

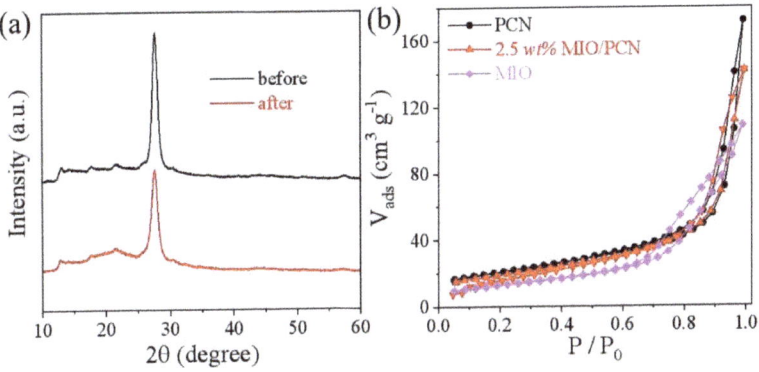

Figure 7. (**a**) XRD pattern of 2.5 wt% MIO/PCN sample before and after reaction, (**b**) low-temperature N_2 adsorption-desorption isotherms of PCN, MIO, and 2.5 wt% MIO/PCN.

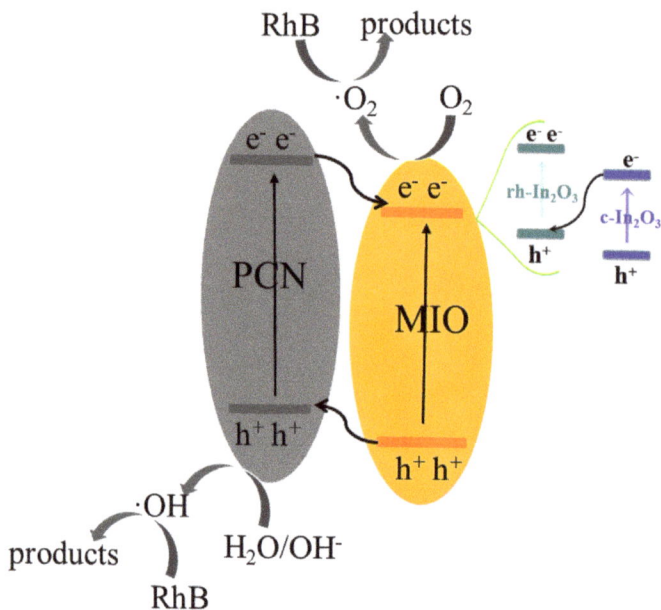

Scheme 2. Possible reaction mechanism in photocatalytic reaction by MIO/PCN.

3. Materials and Methods

Materials: All chemicals were used without further purification.

Indium nitrate hydrate (In(NO$_3$)$_3$·4H$_2$O), 2-aminoterephthalic acid, N,N-dimethyl formamide (DMF), indium chloride tetrahydrate (InCl$_3$·4H$_2$O) and urea were purchased from Sinopharm Chemical Reagent Co., Ltd. (Shanghai, China). N,N-dimethylformamide (DMF) was purchased from Macklin Biochemical Co., Ltd. (Shanghai, China). All reagents were of analytical grade and were used without further purification.

3.1. Preparation of MIL-68(In)-NH$_2$

The precursor MIL-68(In)-NH$_2$ was prepared using a hydrothermal method [26]. Typically, a certain amount of 2-aminoterephthalic acid and In(NO$_3$)$_3$·xH$_2$O were dissolved in DMF and the hydrothermal reaction occurred in an oven. After the reaction ended, the obtained products were washed with methanol and dried overnight at 80 °C.

3.2. Preparation of MIO/PCN and PCN

MIO/PCN samples were obtained using an in situ method [8]. Different amounts of MIL-68(In)-NH$_2$ mixed with urea and calcined at 500 °C for 2 h in a muffle furnace. The products were collected and named x wt% MIO/PCN, where x means the calculated conversion rate of MIO.

Pure PCN was synthesized under the same conditions without the addition of MIL-68(In)-NH$_2$.

3.3. Characterization

XRD patterns were measured on an X-ray diffractometer (D/MAX-2200, Rigaku Company) to examine the crystal structure of samples. FT-IR spectra was used to analyze the functional groups of samples by using an ALPHA-P spectrometer. The morphology and elemental mapping images of samples were characterized by field-emission scanning electron microscopy (SEM, JSM-6700F) and transmission electron microscopy (TEM, JEOL JEM 2100F). The chemical valence of the elements in the samples was obtained using X-ray photoelectron spectroscopy (XPS, Thermo ESCALAB 250XI, standard peak is the C 1s peak at 284.8 eV). Nitrogen adsorption-desorption isotherms and the Brunauer-Emmett-Teller (BET) surface areas were collected at 77 K using Micromeritics ASAP2010 equipment. The visible-light absorption of samples was measured by UV-visible diffuse reflectance spectra (UV-DRS, Cary 500 Scan Spectrophotometer, Varian, Palo Alto, CA, USA). PL was performed on Varian Cary Eclipse (Agilent, Santa Clara, CA, USA) to research the recombination of charge carriers of samples. Electrochemical impedance spectroscopy (EIS) of samples was measured on an electrochemical workstation (Shanghai chenhua) in a standard three-electrode system in which the working electrode was the FTO glass with synthesized samples (10 mg catalyst in 1 mL 1% ethanol and 0.5 mL Nafion with an active area of 1 cm^2); the reference electrode was the Ag/AgCl electrode; and the counter electrode was Pt wire. The electrolyte was 0.4 M Na_2SO_4 aqueous solution.

3.4. Photocatalytic Rhodamine b Degradation

The photocatalytic performance test was conducted under LED lamp (λ = 420 nm) irradiation. In the reaction, 10 mg samples were put into 50 mL 20 ppm rhodamine B (RhB) solution and stirred for 30 min under dark condition. Then the mixture was illuminated under a LED lamp (λ = 420 nm) for 1h. The original solution concentration was labeled C_0. During the irradiation, the suspension (2–3 mL) was taken from the dispersion every 10 min, and the clarified reaction solution (concentration C) could be obtained by filtering with a needle filter. The absorbance of the solution at 664 nm was determined using a Shimazu UV-2600 UV-Vis spectrophotometer. The degradation efficiency of the RhB solution was calculated according to the formula $D = \ln(C/C_0) \times 100\%$. The absorption method replaced C_0 and C according to the Lambert–Beer law.

4. Conclusions

In conclusion, the MIO/PCN composites were synthesized by a facile in situ method that tightly combined PCN and MIO. The successful construction of MIO/PCN was determined by XRD, TEM, and XPS analysis. The best-performing sample was determined to be the 2.5 wt% MIO/PCN composite, which could degrade RhB almost 100% within 50 min under LED light irradiation. The reason for the excellent degradation capability of MIO/PCN was revealed by DRS, PL/TRPL, and EIS analysis. Experimental data showed the phase junction MIO provided a reliable electronic transmission channel for charge transfer and improved the separation efficiency of electron and hole. The active species were identified as $\cdot O_2^-$ and $\cdot OH$ by photodegradation reaction containing scavenger and EPR. Additionally, the durability and stability of the MIO/PCN were upheld to an excellent degree after four cycle tests. Our work provides an avenue for the application of phase junction materials in photocatalysis.

Supplementary Materials: The following supporting information can be downloaded at: https://www.mdpi.com/article/10.3390/ijms232214293/s1.

Author Contributions: X.C. conducted experiments and characterization; Z.L. instructed methodology and characterization; S.T. and L.M. conducted investigation; F.J. performed characterization and analysis; Y.Z. and Y.W. polished the written paper; S.Z. designed the experiments and revised the paper. All authors have read and agreed to the published version of the manuscript.

Funding: Special thanks for the help of Feng Lin and Zhian Lan. This research was funded by the Fundamental Research Funds for Zhejiang Provincial Universities and Research Institutes (2021JZ006), the Natural Science Foundation of Zhejiang Province (LQ22B030010), National Natural Science Foundation of China (22209148), Science Foundation of Donghai Laboratory (DH-2022KF0318), Zhoushan Science and Technology Project (2021C21008, 2020C21023), and the General Projects of Zhejiang Province (Y201942789).

Institutional Review Board Statement: Not applicable.

Informed Consent Statement: Not applicable.

Data Availability Statement: Not applicable.

Conflicts of Interest: The authors declare no conflict of interest.

References

1. Zhang, L.; Song, Y.; Li, Y.; Yin, Y.; Cai, Y. Role of light in methylmercury photodegradation: From irradiation to absorption in the presence of organic ligands. *Sci. Total Environ.* **2022**, *848*, 157550. [CrossRef] [PubMed]
2. Smulek, W.; Bielan, Z.; Pacholak, A.; Zdarta, A.; Zgola-Grzeskowiak, A.; Zielinska-Jurek, A.; Kaczorek, E. Nitrofurazone Removal from Water Enhanced by Coupling Photocatalysis and Biodegradation. *Int. J. Mol. Sci.* **2021**, *22*, 2186. [CrossRef] [PubMed]
3. Liu, X.; Huang, W.-Y.; Zhou, Q.; Chen, X.-R.; Yang, K.; Li, D.; Dionysiou, D.D. Ag-decorated 3D flower-like Bi_2MoO_6/rGO with boosted photocatalytic performance for removal of organic pollutants. *Rare Met.* **2020**, *40*, 1086–1098. [CrossRef]
4. Vinothkumar, K.; Jyothi, M.; Lavanya, C.; Sakar, M.; Valiyaveettil, S.; Balakrishna, R.G. Strongly co-ordinated MOF-PSF matrix for selective adsorption, separation and photodegradation of dyes. *Chem. Eng. J.* **2022**, *428*, 132561. [CrossRef]
5. Zeng, R.; Lian, K.; Su, B.; Lu, L.; Lin, J.; Tang, D.; Lin, S.; Wang, X. Versatile Synthesis of Hollow Metal Sulfides via Reverse Cation Exchange Reactions for Photocatalytic CO_2 Reduction. *Angew. Chem. Int. Ed. Engl.* **2021**, *60*, 25055–25062. [CrossRef] [PubMed]
6. Xu, Y.; Zhang, Y.; Wang, X.; Wang, Z.; Huang, L.; Wu, H.; Ren, J.; Gu, C.; Chen, Z. Enhanced photodegradation of tylosin in the presence of natural montmorillonite: Synergistic effects of adsorption and surface hydroxyl radicals. *Sci. Total Environ.* **2022**, *855*, 158750. [CrossRef]
7. Yu, Z.; Yang, K.; Yu, C.; Lu, K.; Huang, W.; Xu, L.; Zou, L.; Wang, S.; Chen, Z.; Hu, J.; et al. Steering Unit Cell Dipole and Internal Electric Field by Highly Dispersed Er atoms Embedded into NiO for Efficient CO_2 Photoreduction. *Adv. Funct. Mater.* **2022**, *32*, 2111999. [CrossRef]
8. Divakaran, K.; Baishnisha, A.; Balakumar, V.; Perumal, K.N.; Meenakshi, C.; Kannan, R.S. Photocatalytic degradation of tetracycline under visible light using TiO_2@sulfur doped carbon nitride nanocomposite synthesized via in-situ method. *J. Environ. Chem. Eng.* **2021**, *9*, 105560. [CrossRef]
9. Zhang, L.; Zhang, J.; Xia, Y.; Xun, M.; Chen, H.; Liu, X.; Yin, X. Metal-Free Carbon Quantum Dots Implant Graphitic Carbon Nitride: Enhanced Photocatalytic Dye Wastewater Purification with Simultaneous Hydrogen Production. *Int. J. Mol. Sci.* **2020**, *21*, 1054. [CrossRef]
10. Li, S.; Cai, M.; Liu, Y.; Zhang, J.; Wang, C.; Zang, S.; Li, Y.; Zhang, P.; Li, X. In-situ constructing C_3N_5 nanosheets/Bi_2WO_6 nanodots S-scheme heterojunction with enhanced structural defects for efficiently photocatalytic removal of tetracycline and Cr(VI). *Inorg. Chem. Front.* **2022**, *9*, 2479–2497. [CrossRef]
11. Lin, L.; Lin, Z.; Zhang, J.; Cai, X.; Wang, X. Molecular-level insights on the reactive facet of carbon nitride single crystals photocatalysing overall water splitting. *Nat. Catal.* **2020**, *3*, 649–655. [CrossRef]
12. Pei, Z.; Gu, J.; Wang, Y.; Tang, Z.; Liu, Z.; Huang, Y.; Zhao, J.; Chen, Z.; Zhi, C. Component matters: Paving the roadmap toward enhanced electrocatalytic performance of graphitic C_3N_4-based catalysts via atomic tuning. *ACS Nano* **2017**, *11*, 6004–6014. [CrossRef] [PubMed]
13. Fang, Y.; Hou, Y.; Fu, X.; Wang, X. Semiconducting Polymers for Oxygen Evolution Reaction under Light Illumination. *Chem. Rev.* **2022**, *122*, 4204–4256. [CrossRef]
14. Kumar, A.; Raizada, P.; Thakur, V.; Khan, A.A.P.; Singh, N.; Singh, P. An overview on polymeric carbon nitride assisted photocatalytic CO_2 reduction: Strategically manoeuvring solar to fuel conversion efficiency. *Chem. Eng. Sci.* **2021**, *230*, 116219. [CrossRef]
15. Zhan, H.; Zhou, Q.; Li, M.; Zhou, R.; Mao, Y.; Wang, P. Photocatalytic O_2 activation and reactive oxygen species evolution by surface B-N bond for organic pollutants degradation. *Appl. Catal. B Environ.* **2022**, *310*, 121329. [CrossRef]
16. Di, G.; Zhu, Z.; Zhang, H.; Zhu, J.; Qiu, Y.; Yin, D.; Kuppers, S. Visible-light degradation of sulfonamides by Z-scheme ZnO/g-C_3N_4 heterojunctions with amorphous Fe_2O_3 as electron mediator. *J. Colloid. Interface Sci.* **2019**, *538*, 256–266. [CrossRef] [PubMed]
17. Wang, X.; Maeda, K.; Thomas, A.; Takanabe, K.; Antonietti, M. A metal-free polymeric photocatalyst for hydrogen production from water under visible light. *Nat. Mater.* **2009**, *8*, 76–80. [CrossRef]
18. Gao, S.; Liu, Y.; Zhu, J.; Wang, Y.; Han, X.; Xia, X.; Zhao, X. The synthesis of novel FeS_2/g-C_3N_4 nanocomposites for the removal of tetracycline under visible-light irradiation. *Environ. Sci. Water Res. Technol.* **2021**, *7*, 1430–1442. [CrossRef]
19. Wei, X.; Wang, X.; Pu, Y.; Liu, A.; Chen, C.; Zou, W.; Zheng, Y.; Huang, J.; Zhang, Y.; Yang, Y.; et al. Facile ball-milling synthesis of CeO_2/g-C_3N_4 Z-scheme heterojunction for synergistic adsorption and photodegradation of methylene blue: Characteristics, kinetics, models, and mechanisms. *Chem. Eng. J.* **2021**, *420*, 127719. [CrossRef]
20. Zhao, H.; Jian, L.; Gong, M.; Jing, M.; Li, H.; Mao, Q.; Lu, T.; Guo, Y.; Ji, R.; Chi, W.; et al. Transition-Metal-Based Cocatalysts for Photocatalytic Water Splitting. *Small Struct.* **2022**, *3*, 2100229. [CrossRef]
21. Shao, B.; Wang, J.; Liu, Z.; Zeng, G.; Tang, L.; Liang, Q.; He, Q.; Wu, T.; Liu, Y.; Yuan, X. $Ti_3C_2T_x$ MXene decorated black phosphorus nanosheets with improved visible-light photocatalytic activity: Experimental and theoretical studies. *J. Mater. Chem. A* **2020**, *8*, 5171–5185. [CrossRef]

22. Shahiduzzaman, M.; Visal, S.; Kuniyoshi, M.; Kaneko, T.; Umezu, S.; Katsumata, T.; Iwamori, S.; Kakihana, M.; Taima, T.; Isomura, M.; et al. Low-temperature-processed brookite-based TiO$_2$ heterophase junction enhances performance of planar perovskite solar cells. *Nano Lett.* **2018**, *19*, 598–604. [CrossRef] [PubMed]
23. Zhang, J.; Chen, X.; Bai, Y.; Li, C.; Gao, Y.; Li, R.; Li, C. Boosting photocatalytic water splitting by tuning built-in electric field at phase junction. *J. Mater. Chem. A* **2019**, *7*, 10264–10272. [CrossRef]
24. Gao, Y.; Zhu, J.; An, H.; Yan, P.; Huang, B.; Chen, R.; Fan, F.; Li, C. Directly Probing Charge Separation at Interface of TiO$_2$ Phase Junction. *J. Phys. Chem. Lett.* **2017**, *8*, 1419–1423. [CrossRef]
25. Zhang, J.; Xu, Q.; Feng, Z.; Li, M.; Li, C. Importance of the relationship between surface phases and photocatalytic activity of TiO$_2$. *Angew. Chem.* **2008**, *120*, 1790–1793. [CrossRef]
26. Ai, Z.; Zhao, G.; Zhong, Y.; Shao, Y.; Huang, B.; Wu, Y.; Hao, X. Phase junction CdS: High efficient and stable photocatalyst for hydrogen generation. *Appl. Catal. B Environ.* **2018**, *221*, 179–186. [CrossRef]
27. Liu, F.; Shi, R.; Wang, Z.; Weng, Y.; Che, C.M.; Chen, Y. Direct Z-Scheme Hetero-phase Junction of Black/Red Phosphorus for Photocatalytic Water Splitting. *Angew. Chem. Int. Ed. Engl.* **2019**, *58*, 11791–11795. [CrossRef]
28. Han, L.; Jing, F.; Luo, X.; Zhong, Y.-L.; Wang, K.; Zang, S.-H.; Teng, D.-H.; Liu, Y.; Chen, J.; Yang, C.; et al. Environment friendly and remarkably efficient photocatalytic hydrogen evolution based on metal organic framework derived hexagonal/cubic In$_2$O$_3$ phase-junction. *Appl. Catal. B Environ.* **2021**, *282*, 119602. [CrossRef]
29. Friedmann, D.; Caruso, R. Indium Oxides and Related Indium-based Photocatalysts for Water Treatment: Materials Studied, Photocatalytic Performance, and Special Highlights. *Sol. RRL* **2021**, *5*, 2100086. [CrossRef]
30. Feng, Y.; Yan, T.; Wu, T.; Zhang, N.; Yang, Q.; Sun, M.; Yan, L.; Du, B.; Wei, Q. A label-free photoelectrochemical aptasensing platform base on plasmon Au coupling with MOF-derived In$_2$O$_3$@g-C$_3$N$_4$ nanoarchitectures for tetracycline detection. *Sens. Actuators B Chem.* **2019**, *298*, 12817. [CrossRef]
31. Jin, L.-N.; Liu, Q.; Sun, W.-Y. Size-controlled indium(iii)-benzenedicarboxylate hexagonal rods and their transformation to In$_2$O$_3$ hollow structures. *Cryst. Eng. Comm.* **2013**, *15*, 4775–4784. [CrossRef]
32. Wang, S.; Guan, B.Y.; Lou, X.W.D. Construction of ZnIn$_2$S$_4$-In$_2$O$_3$ Hierarchical Tubular Heterostructures for Efficient CO$_2$ Photoreduction. *J. Am. Chem. Soc.* **2018**, *140*, 5037–5040. [CrossRef] [PubMed]
33. Li, R.; Sun, L.; Zhan, W.; Li, Y.; Wang, X.; Han, X. Engineering an effective noble-metal-free photocatalyst for hydrogen evolution: Hollow hexagonal porous micro-rods assembled from In$_2$O$_3$@carbon core-shell nanoparticles. *J. Mater. Chem. A* **2018**, *6*, 15747–15754. [CrossRef]
34. Sun, D.; Wang, W.; Zhang, N.; Liu, C.; Li, X.; Zhou, J.; Ruan, S. G-C$_3$N$_4$/In$_2$O$_3$ composite for effective formaldehyde detection. *Sens. Actuators B Chem.* **2022**, *358*, 131414. [CrossRef]
35. Xu, M.; Zhao, X.; Jiang, H.; Chen, S.; Huo, P. MOFs-derived C-In$_2$O$_3$/g-C$_3$N$_4$ heterojunction for enhanced photoreduction CO$_2$. *J. Environ. Chem. Eng.* **2021**, *9*, 106469. [CrossRef]
36. Jin, X.; Guan, Q.; Tian, T.; Li, H.; Han, Y.; Hao, F.; Cui, Y.; Li, W.; Zhu, Y.; Zhang, Y. In$_2$O$_3$/boron doped g-C$_3$N$_4$ heterojunction catalysts with remarkably enhanced visible-light photocatalytic efficiencies. *Appl. Surf. Sci.* **2020**, *504*, 144241. [CrossRef]
37. Uddin, A.; Rauf, A.; Wu, T.; Khan, R.; Yu, Y.; Tan, L.; Jiang, F.; Chen, H. In$_2$O$_3$/oxygen doped g-C$_3$N$_4$ towards photocatalytic BPA degradation: Balance of oxygen between metal oxides and doped g-C$_3$N$_4$. *J. Colloid. Interface Sci.* **2021**, *602*, 261–273. [CrossRef]
38. Pan, Z.; Zhao, M.; Zhuzhang, H.; Zhang, G.; Anpo, M.; Wang, X. Gradient Zn-Doped Poly Heptazine Imides Integrated with a van der Waals Homojunction Boosting Visible Light-Driven Water Oxidation Activities. *ACS Catal.* **2021**, *11*, 13463–13471. [CrossRef]
39. Vinoth, S.; Rajaitha, P.; Venkadesh, A.; Devi, K.S.S.; Radhakrishnan, S.; Pandikumar, A. Nickel sulfide-incorporated sulfur-doped graphitic carbon nitride nanohybrid interface for non-enzymatic electrochemical sensing of glucose. *Nanoscale Adv.* **2020**, *2*, 4242–4250. [CrossRef]
40. Zang, S.; Cai, X.; Chen, M.; Teng, D.; Jing, F.; Leng, Z.; Zhou, Y.; Lin, F. Tunable Carrier Transfer of Polymeric Carbon Nitride with Charge-Conducting CoV$_2$O$_6$·2H$_2$O for Photocatalytic O$_2$ Evolution. *Nanomaterials* **2022**, *12*, 1931. [CrossRef]
41. Cai, M.; Li, R.; Wang, F.; Guo, X.; Bai, Q.; Sun, L.; Han, X. Architecture of designed hollow indium oxide microspheres assembled by porous nanosheets with high gas sensing capacity. *J. Alloys Compd.* **2017**, *729*, 222–230. [CrossRef]
42. Zhang, K.; Zhou, M.; Yang, K.; Yu, C.; Mu, P.; Yu, Z.; Lu, K.; Huang, W.; Dai, W. Photocatalytic H$_2$O$_2$ production and removal of Cr (VI) via a novel Lu$_3$NbO$_7$: Yb, Ho/CQDs/AgInS$_2$/In$_2$S$_3$ heterostructure with broad spectral response. *J. Hazard. Mater.* **2022**, *423*, 127172. [CrossRef] [PubMed]
43. Sprynskyy, M.; Szczyglewska, P.; Wojtczak, I.; Nowak, I.; Witkowski, A.; Buszewski, B.; Feliczak-Guzik, A. Diatom Biosilica Doped with Palladium(II) Chloride Nanoparticles as New Efficient Photocatalysts for Methyl Orange Degradation. *Int. J. Mol. Sci.* **2021**, *22*, 6734. [CrossRef]
44. Chen, M.; Xu, J.; Tang, R.; Yuan, S.; Min, Y.; Xu, Q.; Shi, P. Roles of microplastic-derived dissolved organic matter on the photodegradation of organic micropollutants. *J. Hazard. Mater.* **2022**, *440*, 129784. [CrossRef] [PubMed]
45. Adeleke, J.; Theivasanthi, T.; Thiruppathi, M.; Akomolafe, T.; Alabi, A. Photocatalytic degradation of methylene blue by ZnO/NiFe$_2$O$_4$ nanoparticles. *Appl. Surf. Sci.* **2018**, *455*, 195–200. [CrossRef]
46. Yadav, R.; Kumar, R.; Aliyan, A.; Dobal, P.S.; Biradar, S.; Vajtai, R.; Singh, D.P.; Martí, A.A.; Ajayan, P.M. Facile synthesis of highly fluorescent free-standing films comprising graphitic carbon nitride (g-C$_3$N$_4$) nanolayers. *New J. Chem.* **2020**, *44*, 2644–2651. [CrossRef]

47. Li, S.; Cai, M.; Liu, Y.; Wang, C.; Yan, R.; Chen, X. Constructing $Cd_{0.5}Zn_{0.5}S/Bi_2WO_6$ S-scheme heterojunction for boosted photocatalytic antibiotic oxidation and Cr (VI) reduction. *Adv. Powder Mater.* **2023**, *2*, 100073. [CrossRef]
48. Fan, G.; Luo, J.; Guo, L.; Lin, R.; Zheng, X.; Snyder, S.A. Doping Ag/AgCl in zeolitic imidazolate framework-8 (ZIF-8) to enhance the performance of photodegradation of methylene blue. *Chemosphere* **2018**, *209*, 44–52. [CrossRef]
49. Lin, B.; Chen, H.; Zhou, Y.; Luo, X.; Tian, D.; Yan, X.; Duan, R.; Di, J.; Kang, L.; Zhou, A.; et al. 2D/2D atomic double-layer WS_2/Nb_2O_5 shell/core nanosheets with ultrafast interfacial charge transfer for boosting photocatalytic H_2 evolution. *Chin. Chem. Lett.* **2021**, *32*, 3128–3132. [CrossRef]
50. Yan, L.; Gao, H.; Chen, Y. Na-Doped Graphitic Carbon Nitride for Removal of Aqueous Contaminants via Adsorption and Photodegradation. *ACS Appl. Nano Mater.* **2021**, *4*, 7746–7757. [CrossRef]
51. Mu, P.; Zhou, M.; Yang, K.; Chen, X.; Yu, Z.; Lu, K.; Huang, W.; Yu, C.; Dai, W. $Cd_{0.5}Zn_{0.5}S/CoWO_4$ Nanohybrids with a Twinning Homojunction and an Interfacial S-Scheme Heterojunction for Efficient Visible-Light-Induced Photocatalytic CO_2 Reduction. *Inorg. Chem.* **2021**, *60*, 14854–14865. [CrossRef] [PubMed]

Review

Photocatalytic Degradation of Some Typical Antibiotics: Recent Advances and Future Outlooks

Xue Bai [1], Wanyu Chen [2], Bao Wang [3], Tianxiao Sun [4], Bin Wu [4] and Yuheng Wang [1,*]

1. Division of Pharmacy and Optometry, Faculty of Biology, Medicine and Health, School of Health Science, University of Manchester, Oxford Road, Manchester M13 9PT, UK; xue.bai@manchester.ac.uk
2. Faculty of Biology, Medicine and Health, School of Health Science, University of Manchester, Oxford Road, Manchester M13 9PT, UK; 16chenj@wycombeabbey.com
3. State Key Laboratory of Biochemical Engineering, Institute of Process Engineering, Chinese Academy of Sciences, Beijing 100190, China; baowang@ipe.ac.cn
4. Helmholtz-Zentrum Berlin für Materialien und Energie GmbH, Albert-Einstein-Straße 15, 12489 Berlin, Germany; tianxiao.sun@helmholtz-berlin.de (T.S.); bin.wu@helmholtz-berlin.de (B.W.)
* Correspondence: yuheng.wang@manchester.ac.uk

Abstract: The existence of antibiotics in the environment can trigger a number of issues by fostering the widespread development of antimicrobial resistance. Currently, the most popular techniques for removing antibiotic pollutants from water include physical adsorption, flocculation, and chemical oxidation, however, these processes usually leave a significant quantity of chemical reagents and polymer electrolytes in the water, which can lead to difficulty post-treating unmanageable deposits. Furthermore, though cost-effectiveness, efficiency, reaction conditions, and nontoxicity during the degradation of antibiotics are hurdles to overcome, a variety of photocatalysts can be used to degrade pollutant residuals, allowing for a number of potential solutions to these issues. Thus, the urgent need for effective and rapid processes for photocatalytic degradation leads to an increased interest in finding more sustainable catalysts for antibiotic degradation. In this review, we provide an overview of the removal of pharmaceutical antibiotics through photocatalysis, and detail recent progress using different nanostructure-based photocatalysts. We also review the possible sources of antibiotic pollutants released through the ecological chain and the consequences and damages caused by antibiotics in wastewater on the environment and human health. The fundamental dynamic processes of nanomaterials and the degradation mechanisms of antibiotics are then discussed, and recent studies regarding different photocatalytic materials for the degradation of some typical and commonly used antibiotics are comprehensively summarized. Finally, major challenges and future opportunities for the photocatalytic degradation of commonly used antibiotics are highlighted.

Keywords: antibiotics; photocatalytic degradation; degradation mechanism; photocatalysts

1. Introduction

Antibiotics are chemotherapeutic agents that cure bacterial infections [1]. Currently, antibiotics in the environment are attracting increased attention, prompting a widespread search for possible methods of containment [1,2]. This issue results in the generation of antibiotic-resistant genes and antibiotic-resistant bacteria, which expedite the spread of antibiotic resistance, creating a threat to human health and ecological systems [2–4]. Thus, in the 21st century, the threat to the integrity of our water resources from antibiotic pollutants is deemed to be one of the most serious environmental problems worldwide, not only because of environmental damage, but due to the potential harm to human health [5].

During the past decade, many strategies have been adopted to address the problem of wastewater antibiotics [6]. Wastewater treatment is usually considered the main method for managing these antibiotics, since wastewater collects discharge from hospitals, industry, and agriculture [7]. However, many more studies have confirmed that conventional treatments are not highly capable of removing these pollutant compounds, which are predominantly water-soluble, and are neither volatile nor biodegradable [8]. Biotic elimination and non-biotic processes, including sorption, hydrolysis, biodegradation by bacteria, and oxidation, as well as reduction, have attracted a great deal of attention [9]. Yang et al. studied the adsorption, desorption, and biodegradation performance of sulfonamide antibiotics in the existence of activated sludge with and without NaN_3 biocide [10]. The experimental results showed that the antibiotics were eliminated by sorption and biodegradation via the activated sludge. Liu et al. investigated four antibiotics including norfloxacin, ofloxacin, roxithromycin, and azithromycin as target antibiotics, and adopted UV_{254} photolysis, ozonation, and UV/O_3 approaches to conduct disposal treatments of nanofiltration, realizing the highest efficiency (>87%) in eliminating antibiotics [11]. Nevertheless, the application of these methods was highly restricted due to the high cost, low stability, and poor recycling ability. Therefore, scientists have been seeking novel methods for degrading antibiotics in wastewater, making the exploration of high-efficiency degradation techniques a popular pursuit for environmental and chemistry researchers [4].

As one of the most promising strategies for degrading antibiotic pollutants, photocatalysis has received much attention due to its low cost, efficiency, and environmental friendliness while degrading antibiotics under sunlight and ambient conditions [12,13]. Most antibiotics are resistant to decomposition owing to their robust molecular structures, thus, the development, design, and fabrication of appropriate photocatalysts with high photocatalytic activities are urgently needed [14]. Though a few catalytic processes have been discussed in the literature so far, there have not been enough examples focusing on the use of appropriate photocatalysts that possess longer wavelength absorption for the photocatalytic degradation of antibiotics, which would help to inform readers about this research field.

The photodegradation of antibiotic pollutants has been reviewed recently [5,15]. However, knowledge of the critical degradation mechanisms and underlying reaction pathways of some typical photodegradation reaction catalysts for antibiotics requires deeper discussion. Furthermore, a comprehensive overview on the possible sources and dangers of the antibiotic pollutants released through the ecological chain, particularly regarding the consequences and damages caused by antibiotic residuals on the environment and human health, is still missing. Additionally, the overall introduction of some commonly-used photocatalytic nanomaterials and their application in the degradation of some typical antibiotics is essential to confirm their practical superiority and effectiveness as photodegradation catalysts.

This review firstly summarizes the effects of antibiotics on living organisms and the environment as well as the basic mechanism of the photocatalytic degradation of antibiotics. Then, commonly used photocatalytic materials for antibiotic degradation are reviewed. Finally, the recent advances in the use of various photocatalytic materials for the degradation of antibiotics are discussed.

2. Consequences of Antibiotics in Wastewater on the Environment and Human Health

Pharmaceuticals can largely improve humans' health and quality of life when used to treat contagious diseases, however, the misuse of drugs, especially antibiotics, has severe damage on the environment and human health [6,16]. Some results reported remarkable changes in sex ratio and fecundity of daphnia manga when exposed to antibiotics such as sulfamethoxazole and trimethoprim [17]. Meanwhile, a decrease in desire and sexual motivation was observed in experiments on male rats given cimetidine [18].

In some countries, antibiotics are not only employed for animal treatment but also to accelerate animal growth and increase production. Thus, antibiotics might be released from animal waste due to incomplete digestion, and that waste may then be used as fertilizer in agriculture or dumped into wastewater, generating a possible pathway to human harm from food or drink exposure, as shown in Figure 1 [19]. A recent study reported that chlortetracycline antibiotic was, to some extent, uptaken by onions, cabbage, and corn [20]. However, those vegetables did not uptake tyrosine antibiotic, probably due to its large molecular size. Thus, continuous release of these antibiotic pollutants into water environments and organisms has a severely negative impact on the environment by causing genetic exchange and activating drug-resistant bacteria. In particular, most antibiotic pollutants, even under low concentrations, may result in a severe risk to the ecosystem and human health [21,22].

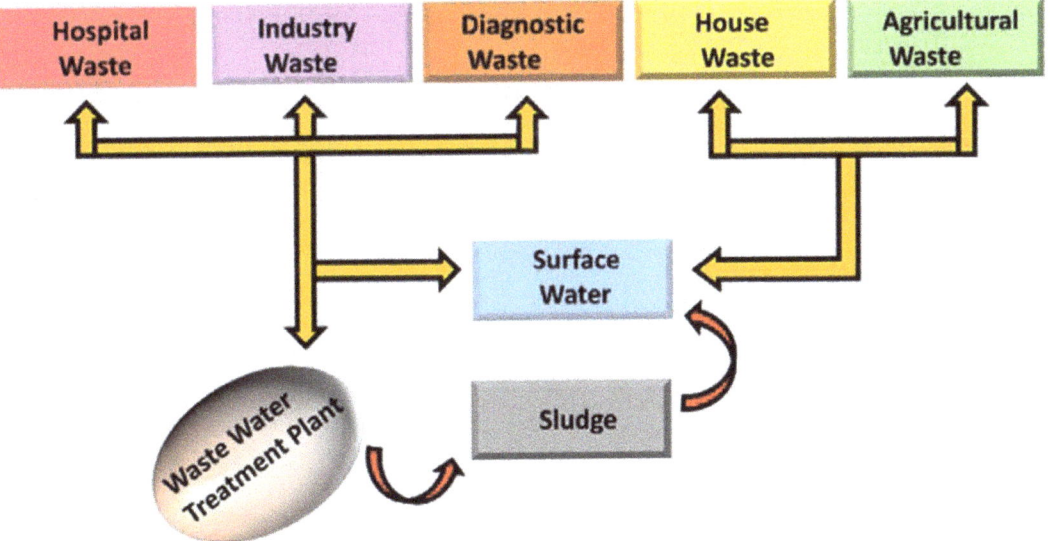

Figure 1. Sources of different emerging antibiotics pollutants in daily life. (Reproduced with permission from [19], copyright 2021, Elsevier).

On one hand, in terms of micro-organisms, the release of antibiotics into the environment could lead to chromosomal mutations of native bacteria, triggering the development of antibiotic-resistant bacterial strains, which may cause environmental threats such as toxicological effects on non-intended pathogens, alteration of structures, and dissemination of algal communities [23–25]. On the other hand, consumption of water or agricultural and sideline products containing antibiotic pollutants may induce symptoms in humans including, but not limited to, vomiting, tremors, nausea, headache, diarrhea, and nervousness [26]. Furthermore, problems such as restraining spinach growth, physiological teratogenesis, and human gene toxicity have also been reported due to the presence of antibiotic pollutants in water or food [3,27].

3. Principle and Fundamental Mechanism of Photocatalytic Degradation of Antibiotics

The steps involved in the photocatalytic degradation of antibiotics are demonstrated in Figure 2. The predominant mechanisms for antibiotic photocatalytic degradation can be summarized as three main steps: photon absorption, excitation, and reaction [5,7]. In detail, once a photocatalyst absorbs photons with an energy higher than its band gap, the electrons in the valance band (VB) can be excited and jump up into the conduction band (CB), where a hole (h^+_{VB}) is produced (Equation (1)) [6,26,28,29]. Subsequently, the photogenerated electrons and holes are efficiently separated and migrate to the surface of the photocatalyst, triggering secondary reactions with the adsorbed materials. Typically, photogenerated holes can also attack those antibiotics directly (Equation (2)), theoretically leading to significant degradation of those toxic antibiotics. In addition, two types of systematic theories about degradation pathways were proposed and recognized by researchers in this field [26]. One is a reductive pathway that happens if the CB potential of the semiconductor is negative compared to that of the $O_2/\bullet O_2^-$ redox potential (−0.13 eV vs. reference hydrogen electrode (RHE)), wherein the photoexcited electrons can react with electron acceptors such as O_2 deposits on the catalyst surface or dissolved in water, thereby reducing it to form superoxide radical anion $\bullet O_2^-$ (Equation (3)) [6,26]. In contrast, another pathway referring to the oxidative pathway was initiated when the holes migrated to the photocatalyst surface, accompanied by hydroxyl radical ($\bullet$OH) generation upon the oxidation of H_2O/OH^- depending on the alkalinity or acidity of the media (Equation (4)) [6,26]. After being excited, hydrogen ions could recombine with the electrons and generate heat energy (Equation (5)), which would decrease the efficiency of the photodegradation. It is noted that the standard redox potential of photocatalysts should be higher than that of $\bullet OH/OH^-$ (+1.99 eV vs. RHE) in this case [6,26]. Then, both of these reactive radicals ($\bullet$OH and $\bullet O_2^-$) are highly active oxidizing agents in the photocatalytic process [30]. They can effectively mineralize any antibiotics and their intermediates to form water and carbon dioxide under prolonged exposure to high-energy UV irradiation, and eventually decompose into CO_2 and H_2O (Equation (6)) [28,29,31,32]. Many studies demonstrated that both pathways (reductive and oxidative) should synergistically occur to largely prevent the accumulation of electrons in the CB and significantly decrease the possibility of the recombination of electrons and positive holes compared to the pathway of direct interaction between photogenerated holes and antibiotics [33].

$$\text{Photocatalyst} + h\nu \rightarrow \text{photocatalyst} + h^+ + e^- \tag{1}$$

$$h^+ + \text{antibiotics} \rightarrow H_2O + CO_2 + \text{degradation products} \tag{2}$$

$$O_2 + e^- \rightarrow \bullet O_2^- \tag{3}$$

$$H_2O/OH^- + h^+ \rightarrow \bullet OH + H^+ \tag{4}$$

$$H^+ + e^- \rightarrow \text{energy} \tag{5}$$

$$\text{Antibiotics} + \bullet OH \text{ or } \bullet O_2^- \rightarrow CO_2 + H_2O + \text{degradation products} \tag{6}$$

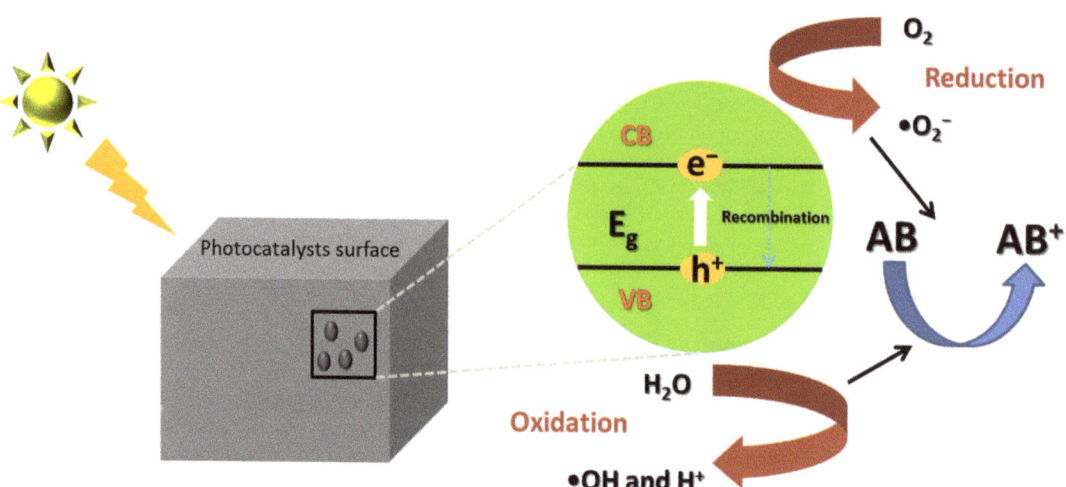

Figure 2. General photocatalytic mechanism on the degradation of antibiotics by the formation of photo-induced charge carriers (e^-/h^+) on the photocatalysts' surface.

Considering the prediction of application and efficiency of a type of photocatalytic material, optical bandgap (E_g) is a very important factor for evaluating photoabsorption ability and photocatalytic efficiency. Mehrorang et al. put forward a method and criterion for bandgap measurement, and divided the concept of the bandgap into the two categories of photonic and electrochemical bandgap facing polyfluorene co-polymers as photocatalysts [34,35]. In addition, They concluded that, the prevention of charge recombination would accordingly lead to a higher lifetime of the active holes, thereby upgrading their antibiotic degradation activity. This proved to be a great strategy for enhancing the activity of photocatalysts under visible light, relating to the interfacial charge transfer from a separate energy surface to a molecular continuous surface from solids [34,35].

All in all, the mechanism of photocatalysis for the degradation of antibiotics can be divided into five main steps: (1) transfer of antibiotics in the fluid phase to the surface; (2) adsorption of the antibiotics; (3) reaction in the adsorbed phase; (4) desorption of the products; and (5) removal of products from the interface region [36,37]. However, photocatalytic degradation suffers the problem of electron-hole recombination in the photocatalyst when the electrons that had been excited to CB rapidly recombine with the separated holes in the VB before creating free radicals [37]. Although this depends on many flexible options such as tuned experimental conditions, the adoption of specific photocatalysts with a low CB–VB bandgap energy and photocatalyst modifications are considered as solutions for these challenges [38,39].

4. Common Photocatalytic Materials for Antibiotic Degradation

4.1. Semiconducting Metal Oxides-Based Photocatalysts

Metal oxide semiconductors have been utilized as pristine photocatalysts or as hybrids, or have been coupled/doped with other materials to facilitate the degradation of organic pollutants such as pesticides, dyes, and polycyclic aromatic hydrocarbons [40]. More importantly, the application of metal oxide-based photocatalysts for antibiotic degradation has recently drawn more interest and attention from researchers due to their good light absorption under UV, visible light, or both, combined with their biocompatibility, safety, and stability when exposed to different conditions [3,41,42]. Generally, metal oxides encounter some challenges regarding ineffectiveness or non-absorbance of photocatalytic activity because of their wide band gap (Figure 3) and faster electron–hole pair recombination [43]. For example, TiO_2 is the most popular metal oxide for photocatalysis because of its good optical and electronic properties, chemical stability and reusability, non-toxicity, and low cost [44]. Additionally, ZnO is another semiconducting material that has a better quantum efficiency and higher photocatalytic efficiency compared to TiO_2, particularly if used for photocatalytic antibiotic degradation at a neutral pH, however, the high recombination rate of the photogenerated electron–hole pairs limits the utilization of ZnO without any functionalization [45]. Several studies demonstrated that doping with metals like Ag and Fe, or non-metals like N and C, into ZnO enhanced the activity of photocatalytic antibiotic degradation [46,47]. WO_3 is another promising metal oxide that has received remarkable attention due to its abundance, cost-effectiveness, and non-toxicity [48,49]. Furthermore, $W_{18}O_{49}$ was also considered a superior photocatalyst with a higher photocatalytic degradation efficiency compared to WO_3 [3,50]. Nevertheless, it is prone to oxidization to WO_3 in spite of its superior photocatalytic performance. Thus, the construction of a hybrid of $W_{18}O_{49}$ and other metal oxides can overcome this oxidization barrier [3,51]. There are many other metal oxides that play important roles in photocatalytic materials for antibiotic degradation, and we may introduce more in the next section.

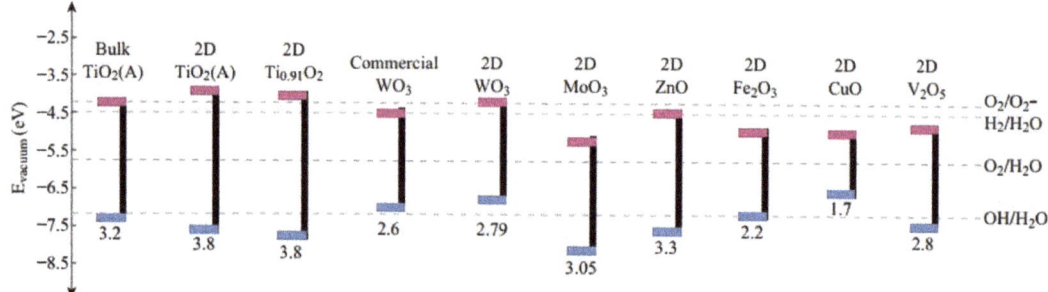

Figure 3. Band energy gaps of selected semiconductors. (Reproduced with permission from [43], copyright 2017, Springer Nature).

4.2. Bismuth-Based Photocatalysts

Bismuth, possessing an atomic electron configuration of $6s^2 6p^3$, is a metallic element from the fifth group of the sixth period in the periodic table, and is usually present in the form of Bi^{3+} [26,52]. A class of novel processes along with bismuth-based catalysts have been developed for antibiotic degradation, as shown in Figure 4 [26]. Bi oxides display a narrower bandgap due to the overlap of O 2p and Bi 6s orbitals in the valence band and lone-pair distortion of the Bi 6s orbital, resulting in the mobility of photoexcited charges, enhancing the visible light response performance [53]. Interestingly, the Bi^{5+} valence state from the oxidation of Bi^{3+} has good absorption of visible light once the 6s orbital is empty [54]. Basically, Bi-based photocatalysts mostly have a bandgap of less than 3 eV. There are some typical Bi-based photocatalysts attracting more attention recently, such as Bi_2O_3 and $BiVO_4$ [55]. Bi_2O_3 is one of the most common photocatalysts, showing excellent photocatalytic performance on water-splitting and water treatment from organic wastes [56]. Bi_2O_3 has a bandgap ranging from 2.1 to 2.8 eV, making its utilization for visible light absorption more efficient. Bi_2O_3 has five different configurations: α, β, γ, δ, and ω-Bi_2O_3 [26]. Additionally, $BiVO_4$, with superior physicochemical properties like ferroelasticity and ionic conductivity, has a theoretical bandgap of 2.047 eV, which maximizes its visible light utilization [57]. $BiVO_4$ was widely used in photocatalytic reactions for organic waste treatment and water splitting in past years [58]. Although it has been confirmed that bismuth-based photocatalysts have good photocatalytic performance and use visible light efficiently, it should be noted that some parameters, such as stability and solubility, need to be emphasized [59].

4.3. Silver-Based Photocatalysts

The application of photocatalytic degradation of silver-based photocatalysts such as AgX (X = Cl, Br, I), Ag_2O, Ag_3PO_4, and Ag_2CO_3 have been reported by various researchers [60–62]. In the case of pristine Ag_2CO_3, the challenge is that pristine Ag_2CO_3 is unstable and photocorrosive due to its possible transformation from Ag^+ to metallic Ag on account of accepted photoelectrons during the photocatalytic processes [63]. Moreover, pristine Ag_2O also exhibits poor stability and rapid electron-hole recombination [64]. Their superior photocatalytic performance on antibiotic degradation depends not only on the reduced electron-hole recombination but may also be from broad and strong absorption ranges in the visible region due to the localized surface plasmon resonance effects induced by Ag nanoparticles [3,7,65].

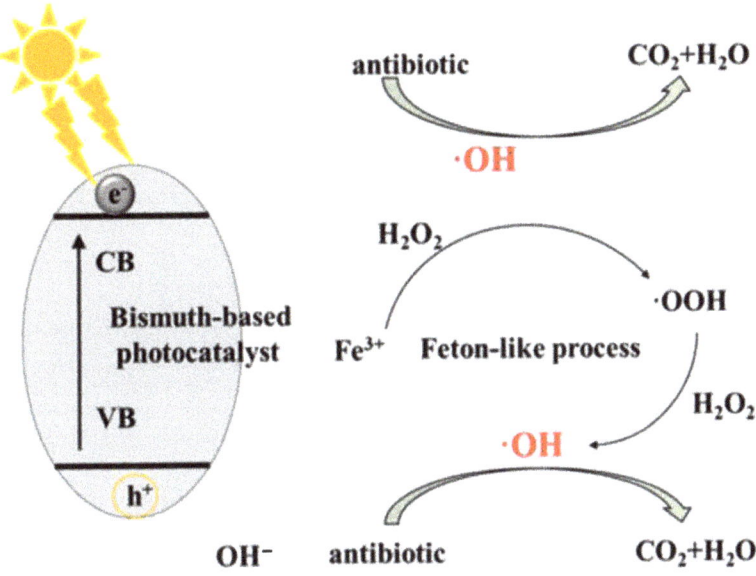

Figure 4. The mechanism of coupled processes with bismuth-based compounds for antibiotic degradation. (Reproduced with permission from [26], copyright 2021, Elsevier).

4.4. Metal-Organic Frameworks (MOFs)-Based Photocatalysts

Metal-organic frameworks (MOFs) are a new class of coordination polymers with periodic network structures formed by the self-assembly between metal ions/metal clusters and organic ligands [66]. By modifying linkers employing functional groups, highly porous structures with remarkable surface areas could be obtained with tuned surface structures [67]. MOFs were first discovered in the mid-1990′s by Omar Yaghi, and the invention of novel MOFs promised long-lived influence in the areas of chemistry, physics, biology, and the material sciences, particularly used extensively in photocatalysis due to their high surface area, adjustable porosity and pore volume [67–69]. Thus, MOFs promise to be highly-effective materials for the photocatalytic degradation of antibiotics in a solution [66,70]. Although various MOF-based materials have been utilized to remove antibiotics, the development of more efficient degradation agents remains a key problem facing more active MOF-based photocatalytic degradation materials with more active sites and large surface areas with group functionalization [66,67]. According to the previous report, the organic linker serves as the VB, while the metallic cluster acts as CB. Under exposure to light, MOFs behave like semiconductors, and can thus be deemed as a potential photocatalyst for highly effective degradation of antibiotics due to their superior high thermal and mechanical stability and their excellent structural characteristics [66].

4.5. Graphitic Carbon Nitrides-Based Photocatalysts

Graphitic carbon nitride (g-C_3N_4), a new class of polymeric semiconducting material, is another kind of promising material for photo-driven catalytic applications [26,71]. On one hand, the past report implied that the g-C_3N_4 has a bandgap of around 2.7 eV, and the CB–VB can also meet the requirement of overall water splitting, as demonstrated in Figure 5, which made it more popular for photocatalytic water splitting in the past decades [72].

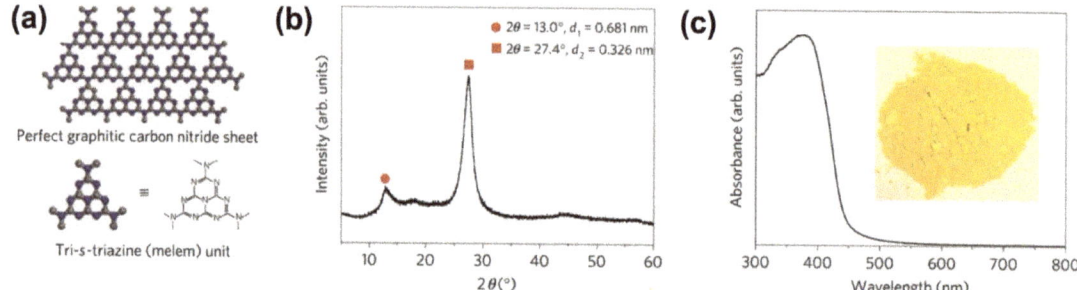

Figure 5. Crystal structure and optical properties of graphitic carbon nitride: (**a**) Schematic diagram of a perfect graphitic carbon nitride sheet constructed from melem units, (**b**) Experimental XRD pattern of the polymeric carbon nitride, revealing a graphitic structure with an interplanar stacking distance of aromatic units of 0.326 nm and (**c**) Ultraviolet-visible diffuse reflectance spectrum of the polymeric carbon nitride. Inset: Photograph of the photocatalyst. (Reproduced with permission from [72], copyright 2009, Springer Nature).

On the other hand, g-C_3N_4 can also be used for the photocatalytic degradation of antibiotic pollutants under visible light. However, pure g-C_3N_4 exhibited a low degradation rate due to negative position, resulting in a weak oxidation ability [26,73,74]. Therefore, surface modification is necessary to overcome these limitations [26]. Researchers found that doping with noble metal ions optimized the photocatalytic performance according to several reports in the literature as a result of the higher separation of the photoproduced electrons and holes due to the excellent capacity of electron capture by the noble metallic ions [26,75]. A great deal of research is ongoing regarding g-C_3N_4 modification for fabricating and designing nanomaterials with different properties in order to obtain the best possible photocatalytic performance for the removal of antibiotics [76].

5. Recent Advances in Photocatalytic Degradation of Antibiotics

5.1. Photocatalytic Degradation of Ciprofloxacin

Ciprofloxacin is a second-generation fluoroquinolone antibiotic used to kill bacteria to prevent severe infection [77]. The chemical structure of ciprofloxacin is exhibited in Figure 6a [1]. Notably, ciprofloxacin dominates 73% of the total consumption, with a daily dose between 0.39 and 1.8 per 1000 patients, and has a broad antimicrobial spectrum that has impact on the DNA gyrase and topoisomerase IV of various Gram-positive and Gram-negative bacteria, thus preventing cell replication [78,79]. It has been considered a good therapy for treating digestive infections, complicated urinary tract infections, sexually transmitted diseases, pulmonary diseases, and skin infections. However, the presence of ciprofloxacin limits photosynthetic pathways and even leads to morphological deformities in higher plants. It also leads to severe damage to human health [78]. In the past, various studies reported the use of modified photocatalysts to meet the demand for higher photocatalytic efficiency in the degradation of ciprofloxacin [80].

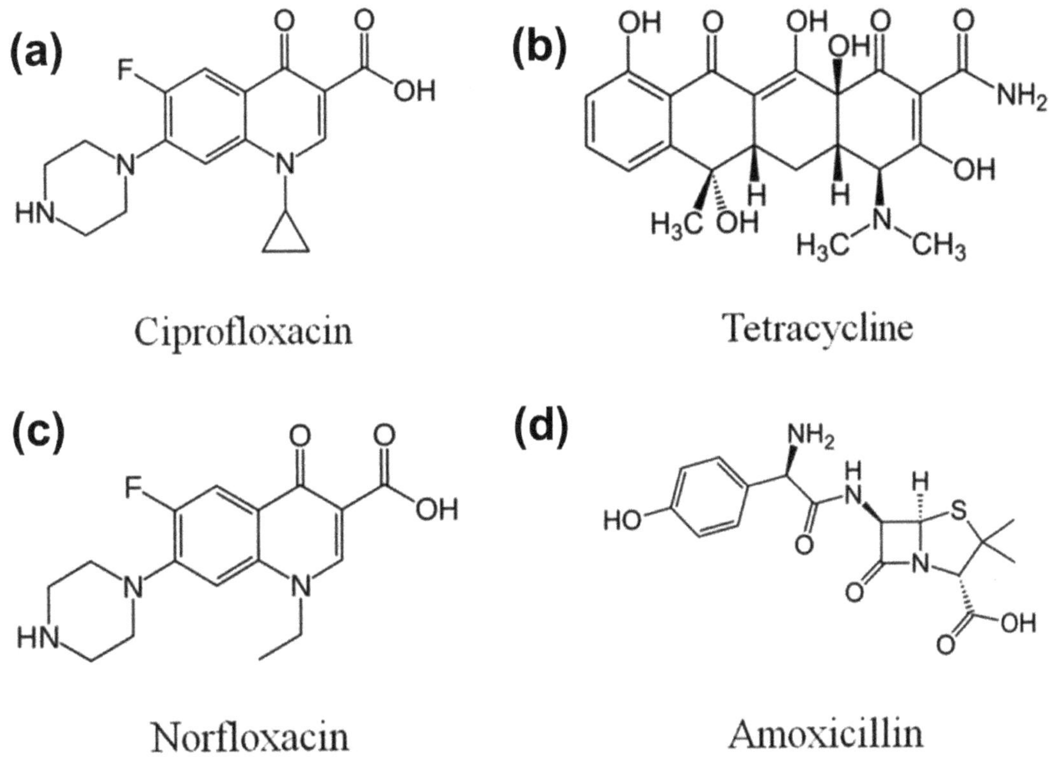

(a) Ciprofloxacin

(b) Tetracycline

(c) Norfloxacin

(d) Amoxicillin

Figure 6. The illustration of commonly investigated antibiotics in photocatalytic processes. (**a**) Ciprofloxacin. (**b**) Tetracycline. (**c**) Norfloxacin. (**d**) Amoxicillin. (Reproduced with permission from [1], copyright 2021, Elsevier).

Yu et al. [80] prepared Zn-doped Cu_2O particles by a solvothermal method to achieve photocatalytic degradation of ciprofloxacin. The photocatalytic results demonstrated that Zn-doped Cu_2O has better photocatalytic performance and reusability compared to the undoped Cu_2O. 94.6% of ciprofloxacin was degraded in presence of Zn-doped Cu_2O, even after 5 cycles, the degradation percentage still remains 91% due to the significantly enhanced absorption intensity in the visible light range, and the increased band gap than that of the undoped Cu_2O (Figure 7a,b). In addition, a novel Z-scheme CeO_2–Ag/AgBr photocatalyst was fabricated by Malakootian et al. [77] using in situ interspersals of AgBr on CeO_2 for subsequent photoreduction process. The results also exhibited largely enhanced photocatalytic activity for the photodegradation of ciprofloxacin under visible light irradiation due to the faster interfacial charge transfer process and the largely enhanced separation of the photogenerated electron-hole pairs. Furthermore, Pattnaik et al. [81] adopted exfoliated graphitic carbon nitride into photocatalytic degradation of ciprofloxacin under solar irradiation and catalytic data have shown that photocatalytic activities of g-C_3N_4 have enhanced after its exfoliation because of its efficient charge separation, low recombination of photogenerated charge carriers and high surface area. They found that 1 g/L exfoliated nano g-C_3N_4 can degrade up to 78% of a 20 ppm solution exposed to solar light for lasted 1 h (Figure 7c,d).

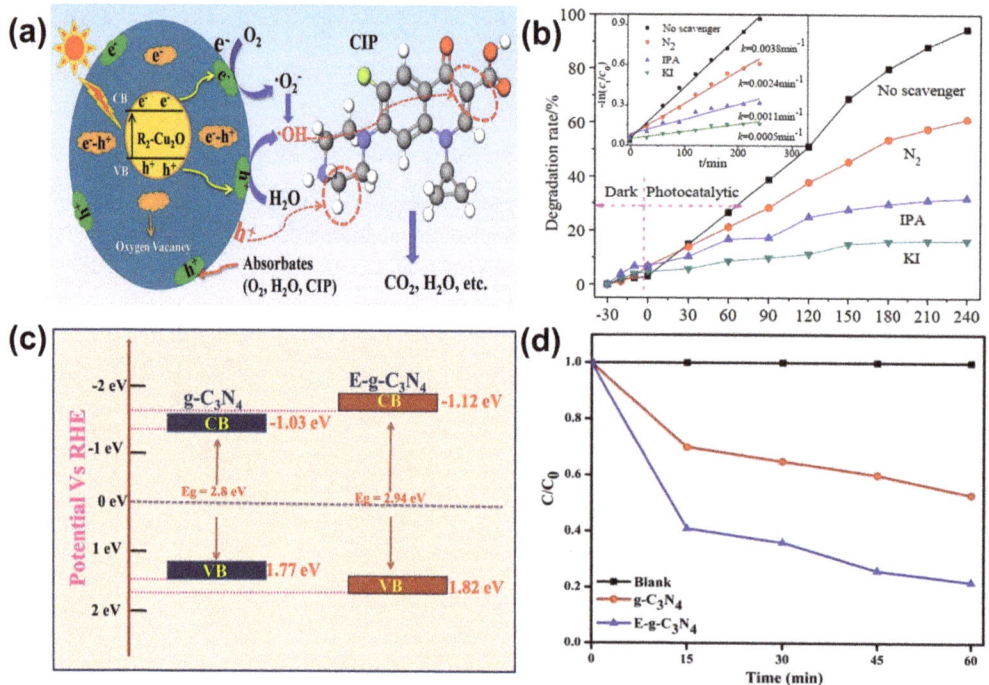

Figure 7. (a) The schematic diagram of the photocatalytic mechanism for the Zn-doped Cu_2O (b) Photocatalytic activity of R_2-Cu_2O with different scavenges, inset is degradation rate of ciprofloxacin. (Reproduced with permission from [80], copyright 2019, Elsevier) (c) Diagram of band gap structure of bulk and exfoliated g-C_3N_4 (d) Photocatalytic degradation of ciprofloxacin over g-C_3N_4 as well as exfoliated g-C_3N_4. (Reproduced with permission from [81], copyright 2019, Springer Nature).

5.2. Photocatalytic Degradation of Tetracycline

Tetracyclines are a series of broad-spectrum antibiotics that were first adopted in 1940, and their structures are shown in Figure 6b [1,82]. All tetracyclines have anti-inflammatory and immunosuppressive effects, and were previously used to treat rheumatism [83]. Due to the additional effect of tetracyclines against lipases and collagenases, these antibiotics were also initially used for the intrapleural treatment of malignant effusions [84,85]. Although tetracycline plays a significant role in medicine, the existence of tetracycline in aquatic media is of great concern because of its ecological impact, including carcinogenicity and toxicity to the environment [84,85]. A number of studies related to the removal or degradation of tetracycline through the use of different photocatalytic materials have been reported in past years [86]. For example, a novel TiO_2/g-C_3N_4 core-shell quantum heterojunction prepared by a feasible strategy of polymerizing the quantum trick graphitic carbon nitride (g-C_3N_4) onto the surface of anatase titanium dioxide nanosheets was put forward by Wang et al. to be employed as a tetracycline degradation photocatalyst, and this catalyst exhibited the highest tetracycline degradation rate: 2.2 mg/min, which is 36% higher than that of the TiO_2/g-C_3N_4 mixture, 2 times higher than that of TiO_2, and 2.3 times higher than that of bulk g-C_3N_4 (Figure 8a,b) [87]. Moreover, Wang et al. synthesized a novel C–N–S tri-doped TiO_2 using a facile and cost-effective sol–gel method with titanium butoxide as titanium precursor and thiourea as the dopant source, which can be used for photocatalytic degradation of tetracycline under visible light [86]. The catalytic results exhibited the highest photocatalytic degradation efficiency of tetracycline under visible light irradiation which is associated with the synergistic effects of tetracycline adsorption due to its high surface area, narrow band gap causing C–N–S tri-doping, presence of carbonaceous species functioning as a photosensitizer, and well-organized anatase phase. Additionally, Chen et al. synthesized a novel heterostructured photocatalyst AgI/BiVO$_4$ by an in situ precipitation procedure, and the results exhibited excellent photoactivity for tetracycline decomposition under visible light irradiation, the tetracycline molecules were apparently eliminated (94.91%) within 60 min, and degradation efficiency was remarkably superior to those of bare BiVO$_4$ (62.68%) and AgI (75.43%) under same experimental conditions (Figure 8c,d) [88].

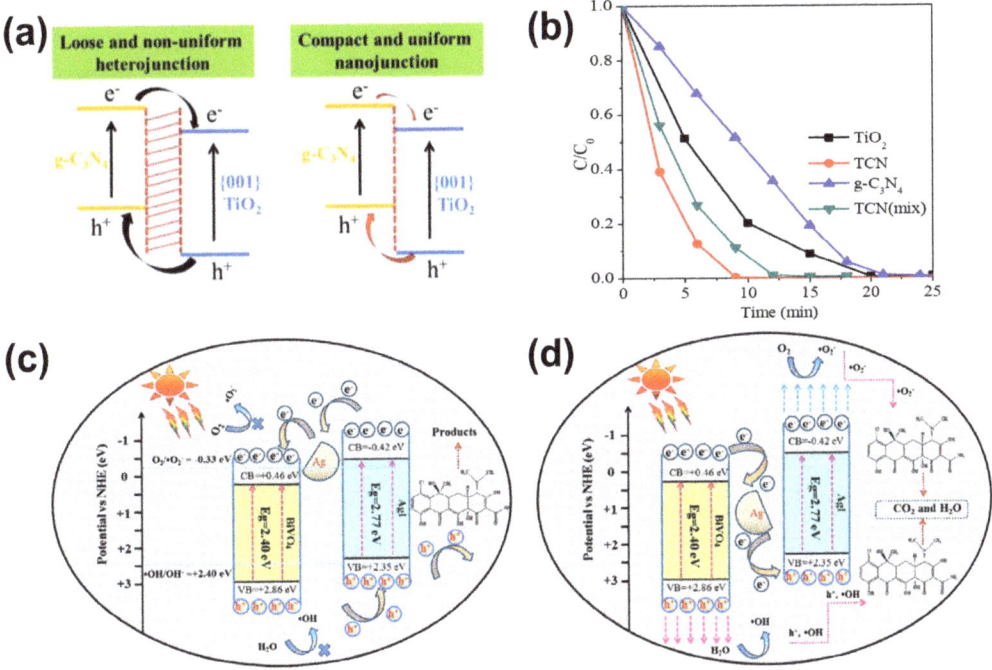

Figure 8. (**a**) Proposed heterojunction differences between TCN and TCN(mix) (**b**) Photocatalytic degradation efficiencies of tetracycline by employing TiO$_2$, TCN, TCN (mix), and g-C$_3$N$_4$ as the photocatalysts under the xenon lamp irradiation. (Reproduced with permission from [87], copyright 2017, Elsevier). Schematic Illustration of the Mechanism for the Photocatalytic Degradation of tetracycline under Visible Light Irradiation over AgI/BiVO$_4$ Nanocomposite: (**c**) Traditional Model and (**d**) Z-Scheme Heterojunction System. (Reproduced with permission from [88], copyright 2016, American Chemical Society).

5.3. Photocatalytic Degradation of Norfloxacin

Norfloxacin is another antibiotic within the fluro-quinolones group, and is widely used for curing urinary tract infections [89]. Figure 6c indicates the chemical structure of norfloxacin [1]. Currently, the presence of norfloxacin in wastewater (especially in hospitals) contains high concentrations and is deemed to be one of the potential pollutants in the aquatic environment. In the past few years, fluoroquinolone antibiotics have triggered tremendous concern due to their widespread use and environmental toxicity [89,90]. There are several reports on the degradation of norfloxacin by photocatalysis using different materials [91]. Sayed et al. [92] prepared a novel immobilized TiO$_2$/Ti film with exposed {001} facets via a facile one-pot hydrothermal route to use in the degradation of norfloxacin from aqueous media, and the experimental results demonstrated excellent photocatalytic performance toward the degradation of norfloxacin in various water matrices, with the observation that •OH is mainly involved in the photocatalytic degradation of norfloxacin by {001} faceted TiO$_2$/Ti film (Figure 9a). Additionally, Tang et al. [93] realized excellent visible-light-driven photocatalytic performance for the degradation of norfloxacin by an as-prepared novel Z-scheme Ag/FeTiO$_3$/Ag/BiFeO$_3$ using a sol–gel method followed by a photo-reduction process. The results showed the photocatalytic degradation extent reaches 96.5% within 150 min when using Ag/FeTiO$_3$/Ag/BiFeO$_3$ at 2.0 wt.% Ag (FeTiO$_3$: BiFeO$_3$ = 1.0:0.5) which can be reused with excellent photocatalytic stability (Figure 9b–d). Moreover, Lv et al. [94] synthesized copper-doped bismuth oxybromide (Cu-doped BiOBr) using a solvothermal method, and assessed their ability to degrade norfloxacin under visible light. The as-prepared Cu-doped BiOBr showed high activity with a photocatalytic degradation constant of 0.64 ×10^{-2} min^{-1} in the photocatalytic degradation of norfloxacin under visible-light irradiation due to its enhanced light-harvesting properties, enhanced charge separation, and interfacial charge transfer, as well as a retention of 95% of its initial activity, even after 5 constant catalytic cycles. Similarly, Bi$_2$WO$_6$, another bismuth-based catalyst, was put forward by Tang et al. [95] and applied to the photodegradation of norfloxacin in a nonionic surfactant Triton-X100 (TX100)/Bi$_2$WO$_6$ dispersion under visible light irradiation. The results found that the degradation of barely insoluble norfloxacin could be strongly enhanced with the addition of TX100. TX100 was adsorbed strongly on the Bi$_2$WO$_6$ surface and promoted norfloxacin photodegradation at the critical micelle concentration (CMC = 0.25 mM).

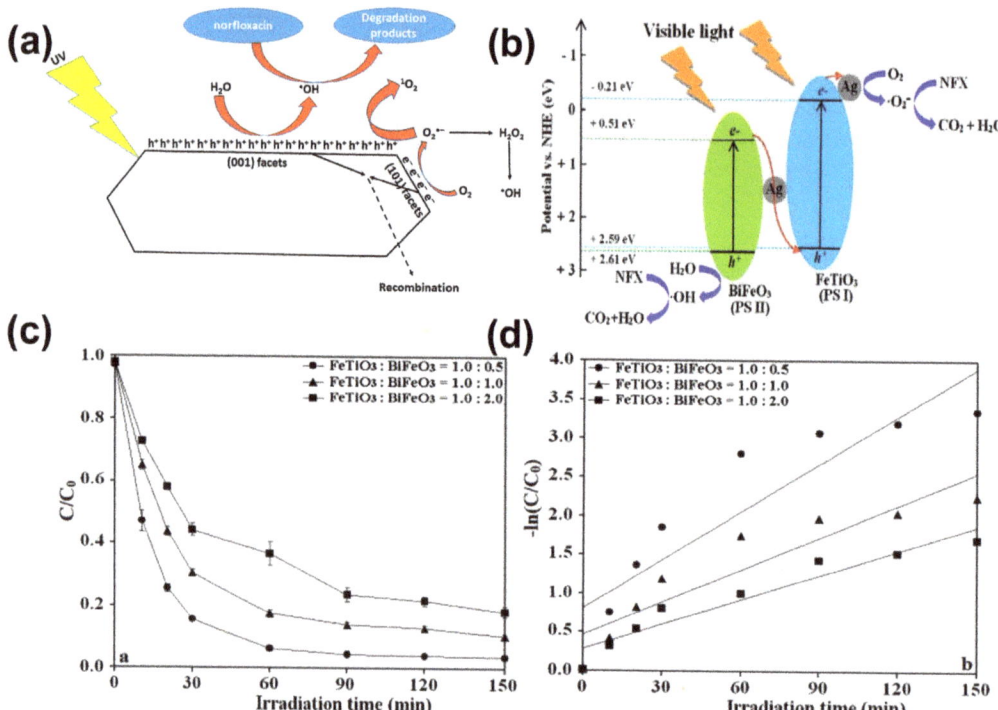

Figure 9. (**a**) Photocatalysis Mechanism of {001} Faceted TiO$_2$/Ti Film. (Reproduced with permission from [92], copyright 2016, American Chemical Society) (**b**) Possible mechanism diagram on Z-scheme Ag/FeTiO$_3$/Ag/BiFeO$_3$ system (**c**) Effect of mass ratio of FeTiO$_3$ and BiFeO$_3$ and (**d**) degradation reaction kinetics on photocatalytic activity (2.0 wt.% Ag; 5.0 mg/L norfloxacin; 1.0 g/L catalyst). (Reproduced with permission from [93], copyright 2018, Elsevier).

5.4. Photocatalytic Degradation of Amoxicillin

Amoxicillin is a penicillin-type antibiotic medicine extensively used for the treatment of various bacterial infections such as dental infections, chest infections, and other infections (ear, throat, and sinus). Figure 6d exhibits the chemical structure of amoxicillin [1]. However, amoxicillin in water or in the ecological environment is considered an emerging pollutant because it can cause several health effects to aquatic life in the presence of solved molecules in water [16]. There are some reports regarding the use of different photocatalytic materials to degrade amoxicillin [96]. For instance, the photocatalytic degradation of amoxicillin by as-prepared titanium dioxide nanoparticles loaded on graphene oxide (GO/TiO_2) by the chemical hydrothermal method was evaluated under UV light by Balarak et al., and the experimental data exhibited that key indexes such as initial pH, GO/TiO_2 dosage, UV intensity, and initial amoxicillin concentration all had a significant impact on amoxicillin degradation (Figure 10a) [97]. The efficiency of amoxicillin degradation collection was measured to be more than 99% at specific conditions, including a pH of 6, a GO/TiO_2 dosage of 0.4 g/L, an amoxicillin concentration of 50 mg/L, and an intensity of 36 W. Additionally, Mirzaei et al. [98] synthesized a new fluorinated graphite carbon nitride photocatalyst with magnetic properties by a gentle hydrothermal method that can be used for the degradation of amoxicillin in water. Compared to the bulk g-C3N4, magnetic fluorinated $Fe_3O_4/g-C_3N_4$ with a high specific surface area (243 m^2g^{-1}) resulted in improved photocatalytic activity regarding amoxicillin degradation and the mineralization of the solution. Furthermore, Huang et al. [99] also prepared novel carbon-rich g-C_3N_4 nanosheets with large surface areas by a facile thermal polymerization method, which displayed superior photocatalytic activity for amoxicillin degradation under solar light (Figure 10b–d). Meanwhile, the catalyst showed high stability and amoxicillin degradation ability under various media conditions, indicating its high applicability for amoxicillin treatment.

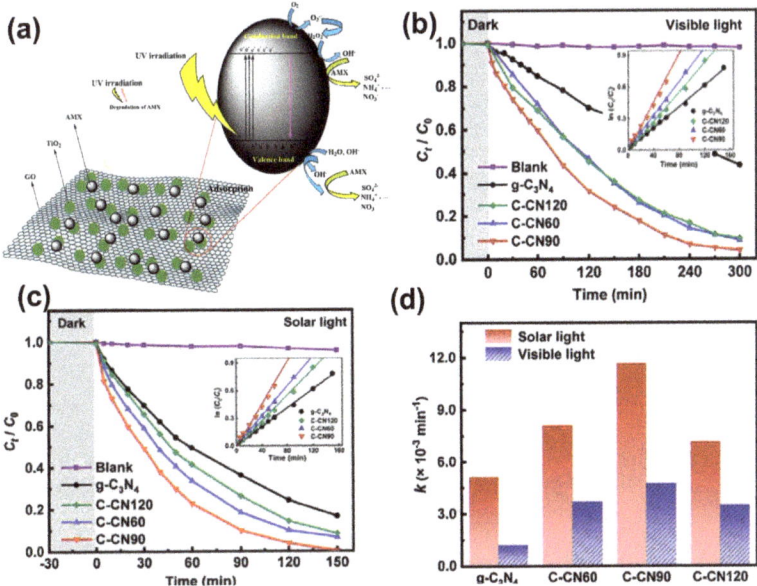

Figure 10. (a) Possible mechanism of amoxicillin degradation at GO/TiO_2 surface. (Reproduced with permission from [97], copyright 2021, Springer Nature). Photocatalytic degradation kinetics of amoxicillin by the synthesized materials under (b) visible light and (c) simulate solar light. (d) amoxicillin degradation rate constants under solar and visible light. (Reproduced with permission from [99], copyright 2021, Elsevier).

Finally, we list a table to compare the different photocatalysts discussed above, and note that the information in Table 1 shows most of the catalysts can easily and effectively remove the antibiotic contaminations within 2 h with relatively high degradation efficiency, which fully proves the superiority of rapid and effective antibiotic removal by photodegradation.

Table 1. Comparison of the photocatalytic activity of different photocatalysts for antibiotic degradation.

Antibiotic	Catalyst	Results	Degradation Mechanism	Ref.
Ciprofloxacin	Zn-doped Cu_2O	94.6% Ciprofloxacin degraded in 240 min	Mechanism by the •OH radical and h^+	[80]
Ciprofloxacin	Exfoliated g-C_3N_4	78% Ciprofloxacin degraded in 60 min	Mechanism by the •O_2^- radical and h^+	[81]
Tetracycline	Heterostructured AgI/$BiVO_4$	94.91% Tetracycline degraded in 60 min	Mechanism by the •OH, •O_2^- radical and h^+	[88]
Tetracycline	Heterostructured TiO_2/g-C_3N_4	100 mg TiO_2/g-C_3N_4 can decompose 2 mg Tetracycline in 9 min (2.2 mg/min)	Mechanism by the •O_2^- radical and h^+	[87]
Tetracycline	C–N–S-TiO_2	>99% Tetracycline degraded in 360 min	Mechanism by the •O_2^- radical and h^+	[86]
Norfloxacin	Cu-doped BiOBr	96.5% Norfloxacin degraded in 150 min	Mechanism by the h^+	[94]
Norfloxacin	Z-scheme Ag/$FeTiO_3$/Ag/$BiFeO_3$	Photocatalytic degradation rate of Norfloxacin is 0.64×10^{-2} min^{-1}	Mechanism by the •OH radical and h^+	[93]
Amoxicillin	Graphene Oxide/TiO_2	91.25% Amoxicillin degraded in 60 min	Mechanism by the h^+	[97]
Amoxicillin	Carbon-rich g-C_3N_4 nanosheets	Photocatalytic degradation rate of Amoxicillin is 0.47×10^{-2} min^{-1}	Mechanism by the •O_2^- radical	[99]

6. Conclusions and Future Perspective

In this review, the photocatalytic degradation of antibiotics was summarized. Firstly, the mechanism of photocatalytic degradation of antibiotics depending on the formation of free radicals and active oxygen species, and the consequences of antibiotics in wastewater on the environment and human health were reviewed. Some widely used antibiotics were then analyzed, and a number of commonly used photocatalysts were introduced. Heteroatom doping is generally used as a strategy to enhance the photocatalytic performance of a photocatalyst, particularly metal atoms as dopants. However, it should be noted that metal dopants could serve as recombination centers at higher concentrations, which can reduce the efficiency of a photocatalyst. Consequently, future research should also focus on other options, including doping with non-metals such as nitrogen, boron, sulfur, and phosphorus. Meanwhile, the formation of a heterojunction with other semiconductors can also play a significant role in the modification of photocatalysts on the degradation of antibiotics due to other semiconductors possibly serving as photosensitizers while simultaneously inhibiting electron-hole recombination. Thus, these methods can achieve visible light-driven photocatalysts with enhanced photocatalytic activity by narrowing the band gap of the photocatalyst or by increasing the activity of charge separation.

Physicochemical properties such as morphology and surface areas are also very critical factors in the performance of catalysts during photodegradation studies. As mentioned above, further studying photocatalysts with different morphologies and surface areas can effectively enhance the performance of catalysts. Furthermore, the degradation pathway also provides a clear introduction to the fate and transformation of antibiotics during the photocatalytic degradation process. Thus, exploring the photocatalytic degradation mechanism at the atomic level is also necessary for accelerating the efficiency of antibiotic degradation.

The utilization of solar radiation and visible light sources to activate photocatalysts during the photodegradation of antibiotics such as ciprofloxacin, tetracycline, norfloxacin, and amoxicillin is still limited. Therefore, the exploration and development of photodegradation induced by UV light sources are still in urgent demand.

Firstly, in the long term, although the removal rate of antibiotics is still being optimized, the removal rate of the chemical oxygen demand is still relatively high during the degradation of antibiotics by photocatalysts, and thus it confirms that the mineralization degree of antibiotics needs to be optimal. There are many intermediates during the process of photodegradation, therefore a deep study into intermediates is also critical for improving the performance of catalysts. Secondly, most experiments involve regular and constant stirring to prevent the agglomeration of materials in media during the degradation of antibiotics by photocatalysts, which requires additional energy consumption. Thirdly, the problems of antibiotics are not only induced by water quality but also by the accumulation of antibiotics in water, which lead to the generation of microbial resistance genes. There is currently still a lack of research on photocatalysts' limited resistance genes. Finally, The recycling ability of a photocatalyst is a significant index for evaluating its cost-effectiveness and feasibility for practical application in the degradation of antibiotics. To minimize any possible waste, the design of photocatalysts with quasi-same photoactivity during each cycle is preferred. Also, it is important to design photocatalysts that are easier to separate and recycle in order not to avoid losing any worthy materials during the photocatalytic reaction. Thus, the separation of photocatalysts from the aqueous phase is crucial from an economic standpoint. It is noted that the operating cost of a photocatalytic reaction mainly originates from its being a single-use photocatalyst, unable to be recycled. Regarding the repetitive usage of photocatalysts, deep research on how to sustainably use recyclable photocatalysts for antibiotic degradation is still urgently needed.

Funding: This work is supported by the National Natural Science Foundation of China (Grant No. 51772296).

Institutional Review Board Statement: Not applicable.

Informed Consent Statement: Not applicable.

Data Availability Statement: No data.

Conflicts of Interest: The authors declare no conflict of interest.

References

1. Abdurahman, M.H.; Abdullah, A.Z.; Shoparwe, N.F. A Comprehensive Review on Sonocatalytic, Photocatalytic, and Sonophotocatalytic Processes for the Degradation of Antibiotics in Water: Synergistic Mechanism and Degradation Pathway. *Chem. Eng. J.* **2021**, *413*, 127412. [CrossRef]
2. Calvete, M.J.F.; Piccirillo, G.; Vinagreiro, C.S.; Pereira, M.M. Hybrid Materials for Heterogeneous Photocatalytic Degradation of Antibiotics. *Coord. Chem. Rev.* **2019**, *395*, 63–85. [CrossRef]
3. Shurbaji, S.; Huong, P.T.; Altahtamouni, T.M. Review on the Visible Light Photocatalysis for the Decomposition of Ciprofloxacin, Norfloxacin, Tetracyclines, and Sulfonamides Antibiotics in Wastewater. *Catalysts* **2021**, *11*, 437. [CrossRef]
4. Li, D.; Shi, W. Recent Developments in Visible-Light Photocatalytic Degradation of Antibiotics. *Chin. J. Catal.* **2016**, *37*, 792–799. [CrossRef]
5. Wei, Z.; Liu, J.; Shangguan, W. A Review on Photocatalysis in Antibiotic Wastewater: Pollutant Degradation and Hydrogen Production. *Chin. J. Catal.* **2020**, *41*, 1440–1450. [CrossRef]
6. Velempini, T.; Prabakaran, E.; Pillay, K. Recent Developments in the Use of Metal Oxides for Photocatalytic Degradation of Pharmaceutical Pollutants in Water—A Review. *Mater. Today Chem.* **2021**, *19*, 100380. [CrossRef]
7. Shehu Imam, S.; Adnan, R.; Mohd Kaus, N.H. Photocatalytic Degradation of Ciprofloxacin in Aqueous Media: A Short Review. *Toxicol. Environ. Chem.* **2018**, *100*, 518–539. [CrossRef]
8. Cuerda-Correa, E.M.; Alexandre-Franco, M.F.; Fernández-González, C. Advanced Oxidation Processes for the Removal of Antibiotics from Water. An Overview. *Water* **2020**, *12*, 102. [CrossRef]
9. Yang, Q.; Gao, Y.; Ke, J.; Show, P.L.; Ge, Y.; Liu, Y.; Guo, R.; Chen, J. Antibiotics: An Overview on the Environmental Occurrence, Toxicity, Degradation, and Removal Methods. *Bioengineered* **2021**, *12*, 7376–7416. [CrossRef]
10. Yang, S.-F.; Lin, C.-F.; Yu-Chen Lin, A.; Andy Hong, P.-K. Sorption and Biodegradation of Sulfonamide Antibiotics by Activated Sludge: Experimental Assessment Using Batch Data Obtained Under Aerobic Conditions. *Water Res.* **2011**, *45*, 3389–3397. [CrossRef]
11. Liu, P.; Zhang, H.; Feng, Y.; Yang, F.; Zhang, J. Removal of Trace Antibiotics from Wastewater: A Systematic Study of Nanofiltration Combined with Ozone-Based Advanced Oxidation Processes. *Chem. Eng. J.* **2014**, *240*, 211–220. [CrossRef]
12. Boxi, S.S.; Paria, S. Visible Light Induced Enhanced Photocatalytic Degradation of Organic Pollutants in Aqueous Media Using Ag Doped Hollow TiO2 Nanospheres. *RSC Adv.* **2015**, *5*, 37657–37668. [CrossRef]
13. Kushwaha, H.S.; Halder, A.; Jain, D.; Vaish, R. Visible Light-Induced Photocatalytic and Antibacterial Activity of Li-Doped Bi0.5Na0.45K0.5TiO3–BaTiO3 Ferroelectric Ceramics. *J. Electron. Mater.* **2015**, *44*, 4334–4342. [CrossRef]
14. Huo, P.; Lu, Z.; Liu, X.; Wu, D.; Liu, X.; Pan, J.; Gao, X.; Guo, W.; Li, H.; Yan, Y. Preparation Photocatalyst of Selected Photodegradation Antibiotics by Molecular Imprinting Technology onto TiO_2/Fly-Ash Cenospheres. *Chem. Eng. J.* **2012**, *189–190*, 75–83. [CrossRef]
15. Yang, X.; Chen, Z.; Zhao, W.; Liu, C.; Qian, X.; Zhang, M.; Wei, G.; Khan, E.; Hau Ng, Y.; Sik Ok, Y. Recent Advances in Photodegradation of Antibiotic Residues in Water. *Chem. Eng. J.* **2021**, *405*, 126806. [CrossRef]

16. Kraemer, S.A.; Ramachandran, A.; Perron, G.G. Antibiotic Pollution in the Environment: From Microbial Ecology to Public Policy. *Microorganisms* **2019**, *7*, 180. [CrossRef]
17. De Liguoro, M.; Fioretto, B.; Poltronieri, C.; Gallina, G. The Toxicity of Sulfamethazine to Daphnia Magna and Its Additivity to Other Veterinary Sulfonamides and Trimethoprim. *Chemosphere* **2009**, *75*, 1519–1524. [CrossRef]
18. Bialy, M.; Bogacki-Rychlik, W.; Przybylski, J.; Zera, T. The Sexual Motivation of Male Rats as a Tool in Animal Models of Human Health Disorders. *Front. Behav. Neurosci.* **2019**, *13*, 257. [CrossRef]
19. Kar, P.; Shukla, K.; Jain, P.; Sathiyan, G.; Gupta, R.K. Semiconductor Based Photocatalysts for Detoxification of Emerging Pharmaceutical Pollutants from Aquatic Systems: A Critical Review. *Nano Mater. Sci.* **2021**, *3*, 25–46. [CrossRef]
20. Manyi-Loh, C.; Mamphweli, S.; Meyer, E.; Okoh, A. Antibiotic Use in Agriculture and Its Consequential Resistance in Environmental Sources: Potential Public Health Implications. *Molecules* **2018**, *23*, 795. [CrossRef]
21. Fletcher, S. Understanding the Contribution of Environmental Factors in the Spread of Antimicrobial Resistance. *Environ. Health Prev. Med.* **2015**, *20*, 243–252. [CrossRef] [PubMed]
22. Larsson, D.G.J.; Flach, C.-F. Antibiotic Resistance in the Environment. *Nat. Rev. Microbiol.* **2022**, *20*, 257–269. [CrossRef] [PubMed]
23. Uddin, T.M.; Chakraborty, A.J.; Khusro, A.; Zidan, B.R.M.; Mitra, S.; Emran, T.B.; Dhama, K.; Ripon, K.H.; Gajdács, M.; Sahibzada, M.U.K.; et al. Antibiotic Resistance in Microbes: History, Mechanisms, Therapeutic Strategies and Future Prospects. *J. Infect. Public Health* **2021**, *14*, 1750–1766. [CrossRef] [PubMed]
24. Davies Julian; Davies Dorothy Origins and Evolution of Antibiotic Resistance. *Microbiol. Mol. Biol. Rev.* **2010**, *74*, 417–433. [CrossRef]
25. Peterson, E.; Kaur, P. Antibiotic Resistance Mechanisms in Bacteria: Relationships Between Resistance Determinants of Antibiotic Producers, Environmental Bacteria, and Clinical Pathogens. *Front. Microbiol.* **2018**, *9*, 2928. [CrossRef]
26. Qin, K.; Zhao, Q.; Yu, H.; Xia, X.; Li, J.; He, S.; Wei, L.; An, T. A Review of Bismuth-Based Photocatalysts for Antibiotic Degradation: Insight into the Photocatalytic Degradation Performance, Pathways and Relevant Mechanisms. *Environ. Res.* **2021**, *199*, 111360. [CrossRef]
27. Founou, L.L.; Founou, R.C.; Essack, S.Y. Antibiotic Resistance in the Food Chain: A Developing Country-Perspective. *Front. Microbiol.* **2016**, *7*, 1881. [CrossRef]
28. Gunawan, M.A.; Hierso, J.-C.; Poinsot, D.; Fokin, A.A.; Fokina, N.A.; Tkachenko, B.A.; Schreiner, P.R. Diamondoids: Functionalization and Subsequent Applications of Perfectly Defined Molecular Cage Hydrocarbons. *New J. Chem.* **2014**, *38*, 28–41. [CrossRef]
29. Li, Y.; Fu, Y.; Zhu, M. Green Synthesis of 3d Tripyramid TiO2 Architectures with Assistance of Aloe Extracts for Highly Efficient Photocatalytic Degradation of Antibiotic Ciprofloxacin. *Appl. Catal. B Environ.* **2020**, *260*, 118149. [CrossRef]
30. Soutsas, K.; Karayannis, V.; Poulios, I.; Riga, A.; Ntampegliotis, K.; Spiliotis, X.; Papapolymerou, G. Decolorization and Degradation of Reactive Azo Dyes via Heterogeneous Photocatalytic Processes. *Desalination* **2010**, *250*, 345–350. [CrossRef]
31. Chen, X.; Yao, J.; Xia, B.; Gan, J.; Gao, N.; Zhang, Z. Influence of pH and DO on the Ofloxacin Degradation in Water by UVA-LED/TiO2 Nanotube Arrays Photocatalytic Fuel Cell: Mechanism, ROSs Contribution and Power Generation. *J. Hazard. Mater.* **2020**, *383*, 121220. [CrossRef] [PubMed]
32. Pinto, M.F.; Olivares, M.; Vivancos, Á.; Guisado-Barrios, G.; Albrecht, M.; Royo, B. (Di)triazolylidene Manganese Complexes in Catalytic Oxidation of Alcohols to Ketones and Aldehydes. *Catal. Sci. Technol.* **2019**, *9*, 2421–2425. [CrossRef]
33. Muruganandham, M.; Swaminathan, M. Solar Photocatalytic Degradation of a Reactive Azo Dye in TiO2-Suspension. *Sol. Energy Mater. Sol. Cells* **2004**, *81*, 439–457. [CrossRef]
34. Taghizadeh, A.; Taghizadeh, M.; Sabzehmeidani, M.M.; Sadeghfar, F.; Ghaedi, M. Chapter 1—Electronic Structure: From Basic Principles to Photocatalysis. In *Interface Science and Technology*; Ghaedi, M., Ed.; Elsevier: Amsterdam, The Netherlands, 2021; Volume 32, pp. 1–53, ISBN 1573-4285.
35. Sadeghfar, F.; Zalipour, Z.; Taghizadeh, M.; Taghizadeh, A.; Ghaedi, M. Chapter 2—Photodegradation Processes. In *Interface Science and Technology*; Ghaedi, M., Ed.; Elsevier: Amsterdam, The Netherlands, 2021; Volume 32, pp. 55–124, ISBN 1573-4285.
36. Sacco, O.; Vaiano, V.; Han, C.; Sannino, D.; Dionysiou, D.D. Photocatalytic Removal of Atrazine Using N-Doped TiO2 Supported on Phosphors. *Appl. Catal. B Environ.* **2015**, *164*, 462–474. [CrossRef]
37. Zhao, G.; Ding, J.; Zhou, F.; Chen, X.; Wei, L.; Gao, Q.; Wang, K.; Zhao, Q. Construction of a Visible-Light-Driven Magnetic Dual Z-Scheme BiVO4/g-C3N4/NiFe2O4 Photocatalyst for Effective Removal of Ofloxacin: Mechanisms and Degradation Pathway. *Chem. Eng. J.* **2021**, *405*, 126704. [CrossRef]
38. Topkaya, E.; Konyar, M.; Yatmaz, H.C.; Öztürk, K. Pure ZnO and Composite ZnO/TiO2 Catalyst Plates: A Comparative Study for the Degradation of Azo Dye, Pesticide and Antibiotic in Aqueous Solutions. *J. Colloid Interface Sci.* **2014**, *430*, 6–11. [CrossRef]
39. Ni, M.; Leung, M.K.H.; Leung, D.Y.C.; Sumathy, K. A Review and Recent Developments in Photocatalytic Water-Splitting Using TiO2 for Hydrogen Production. *Renew. Sustain. Energy Rev.* **2007**, *11*, 401–425. [CrossRef]
40. Djurišić, A.B.; He, Y.; Ng, A.M.C. Visible-Light Photocatalysts: Prospects and Challenges. *APL Mater.* **2020**, *8*, 030903. [CrossRef]
41. Soni, V.; Khosla, A.; Singh, P.; Nguyen, V.-H.; Le, Q.V.; Selvasembian, R.; Hussain, C.M.; Thakur, S.; Raizada, P. Current Perspective in Metal Oxide Based Photocatalysts for Virus Disinfection: A Review. *J. Environ. Manag.* **2022**, *308*, 114617. [CrossRef]
42. Zeng, J.; Li, Z.; Jiang, H.; Wang, X. Progress on Photocatalytic Semiconductor Hybrids for Bacterial Inactivation. *Mater. Horiz.* **2021**, *8*, 2964–3008. [CrossRef]

43. Haque, F.; Daeneke, T.; Kalantar-zadeh, K.; Ou, J.Z. Two-Dimensional Transition Metal Oxide and Chalcogenide-Based Photocatalysts. *Nano-Micro Lett.* **2017**, *10*, 23. [CrossRef] [PubMed]
44. FUJISHIMA, A.; HONDA, K. Electrochemical Photolysis of Water at a Semiconductor Electrode. *Nature* **1972**, *238*, 37–38. [CrossRef] [PubMed]
45. Chen, X.; He, Y.; Zhang, Q.; Li, L.; Hu, D.; Yin, T. Fabrication of Sandwich-Structured ZnO/Reduced Graphite Oxide Composite and Its Photocatalytic Properties. *J. Mater. Sci.* **2010**, *45*, 953–960. [CrossRef]
46. Semeraro, P.; Bettini, S.; Sawalha, S.; Pal, S.; Licciulli, A.; Marzo, F.; Lovergine, N.; Valli, L.; Giancane, G. Photocatalytic Degradation of Tetracycline by ZnO/γ-Fe_2O_3 Paramagnetic Nanocomposite Material. *Nanomaterials* **2020**, *10*, 1458. [CrossRef] [PubMed]
47. Mirzaeifard, Z.; Shariatinia, Z.; Jourshabani, M.; Rezaei Darvishi, S.M. ZnO Photocatalyst Revisited: Effective Photocatalytic Degradation of Emerging Contaminants Using S-Doped ZnO Nanoparticles Under Visible Light Radiation. *Ind. Eng. Chem. Res.* **2020**, *59*, 15894–15911. [CrossRef]
48. Gholami, P.; Khataee, A.; Bhatnagar, A. Photocatalytic Degradation of Antibiotic and Hydrogen Production Using Diatom-Templated 3d Wo_{3-x}@mesoporous Carbon Nanohybrid Under Visible Light Irradiation. *J. Clean. Prod.* **2020**, *275*, 124157. [CrossRef]
49. Nguyen, T.T.; Nam, S.-N.; Son, J.; Oh, J. Tungsten Trioxide (Wo_3)-Assisted Photocatalytic Degradation of Amoxicillin by Simulated Solar Irradiation. *Sci. Rep.* **2019**, *9*, 9349. [CrossRef]
50. Wang, J.; Fang, X.-C.; Liu, Y.; Fu, S.-L.; Zhuo, M.-P.; Yao, M.-D.; Wang, Z.-S.; Chen, W.-F.; Liao, L.-S. $W_{18}O_{49}$/N-Doped Reduced Graphene Oxide Hybrid Architectures for Full-Spectrum Photocatalytic Degradation of Organic Contaminants in Water. *J. Mater. Chem. C* **2021**, *9*, 829–835. [CrossRef]
51. Yang, Y.; Qiu, M.; Chen, F.; Qi, Q.; Yan, G.; Liu, L.; Liu, Y. Charge-Transfer-Mediated Photocatalysis of $W_{18}O_{49}$@CdS Nanotubes to Boost Photocatalytic Hydrogen Production. *Appl. Surf. Sci.* **2021**, *541*, 148415. [CrossRef]
52. Gurunathan, K. Photocatalytic Hydrogen Production Using Transition Metal Ions-Doped γ-Bi_2O_3 Semiconductor Particles. *Int. J. Hydrogen Energy* **2004**, *29*, 933–940. [CrossRef]
53. Chen, P.; Zhang, Q.; Zheng, X.; Tan, C.; Zhuo, M.; Chen, T.; Wang, F.; Liu, H.; Liu, Y.; Feng, Y.; et al. Phosphate-Modified m-Bi_2O_4 Enhances the Absorption and Photocatalytic Activities of Sulfonamide: Mechanism, Reactive Species, and Reactive Sites. *J. Hazard. Mater.* **2020**, *384*, 121443. [CrossRef] [PubMed]
54. Tang, J.; Zou, Z.; Ye, J. Efficient Photocatalysis on $BaBiO_3$ Driven by Visible Light. *J. Phys. Chem. C* **2007**, *111*, 12779–12785. [CrossRef]
55. He, R.; Cao, S.; Zhou, P.; Yu, J. Recent Advances in Visible Light Bi-Based Photocatalysts. *Chin. J. Catal.* **2014**, *35*, 989–1007. [CrossRef]
56. Huang, Y.; Fan, W.; Long, B.; Li, H.; Zhao, F.; Liu, Z.; Tong, Y.; Ji, H. Visible Light Bi_2S_3/Bi_2O_3/$Bi_2O_2CO_3$ Photocatalyst for Effective Degradation of Organic Pollutions. *Appl. Catal. B Environ.* **2016**, *185*, 68–76. [CrossRef]
57. Gotić, M.; Musić, S.; Ivanda, M.; Šoufek, M.; Popović, S. Synthesis and Characterisation of Bismuth(III) Vanadate. *J. Mol. Struct.* **2005**, *744–747*, 535–540. [CrossRef]
58. Wang, Z.; Huang, X.; Wang, X. Recent Progresses in the Design of BiVO4-Based Photocatalysts for Efficient Solar Water Splitting. *Catal. Today* **2019**, *335*, 31–38. [CrossRef]
59. Wang, G.; Sun, Q.; Liu, Y.; Huang, B.; Dai, Y.; Zhang, X.; Qin, X. A Bismuth-Based Metal–Organic Framework as an Efficient Visible-Light-Driven Photocatalyst. *Chem. A Eur. J.* **2015**, *21*, 2364–2367. [CrossRef]
60. Liu, W.; Liu, D.; Wang, K.; Yang, X.; Hu, S.; Hu, L. Fabrication of Z-Scheme Ag_3PO_4/TiO_2 Heterostructures for Enhancing Visible Photocatalytic Activity. *Nanoscale Res. Lett.* **2019**, *14*, 203. [CrossRef]
61. Wang, H.; Li, J.; Huo, P.; Yan, Y.; Guan, Q. Preparation of Ag_2O/Ag_2CO_3/MWNTs Composite Photocatalysts for Enhancement of Ciprofloxacin Degradation. *Appl. Surf. Sci.* **2016**, *366*, 1–8. [CrossRef]
62. Li, W.; Chen, Q.; Lei, X.; Gong, S. Fabrication of Ag/AgBr/Ag_3VO_4 Composites with High Visible Light Photocatalytic Performance. *RSC Adv.* **2019**, *9*, 5100–5109. [CrossRef]
63. Dai, G.; Yu, J.; Liu, G. A New Approach for Photocorrosion Inhibition of Ag_2CO_3 Photocatalyst with Highly Visible-Light-Responsive Reactivity. *J. Phys. Chem. C* **2012**, *116*, 15519–15524. [CrossRef]
64. Tian, L.; Sun, K.; Rui, Y.; Cui, W.; An, W. Facile Synthesis of an Ag@AgBr Nanoparticle-Decorated $K_4Nb_6O_{17}$ Photocatalyst with Improved Photocatalytic Properties. *RSC Adv.* **2018**, *8*, 29309–29320. [CrossRef]
65. Khanam, S.; Rout, S.K. A Photocatalytic Hydrolysis and Degradation of Toxic Dyes by Using Plasmonic Metal–Semiconductor Heterostructures: A Review. *Chemistry* **2022**, *4*, 454–479. [CrossRef]
66. Du, C.; Zhang, Z.; Yu, G.; Wu, H.; Chen, H.; Zhou, L.; Zhang, Y.; Su, Y.; Tan, S.; Yang, L.; et al. A Review of Metal Organic Framework (MOFs)-Based Materials for Antibiotics Removal via Adsorption and Photocatalysis. *Chemosphere* **2021**, *272*, 129501. [CrossRef] [PubMed]
67. Zhang, J.; Xiang, S.; Wu, P.; Wang, D.; Lu, S.; Wang, S.; Gong, F.; Wei, X.; Ye, X.; Ding, P. Recent Advances in Performance Improvement of Metal-Organic Frameworks to Remove Antibiotics: Mechanism and Evaluation. *Sci. Total Environ.* **2022**, *811*, 152351. [CrossRef] [PubMed]
68. Wang, F.; Xue, R.; Ma, Y.; Ge, Y.; Wang, Z.; Qiao, X.; Zhou, P. Study on the Performance of a MOF-808-Based Photocatalyst Prepared by a Microwave-Assisted Method for the Degradation of Antibiotics. *RSC Adv.* **2021**, *11*, 32955–32964. [CrossRef]

69. Gheytanzadeh, M.; Baghban, A.; Habibzadeh, S.; Jabbour, K.; Esmaeili, A.; Mohaddespour, A.; Abida, O. An Insight into Tetracycline Photocatalytic Degradation by MOFs Using the Artificial Intelligence Technique. *Sci. Rep.* **2022**, *12*, 6615. [CrossRef]
70. Li, X.; Zhang, D.; Bai, R.; Mo, R.; Yang, C.; Li, C.; Han, Y. Zr-MOFs Based BiOBr/UiO-66 Nanoplates with Enhanced Photocatalytic Activity for Tetracycline Degradation Under Visible Light Irradiation. *AIP Adv.* **2020**, *10*, 125228. [CrossRef]
71. Dong, G.; Zhang, Y.; Pan, Q.; Qiu, J. A Fantastic Graphitic Carbon Nitride (g-C3n4) Material: Electronic Structure, Photocatalytic and Photoelectronic Properties. *J. Photochem. Photobiol. C Photochem. Rev.* **2014**, *20*, 33–50. [CrossRef]
72. Wang, X.; Maeda, K.; Thomas, A.; Takanabe, K.; Xin, G.; Carlsson, J.M.; Domen, K.; Antonietti, M. A Metal-Free Polymeric Photocatalyst for Hydrogen Production from Water Under Visible Light. *Nat. Mater.* **2009**, *8*, 76–80. [CrossRef]
73. Alaghmandfard, A.; Ghandi, K. A Comprehensive Review of Graphitic Carbon Nitride (g-C3n4)–Metal Oxide-Based Nanocomposites: Potential for Photocatalysis and Sensing. *Nanomaterials* **2022**, *12*, 294. [CrossRef] [PubMed]
74. Li, G.; Wang, B.; Zhang, J.; Wang, R.; Liu, H. Er-Doped g-C3n4 for Photodegradation of Tetracycline and Tylosin: High Photocatalytic Activity and Low Leaching Toxicity. *Chem. Eng. J.* **2020**, *391*, 123500. [CrossRef]
75. Pattanayak, D.S.; Pal, D.; Mishra, J.; Thakur, C. Noble Metal–Free Doped Graphitic Carbon Nitride (g-C3n4) for Efficient Photodegradation of Antibiotics: Progress, Limitations, and Future Directions. *Environ. Sci. Pollut. Res.* **2022**, 1–13. [CrossRef] [PubMed]
76. Dong, J.; Zhang, Y.; Hussain, M.I.; Zhou, W.; Chen, Y.; Wang, L.-N. G-C3n4: Properties, Pore Modifications, and Photocatalytic Applications. *Nanomaterials* **2022**, *12*, 121. [CrossRef] [PubMed]
77. Malakootian, M.; Nasiri, A.; Amiri Gharaghani, M. Photocatalytic Degradation of Ciprofloxacin Antibiotic by TiO2 Nanoparticles Immobilized on a Glass Plate. *Chem. Eng. Commun.* **2020**, *207*, 56–72. [CrossRef]
78. Campoli-Richards, D.M.; Monk, J.P.; Price, A.; Benfield, P.; Todd, P.A.; Ward, A. Ciprofloxacin. *Drugs* **1988**, *35*, 373–447. [CrossRef]
79. Fief, C.A.; Hoang, K.G.; Phipps, S.D.; Wallace, J.L.; Deweese, J.E. Examining the Impact of Antimicrobial Fluoroquinolones on Human DNA Topoisomerase IIα and IIβ. *ACS Omega* **2019**, *4*, 4049–4055. [CrossRef]
80. Yu, X.; Zhang, J.; Zhang, J.; Niu, J.; Zhao, J.; Wei, Y.; Yao, B. Photocatalytic Degradation of Ciprofloxacin Using Zn-Doped Cu2O Particles: Analysis of Degradation Pathways and Intermediates. *Chem. Eng. J.* **2019**, *374*, 316–327. [CrossRef]
81. Pattnaik, S.P.; Behera, A.; Martha, S.; Acharya, R.; Parida, K. Facile Synthesis of Exfoliated Graphitic Carbon Nitride for Photocatalytic Degradation of Ciprofloxacin Under Solar Irradiation. *J. Mater. Sci.* **2019**, *54*, 5726–5742. [CrossRef]
82. Chopra Ian; Roberts Marilyn Tetracycline Antibiotics: Mode of Action, Applications, Molecular Biology, and Epidemiology of Bacterial Resistance. *Microbiol. Mol. Biol. Rev.* **2001**, *65*, 232–260. [CrossRef]
83. Florou, D.T.; Mavropoulos, A.; Dardiotis, E.; Tsimourtou, V.; Siokas, V.; Aloizou, A.-M.; Liaskos, C.; Tsigalou, C.; Katsiari, C.; Sakkas, L.I.; et al. Tetracyclines Diminish In Vitro IFN-γ and IL-17-Producing Adaptive and Innate Immune Cells in Multiple Sclerosis. *Front. Immunol.* **2021**, *12*, 4099. [CrossRef] [PubMed]
84. Tettey, M.; Sereboe, L.; Edwin, F.; Frimpong-Boateng, K. Tetracycline Pleurodesis for Malignant Pleural Effusion-A Review of 38 Cases. *Ghana Med. J.* **2005**, *39*, 128–131.
85. Daghrir, R.; Drogui, P. Tetracycline Antibiotics in the Environment: A Review. *Environ. Chem. Lett.* **2013**, *11*, 209–227. [CrossRef]
86. Wang, P.; Yap, P.-S.; Lim, T.-T. C–N–S Tridoped TiO2 for Photocatalytic Degradation of Tetracycline Under Visible-Light Irradiation. *Appl. Catal. A Gen.* **2011**, *399*, 252–261. [CrossRef]
87. Wang, W.; Fang, J.; Shao, S.; Lai, M.; Lu, C. Compact and Uniform TiO2@g-C3n4 Core-Shell Quantum Heterojunction for Photocatalytic Degradation of Tetracycline Antibiotics. *Appl. Catal. B Environ.* **2017**, *217*, 57–64. [CrossRef]
88. Chen, F.; Yang, Q.; Sun, J.; Yao, F.; Wang, S.; Wang, Y.; Wang, X.; Li, X.; Niu, C.; Wang, D.; et al. Enhanced Photocatalytic Degradation of Tetracycline by AgI/BiVO4 Heterojunction Under Visible-Light Irradiation: Mineralization Efficiency and Mechanism. *ACS Appl. Mater. Interfaces* **2016**, *8*, 32887–32900. [CrossRef]
89. Ezelarab, H.A.A.; Abbas, S.H.; Hassan, H.A.; Abuo-Rahma, G.E.-D.A. Recent Updates of Fluoroquinolones as Antibacterial Agents. *Arch. Der. Pharm.* **2018**, *351*, 1800141. [CrossRef]
90. Janecko, N.; Pokludova, L.; Blahova, J.; Svobodova, Z.; Literak, I. Implications of Fluoroquinolone Contamination for the Aquatic Environment—A Review. *Environ. Toxicol. Chem.* **2016**, *35*, 2647–2656. [CrossRef]
91. Pretali, L.; Fasani, E.; Sturini, M. Current Advances on the Photocatalytic Degradation of Fluoroquinolones: Photoreaction Mechanism and Environmental Application. *Photochem. Photobiol. Sci.* **2022**, *21*, 899–912. [CrossRef]
92. Sayed, M.; Shah, L.A.; Khan, J.A.; Shah, N.S.; Nisar, J.; Khan, H.M.; Zhang, P.; Khan, A.R. Efficient Photocatalytic Degradation of Norfloxacin in Aqueous Media by Hydrothermally Synthesized Immobilized TiO2/Ti Films with Exposed {001} Facets. *J. Phys. Chem. A* **2016**, *120*, 9916–9931. [CrossRef]
93. Tang, J.; Wang, R.; Liu, M.; Zhang, Z.; Song, Y.; Xue, S.; Zhao, Z.; Dionysiou, D.D. Construction of Novel Z-Scheme Ag/FeTiO3/Ag/BiFeO3 Photocatalyst with Enhanced Visible-Light-Driven Photocatalytic Performance for Degradation of Norfloxacin. *Chem. Eng. J.* **2018**, *351*, 1056–1066. [CrossRef]
94. Lv, X.; Yan, D.Y.S.; Lam, F.L.-Y.; Ng, Y.H.; Yin, S.; An, A.K. Solvothermal Synthesis of Copper-Doped BiOBr Microflowers with Enhanced Adsorption and Visible-Light Driven Photocatalytic Degradation of Norfloxacin. *Chem. Eng. J.* **2020**, *401*, 126012. [CrossRef]
95. Tang, L.; Wang, J.; Zeng, G.; Liu, Y.; Deng, Y.; Zhou, Y.; Tang, J.; Wang, J.; Guo, Z. Enhanced Photocatalytic Degradation of Norfloxacin in Aqueous Bi2WO6 Dispersions Containing Nonionic Surfactant Under Visible Light Irradiation. *J. Hazard. Mater.* **2016**, *306*, 295–304. [CrossRef] [PubMed]

96. Lalliansanga; Tiwari, D.; Lee, S.-M.; Kim, D.-J. Photocatalytic Degradation of Amoxicillin and Tetracycline by Template Synthesized Nano-Structured Ce3+@TiO2 Thin Film Catalyst. *Environ. Res.* **2022**, *210*, 112914. [CrossRef] [PubMed]
97. Balarak, D.; Mengelizadeh, N.; Rajiv, P.; Chandrika, K. Photocatalytic Degradation of Amoxicillin from Aqueous Solutions by Titanium Dioxide Nanoparticles Loaded on Graphene Oxide. *Environ. Sci. Pollut. Res.* **2021**, *28*, 49743–49754. [CrossRef]
98. Mirzaei, A.; Chen, Z.; Haghighat, F.; Yerushalmi, L. Magnetic Fluorinated Mesoporous g-C3n4 for Photocatalytic Degradation of Amoxicillin: Transformation Mechanism and Toxicity Assessment. *Appl. Catal. B Environ.* **2019**, *242*, 337–348. [CrossRef]
99. Huang, D.; Sun, X.; Liu, Y.; Ji, H.; Liu, W.; Wang, C.-C.; Ma, W.; Cai, Z. A Carbon-Rich g-C3n4 with Promoted Charge Separation for Highly Efficient Photocatalytic Degradation of Amoxicillin. *Chin. Chem. Lett.* **2021**, *32*, 2787–2791. [CrossRef]

Article

Nitrogen-Doped Bismuth Nanosheet as an Efficient Electrocatalyst to CO$_2$ Reduction for Production of Formate

Sanxiu Li [1,†], Yufei Kang [1,†], Chenyang Mo [1], Yage Peng [1], Haijun Ma [2] and Juan Peng [1,*]

1. State Key Laboratory of High-Efficiency Utilization of Coal and Green Chemical Engineering, College of Chemistry and Chemical Engineering, Ningxia University, Yinchuan 750021, China
2. Key Laboratory of Ministry of Education for Protection and Utilization of Special Biological Resources in Western China, School of Life Sciences, Ningxia University, Yinchuan 750021, China
* Correspondence: pengjuan@nxu.edu.cn
† These authors contributed equally to this work.

Abstract: Electrochemical CO$_2$ reduction (CO$_2$RR) to produce high value-added chemicals or fuels is a promising technology to address the greenhouse effect and energy challenges. Formate is a desirable product of CO$_2$RR with great economic value. Here, nitrogen-doped bismuth nanosheets (N-BiNSs) were prepared by a facile one-step method. The N-BiNSs were used as efficient electrocatalysts for CO$_2$RR with selective formate production. The N-BiNSs exhibited a high formate Faradic efficiency (FE$_{formate}$) of 95.25% at −0.95 V (vs. RHE) with a stable current density of 33.63 mA cm^{-2} in 0.5 M KHCO$_3$. Moreover, the N-BiNSs for CO$_2$RR yielded a large current density (300 mA cm^{-2}) for formate production in a flow-cell measurement, achieving the commercial requirement. The FE$_{formate}$ of 90% can maintain stability for 14 h of electrolysis. Nitrogen doping could induce charge transfer from the N atom to the Bi atom, thus modulating the electronic structure of N-Bi nanosheets. DFT results demonstrated the N-BiNSs reduced the adsorption energy of the *OCHO intermediate and promoted the mass transfer of charges, thereby improving the CO$_2$RR with high FE$_{formate}$. This study provides a valuable strategy to enhance the catalytic performance of bismuth-based catalysts for CO$_2$RR by using a nitrogen-doping strategy.

Keywords: electrocatalysis; electronic structure; flow cell; large current density

1. Introduction

With rapid economic development and the extensive use of fossil fuels, carbon dioxide (CO$_2$) in the atmosphere continues to increase. Excessive amounts of CO$_2$ emissions will cause severe global warming and sea levels to rise. Therefore, the necessary measures to reduce the environmental impact of CO$_2$ are of great importance. Among the various existing CO$_2$ conversion technologies, the electrochemical CO$_2$ reduction reaction (CO$_2$RR) is a prospective strategy to produce fuels or value-added chemicals. Electrochemical CO$_2$RR can convert CO$_2$ into CO [1], formate [2], methanol [3], ethanol [4,5], and hydrocarbons [6–8] under room temperature and atmospheric pressure. Among the CO$_2$ reduction products, formate has attracted great interest because of its commercial value. Mainly, formate is an essential feedstock for the pharmaceutical and chemical industries. Therefore, it is significant to develop an efficient way to produce formate.

The practical application of electrochemical CO$_2$RR to formate production is constrained owing to the following factors, such as high overpotential, colossal cost, low selectivity, and poor stability [9]. In addition, since electrochemical CO$_2$RR usually occurs in an aqueous solution, it will be accompanied by a hydrogen evolution reaction (HER), resulting in low Faradaic efficiency (FE) of the product. So far, many metal-based catalysts, including Pb [10], In [11], Sb [12,13], Sn [14], Co [15], Cu [16,17], Bi [18], and Pd [19] were designed and developed to improve the efficiency and selectivity of CO$_2$RR to produce the formate in an aqueous solution. Bi-based catalysts are popular because of the selective

production of formate, low cost, and ability to inhibit the occurrence of adverse reaction HER in an aqueous solution [20]. In addition, Bi-based catalysts tend to stabilize *OCHO, an essential intermediate in formate formation [21]. However, the catalytic efficiency and selectivity of Bi-based nanomaterials are far from meeting the requirements of CO_2RR [22]. Therefore, there is an urgent need to develop effective and straightforward pathways to improve catalytic efficiency and selectivity. As demonstrated recently, plenty of studies have exhibited that the properties of CO_2RR were improved by changing the structure of the catalyst, such as size and morphology [23–25]. Two-dimensional (2D) nanostructures with high conductivity and abundant active sites have been extensively studied as potential electrocatalysts, which hold a key to enhanced CO_2RR activity [26,27]. Meanwhile, the FE of target products was improved via surface modification, doping hetero-atoms, and defects [28,29]. Especially for doping hetero-atoms, such as N, S, or O, this tactic would optimize the electronic structure of the catalyst and provide more active sites. For instance, Wu et al. [30] used this strategy to prepare the nitrogen-doped Sn(S) nanosheets. It can reach the highest FE of formate of 93.3% at −0.7 V versus reversible hydrogen electrode (vs. RHE). Liu et al. [31] prepared a Bi_2O_3-NGQDs catalyst and achieved the FE of formate was nearly 100% at −0.71 V (vs. RHE). Besides, Kang et al. [32] prepared nitrogen-doped SnO_2/C material that could enhance electrocatalytic CO_2RR activity, showing high FE (90%) for formate at –0.65 V vs. RHE. Therefore, doping hetero-atoms can increase the amounts of active sites and optimize the electronic structure to achieve high formate current density and selectivity.

Herein, we developed a one-step activation and nitrogen-doping combination method using bismuth citrate as the Bi precursor, and the combination of $Ca(OH)_2$ and NH_4Cl as an activator and a nitrogen source to prepare the nitrogen-doped bismuth nanosheets (N-BiNSs). The N-BiNSs can be used to electrocatalyze CO_2RR to produce formate with high efficiency. N-doping improved the electrical conductivity, which will make up for the poor conductivity of bismuth-based materials. The theoretical study shows that the outstanding selectivity could be attributed to the change in the electronic structure of bismuth after doping N atoms. At the same time, N-BiNSs reduced the adsorption energy of *OCHO intermediate and promoted the mass transfer of charge. The optimal adsorption *OCHO intermediate promoted formate formation while inhibiting the CO product pathway, thereby enhancing the selectivity of CO_2RR for formate.

2. Results and Discussion
2.1. Morphology and Structure Analysis

In this work, nitrogen-doped bismuth nanosheets (N-BiNSs) were prepared by a simple and rapid one-pot method. A series of N-BiNSs-X was successfully constructed by adjusting the amount of $Ca(OH)_2$ and NH_4Cl in different proportions to provide an activating agent and a nitrogen source (Figure 1). To study the phase composition of N-BiNSs, the catalyst was determined by X-ray diffraction (XRD) measurement. It can be drawn from the XRD pattern in Figure 2a, the N-BiNSs exhibited different peaks at 27.12°, 37.95°, and 39.62°, which were indexed to the (012), (104), and (110) planes of tripartite crystal system Bi (PDF#85-1329). Compared with BiNSs catalyst, the N-BiNSs had stronger (110) and (012) crystal planes. The morphologies of the BiNSs and N-BiNSs were studied by scanning electron microscopy (SEM), as shown in Supplementary Figure S1a,c and Figure 2b. It can be seen that the BiNSs showed the stacked flakiness morphology of bulk Bi (Supplementary Figure S1c), while the N-BiNSs showed a uniform 2D nanosheet structure with a large thin layer area (Figure 2b). This 2D nanostructure was very beneficial for increasing surface area and abundant active sites. Additionally, the transmission electron microscopy (TEM) observations (Figure 2c) revealed the ultrathin feature of N-BiNSs. These nanosheets were ubiquitously present. Besides, the high-resolution transmission microscopic (HRTEM) images of N-BiNSs display a lattice spacing of about 0.272 nm that corresponds to the (110) crystal plane of the tripartite crystal system Bi. The lattice band gap of the material was very uniform. As shown in Figure 2e, the high-angle annular dark field scanning TEM (HAADF-STEM) images also exhibited the morphology of nanosheets. Moreover, the N-BiNSs structure was confirmed again by the dispersive energy X-ray (EDX) elemental mapping. The Bi (red) and N (green) elements were well distributed on the whole surface of N-BiNSs (Figure 2f–h). Meanwhile, the EDS elemental mapping and line scan (Supplementary Figure S1b) were first recorded to demonstrate Bi and N distribution on N-Bi nanosheets accompanied by a disparate atomic ratio of 97.85% (Bi) and 2.85% (N). The above results suggested that nitrogen had been triumphantly incorporated into the BiNSs catalyst.

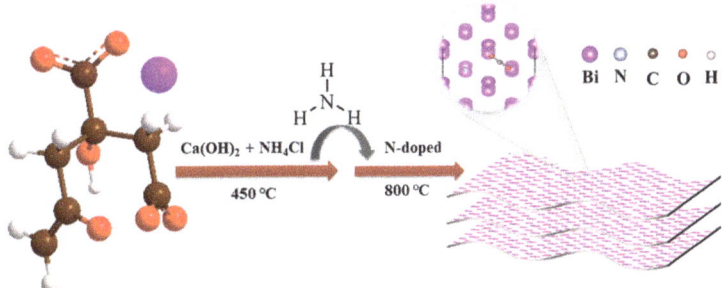

Figure 1. The illustration scheme of N-BiNSs preparation.

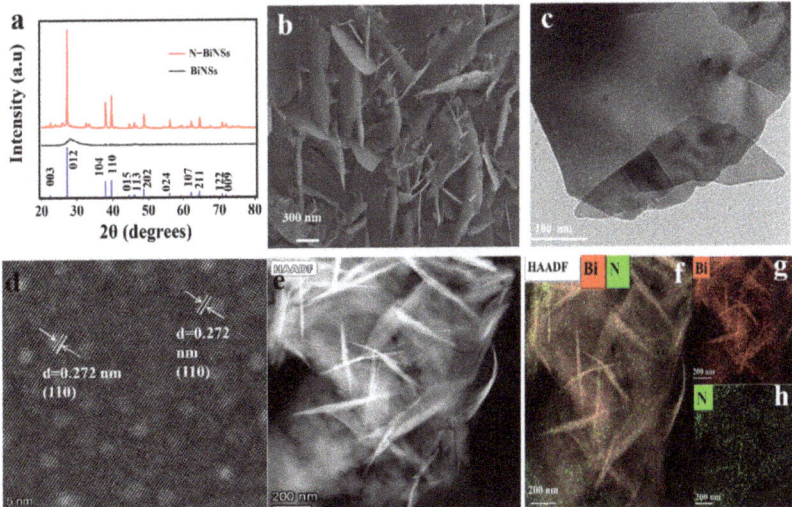

Figure 2. (a) XRD pattern of BiNSs and N-BiNSs. Characterization of N-BiNSs: (b) SEM image, (c) TEM image, (d) HR-TEM image, (e) HAADF-STEM and (f–h) EDS mapping (green and red represent N, Bi element, respectively).

X-ray photoelectron spectroscopy (XPS) was also used to characterize the surface chemical constituents of the specimens and the valence state of the nanomaterials. The position of the C(sp2) peak in the C1s spectrum was taken as the reference value of 284.8 eV, and the obtained XPS spectrum was calibrated. Figure 3a,b reveals the detailed Bi 4f spectra by comparing the XPS spectra of BiNSs and N-BiNSs. For the BiNSs catalyst, binding energies at 165 eV and 159.6 eV were belong to the Bi^{3+} $4f_{5/2}$ and $4f_{7/2}$ [23], respectively. However, binding energies for Bi^{3+} $4f_{5/2}$ and $4f_{7/2}$ in N-BiNSs shifted to 164.5 and 159.11 eV, respectively. These could be caused by the charge transfer from the N to the Bi atom, thus optimizing the electronic structure of N-BiNSs. Furthermore, the N1s peaks of N-BiNSs can be split into three peaks (Figure 3c), pyridinic-N (397.6 eV), N-oxidized (404.6 eV), and Nitrate (405.4 eV) [12,33–35], respectively. XPS analyses showed that the atomic ratio of N was 0.41% and pyridinic N accounted for the largest proportion, it indicated that the N element was doped successfully in BiNSs.

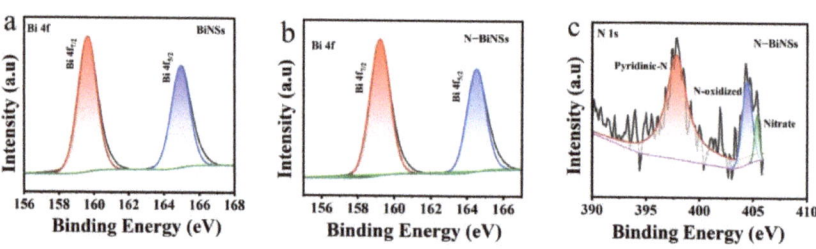

Figure 3. (a) XPS Bi 4f spectrum of BiNSs and (b) N-BiNSs; (c) XPS N 1s spectrum of N-BiNSs.

2.2. Electrocatalytic Perfomance

To estimate the CO_2 reduction performance of BiNSs, N-BiNSs, N-BiNSs-1, and N-BiNSs-2, the catalytic reactions are applied in a proton exchange membrane (PEM) separated two-compartment cell. Linear sweep voltammetry (LSV) curves (Supplementary Figure S3) of N-BiNSs were measured in both CO_2-saturated and Ar-saturated 0.5 M $KHCO_3$ electrolytes. The N-BiNSs catalyst exhibited a larger current in the CO_2-saturated electrolyte than in N_2. The current density increases sharply from −0.6 V (vs. RHE), reaching about 30 mA cm^{-2} at −0.8 V (vs. RHE) to about 50 mA cm^{-2} at −1.2 V (vs. RHE). The polarization curves stated that the catalyst has higher activity to CO_2RR. Meanwhile, the LSV curves (Figure 4a) were tested in the CO_2-saturated 0.5 M $KHCO_3$ electrolyte. Strikingly, for the N-BiNSs catalyst, the current density in the CO_2-saturated atmosphere was higher than the BiNSs catalysts, indicating the introduction of N could improve the electrocatalytic activity of Bi on CO_2RR. Moreover, comparing the three samples with different N doping amounts, the N-BiNSs show the highest current with the optimized Nitrogen amount of 2.85%.

To evaluate the selectivity of CO_2RR, electrolysis was performed in a CO_2-saturated 0.5 M $KHCO_3$ aqueous solution at various applied potentials. The gaseous and liquid products were quantitatively analyzed via gas chromatography (GC) and ion chromatography (IC) (Supplementary Figure S2), respectively. As shown in Figure 4b, the FE$_{formate}$ of N-BiNSs was higher than BiNSs, N-BiNSs-1, and N-BiNSs-2 at all applied potentials. To investigate the catalytic activity of N-BiNSs, the CO_2RR was carried out at different potentials between −0.4 V and −1.3 V (vs. RHE). As the applied potential changes, the FE of the CO_2RR products was displayed in Figure 4c. The N-BiNSs also showed that formate was the main product of CO_2RR. The FE value of CO was lower than 20% in the whole potential window, and the FE value of H_2 decreased significantly from 60% at −0.5 V to 2–3% at −1.0 V (vs. RHE), the FE$_{formate}$ at −0.95 V (vs. RHE) was 95.25%. In contrast, for the BiNSs without nitrogen element, as shown in Supplementary Figure S4a, the FE values of formate, CO, and H_2 at −0.95 V (vs. RHE) were 81.54%, 3.09%, and 10.56%, respectively. This indicated that the nitrogen-doped bismuth catalyst was beneficial to improve the selectivity of formate. Meanwhile, in the N-BiNSs-1 catalyst with a small amount of nitrogen, the FE$_{formate}$ could reach 88.94% at −0.95 V (vs. RHE) in Supplementary Figure S4b. For the catalyst N-BiNSs-2 with a higher amount of doping nitrogen, the maximum FE of formate in the H-shaped cell reached 78.63% in Supplementary Figure S4c. It could be concluded that varying N doping amounts lead to the formation of different catalyst morphologies, resulting in different activities. Supplementary Figure S4d showed the constant potential electrolysis of CO_2 under a series of potentials. The stable current density showed that the N-BiNSs catalyst had good electrochemical stability in the CO_2RR test. Furthermore, the formate partial current densities (j$_{formate}$) of the N-BiNSs were measured in the whole potential region in Figure 4d, with a maximum value of j$_{formate}$ = 45 mA cm^{-2} at −1.2 V (vs. RHE). To better understand the activity and kinetics of N-BiNSs materials for CO$_2$RR, the Tafel diagram of the catalyst was obtained in the low current density area. In Supplementary Figure S5a, the slope value obtained by the N-BiNSs catalyst was 184.13 mV dec^{-1}, indicating that the electron transfer rate of this catalyst was relatively fast, thus facilitating the adsorption and desorption of CO$_2$* intermediate on the N-BiNSs catalyst surface [36].

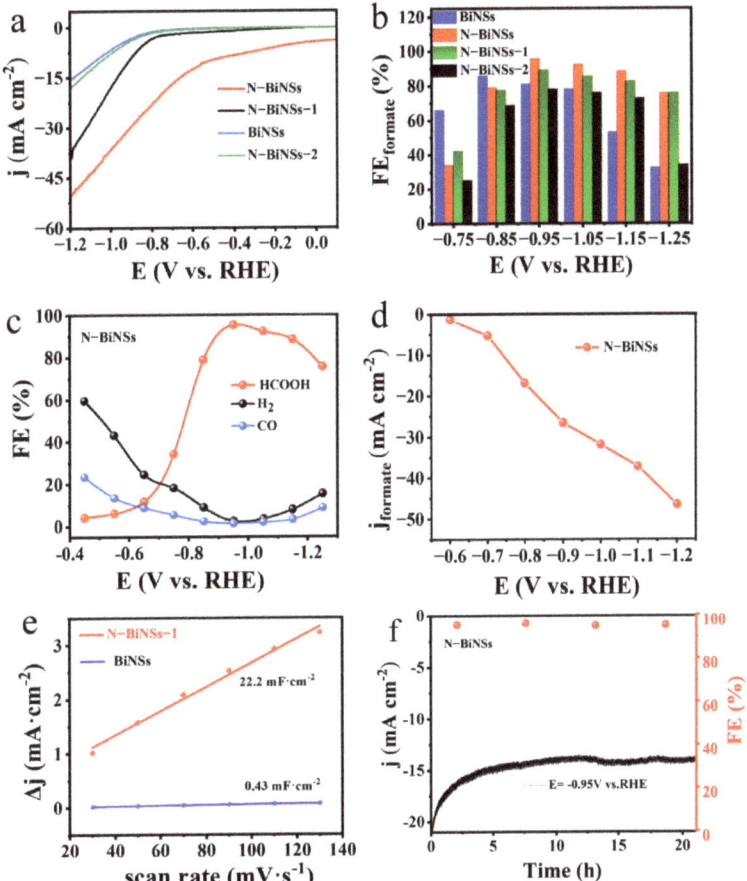

Figure 4. CO$_2$RR performances of BiNSs, N-BiNSs, N-BiNSs-1, and N-BiNSs-2: (**a**) Linear sweep voltammetry curves in 0.5 M KHCO$_3$ aqueous solutions with saturated gases CO$_2$, sweeping speed of 5 mV s^{-1}; (**b**) the product FE$_{formate}$ at different applied potentials; (**c**) corresponding FE of N-BiNSs at various potentials; (**d**) the N-BiNSs of j$_{formate}$ recorded at different potentials in 0.5 M KHCO$_3$; (**e**) the BiNSs and N-BiNSs the relationship between charge current density difference (ΔJ) and scanning rate; (**f**) stability test of N-BiNSs in the H-type electrolytic cell and Faraday efficiency test of formate production.

More importantly, the CO$_2$RR activity was also related to the electrochemical active surface area (ECSA) of the catalyst. To evaluate the ECSA of BiNSs, N-BiNSs, N-BiNSs-1, and N-BiNSs-2, the double-layer capacitance (Cdl) was calculated. According to the cyclic voltammograms (CVs) of BiNSs, N-BiNSs, N-BiNSs-1, and N-BiNSs-2 at different sweep speeds in the potential region of 0.12 V–0.22 V (vs. RHE) (Supplementary Figure S6a–d). It can be seen from Figure 4e and Supplementary Figure S5c that the capacitance values of BiNSs, N-BiNSs, N-BiNSs-1, and N-BiNSs-2 were 0.43 mF cm^{-2}, 22.2 mF cm^{-2},

0.56 mF cm^{-2}, and 0.075 mF cm^{-2}, respectively. These results indicated that the Cdl of N-BiNSs electrocatalyst was highest, which can provide abundant catalytic active sites for increasing the electrocatalytic performance of CO_2RR. At the same time, we measured the impedance of some different catalysts at open circuit voltage and obtained the Nyquist diagram (Supplementary Figure S5b) to explore the kinetic processes among the catalyst interfaces. The N-BiNSs material corresponds to the most minor semicircle among the three. The results showed that the interfacial charge could be transferred rapidly to improve the catalytic activity in the reaction process, which was consistent with our inferred results.

We further studied the long-term durability of the material for CO_2RR, as shown in Figure 4f. The material was tested for electrolysis at −0.95 V (vs. RHE) for about 20 h, and the FE of formate remained very stable, basically around 95%, indicating that the N-BiNSs material had significant durability to CO_2RR. A comparison of the performance of N-Bi nanosheets in electrocatalytic CO_2RR with other representative electrocatalysts is in the literature (Supplementary Table S1). It was noteworthy that the N-BiNSs catalyst showed the morphology of the nanosheets was maintained and became thinner after a period of electrolysis process (Supplementary Figure S7a). The crystal shape of the nanosheets remained after prolonged electrolysis (Supplementary Figure S7b). It was proved that the catalyst had excellent morphology stability.

To eliminate CO_2 mass transfer constraints in the H-cell and achieve a commercially viable high current density (≈200 mA cm^{-2}), a flow cell reactor was assembled using catalysts coated on the gas diffusion layer, carbon paper of 2 × 3 cm size, and commercial mercury oxide anion exchange membrane (Supplementary Figure S8). In this unit, carbon dioxide gas can react at the gas-liquid-solid three-phase boundary. Peristaltic pumps and gas-liquid mixed flow pumps are installed on the cathode and anode to remove liquid or gas products, keeping the pH of the electrolyte constant and fully contacting the electrode surface. Then, we systematically evaluated the CO_2RR performance of the N-BiNSs catalyst on a carbon-based gas diffusion layer (C-GDL) substrate, as shown in Figure 5a,b. 1.0 M KOH aqueous solution was used as a flow-electrolyte. The alkaline electrolyte not only effectively inhibits HER, but also effectively reduces the activation energy barrier. Through the LSV curve, it can be displayed very clearly that the current density of the catalyst has met the commercial requirements, as shown in Figure 5a. Notably, the catalyst was maintained for more than 14 h at a high current density of ~300 mA cm^{-2} at a potential of −1.2 V (vs. RHE) in Figure 5b. The catalyst had significant durability at high current density, with the average selectivity of the catalyst being about 89.30%. After 14 h of electrolysis, due to the existing structure of the flow tank, flooding, seepage, and other problems occurred. Therefore, the optimization of the flow tank structure is still an urgent problem to be solved.

The two-electrode electrolyzer was assembled by utilizing the N-BiNSs loaded carbon paper as a cathode for CO_2RR and IrO_2 loaded on carbon paper as an anode for the oxygen evolution reaction (OER). As shown in Figure 5c, the polarization curve of the N-BiNSs | | IrO_2 cell exhibits electrocatalytic performance, with a current density of 6.19 mA cm^{-2} at 3.0 V, capable of delivering 10 mA cm^{-2} at 3.4V. It was worth noting that the initial current density was maintained at 6 mA cm^{-2}. After 11 h of continuous reaction, the current density was maintained at 5.4 mA cm^{-2}, the current density decreased by about 9.3%, and the stability was pretty, as shown in Figure 5d. The OER electrocatalyst IrO_2 may also be replaced by non-noble metal-based materials [37].

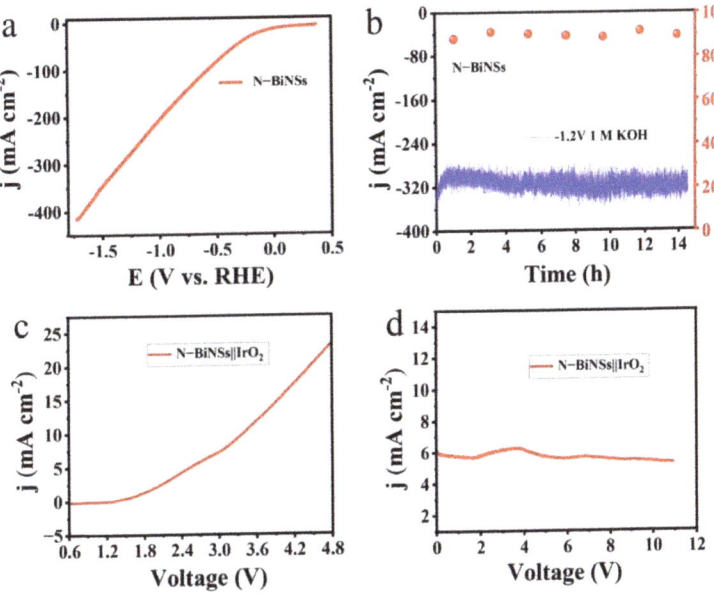

Figure 5. (a) The LSV curve in a flow cell under CO_2 atmosphere; (b) Stability test of the material at −1.2 V vs. RHE and corresponding Faraday efficiency test of formate production of N-BiNSs; (c) Polarization curve of N-BiNSs||IrO$_2$ couple in the two-electrode system; (d) Chronoamperometry measurement of N-BiNSs||IrO$_2$ couple at 3 V.

2.3. Catalytic Mechanisms Revealed by DFT Computations

All the above electrochemical performance studies proved that the N-BiNSs catalysts had excellent activity and selectivity compared with pure BiNSs. In order to further explore the reasons why the catalyst improves CO_2RR activity and selectivity, and determine the reaction mechanism of formate, the density functional theory (DFT) calculation was used to simulate and compare the CO_2RR pathway on N-BiNSs and BiNSs surfaces. Figure 6a,b depicts the optimized adsorption geometries and their energy distributions for *OCHO (intermediate to formate) on N-BiNSs and BiNSs surfaces, respectively. As shown in Figure 6b, the total reaction pathway for the electrochemical reduction of CO_2 to formate was two protons, and two electrons were transferred through the *OCHO intermediate and adsorbed by HCOOH (aq) [38]. Both N-BiNSs and BiNSs displayed a great energy barrier for *OCHO formation, confirming the original proton-coupled electron transfer is the potential limiting procedure [31] and that the optimally adsorbed *OCHO intermediate promotes formate production. Obviously, on the N-BiNSs (012) surface, the calculated Gibbs free energy ΔG for the formation of *OCHO was +0.36 eV, and the ΔG for the formation of *OCHO from BiNSs was +0.65 eV. Thus, the N-Bi nanosheets had a more negative Gibbs free energy than that of the Bi nanosheets site, indicating that *OCHO formation and protonation were more spontaneous.

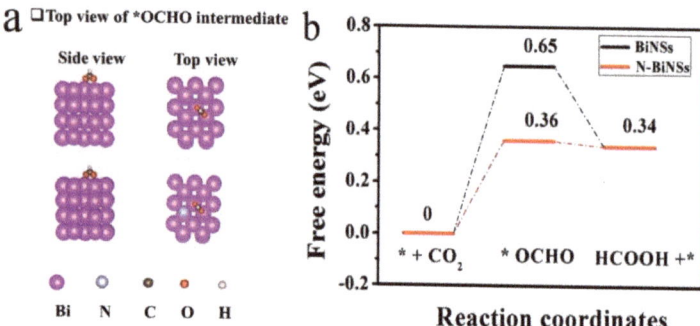

Figure 6. (a) Side and Top views of BiNSs (012) and N-BiNSs (012) configurations (the red, white, blue, purple and brown spheres represent: O, H, N, Bi, and C atoms, respectively); (b) Calculated free-energy diagram for *OCHO generation of N-BiNSs and BiNSs for the electrochemical reduction to formate process (where * represents the active site of the catalyst).

3. Materials and Methods

3.1. Synthesis of Nitrogen-Doped Bismuth Nanosheets (N-BiNSs)

The synthesis method was improved according to the literature [39], and the synthesis of the specific process was as follows. Firstly, 5 g of $C_6H_5BiO_7$, 3.7 g of $Ca(OH)_2$, and 5.4 g of NH_4Cl were weighed and placed in a sample vial the reagents were thoroughly mixed. Then the mixed reagents were transferred to a porcelain boat and placed in a tube furnace. Subsequently, the tube furnace was vacuumed, and N_2 was pumped into it. The tube furnace was activated at 450 °C for 1 h in the N_2 atmosphere, and then the temperature was increased to 800 °C for 2 h. During the whole synthesis process, the heating rate was 5 °C min^{-1} in the tube furnace. At the end of the reaction, the samples were cooled to room temperature, the samples were first repeatedly cleared with 2.0 M hydrochloric acid (HCl), and the inorganic salts of the samples were removed entirely. The samples were repeatedly cleared with deionized water until the pH value was neutral. Finally, the samples were dried under vacuum at 60 °C for 24 h. The samples were named nitrogen-doped bismuth nanosheets (N-BiNSs). For comparison, samples prepared at different $Ca(OH)_2$ and NH_4Cl were denoted as N-BiNSs-1 and N-BiNSs-2. The N-BiNSs-1 catalyst was prepared by adding $Ca(OH)_2$ (1.85 g) and NH_4Cl (2.7 g), and the N-BiNSs-2 was prepared by adding $Ca(OH)_2$ (5.55 g) and NH_4Cl (8.1 g). Moreover, the bismuth nanosheets (BiNSs) sample was prepared on the basis of the same procedure as introduced above, but without added $Ca(OH)_2$ and NH_4Cl.

3.2. Preparation of Working Electrode

Generally, 10 mg of as-prepared N-BiNSs-X were mixed with 600 µL of deionized water, 350 µL ethanol, and 50 µL of Nafion D-521 dispersion within a 2 mL vial in an ultrasonic bath for 2 h to become a homogeneous catalyst ink suspension; 10 µL of the catalyst ink was dropped onto a carbon paper (1 mg cm^{-2}), and the catalyst-covered electrode was dried in a desiccator before use.

3.3. Electrochemical Measurements

In this work, all electrochemical properties were tested at an electrochemical workstation (Shanghai Chenhua, CHI760E). The electrocatalytic CO_2RR was performed in an H-type electrolytic cell with the two chambers isolated by a Nafion 117 proton exchange membrane to stop reoxidation of the cathode generated to the anode. The prepared catalyst, the Ag/AgCl(saturated KCl), and the platinum sheet were used as the working electrodes, reference electrodes, and counter electrodes, respectively. All potential values were measured with Ag/AgCl and then converted to RHE. All electrode potentials were converted to electrode potentials relative to RHE by the Nernst equation: E (vs. RHE) = E (vs. Ag/AgCl) + 0.0591 × pH + 0.222 V. Each electrode chamber contained 35 mL of 0.5 M KHCO$_3$ electrolyte. Before the test, the above electrolyte was continuously bubbled with high-purity CO$_2$ or N$_2$ for at least 30 min to saturate the electrolyte with N$_2$ (pH = 8.5) or CO$_2$ (pH = 7.2). The electrochemical reduction of CO$_2$ on BiNSs and N-BiNSs-X electrodes were performed in N$_2$-saturated 0.5 M KHCO$_3$ or CO$_2$-saturated 0.5 M KHCO$_3$ at ambient temperature and pressure. Double-layer capacitance (Cdl) was obtained from cyclic voltammetry (CV) curves surveyed at unlike scan rates (10, 30, 50, 70, 90, 110, and 130 mV s^{-1}) in the scope of +0.12 V to +0.22 V (vs. RHE). Electrochemical impedance spectroscopy (EIS) was employed to understand the charge mass transfer resistance in the electrocatalytic CO$_2$RR course. The frequency range was 10 kHz–1.0 Hz and the amplitude was 5 mV s^{-1}. The durability of the catalyst was obtained by using the current-time curve method (i-t). At the end of each potential test, the liquid products generated at each potential were collected and detected by ion chromatography.

3.4. Product Analysis

Faraday efficiency test method: Control potential electrolytic Coulomb method was used for the CO$_2$ saturated solution, and the electrolytic reduction products were analyzed and calculated 0.5 h later. The flow rate of CO$_2$ was mastered at 20 mL min^{-1} during electrolysis. The liquid products after electrolysis were detected by ion chromatography (AS-DV, Thermo Scientific, New York City, MA, USA). The Faraday efficiency (FE) of liquid-phase products was calculated as follows:

$$FE = \frac{NnF}{Q} \times 100\%$$

where N is the number of electrons transferred, n is the total mole fraction of the gas measured by gas chromatography, F is the Faraday constant (96,485 C mol^{-1}), and Q is the total electric charge passed through the electrode.

The gaseous products were quantified by gas chromatography (GC7900, Tianmei, China), and the H$_2$ and CO were detected by a thermal conductivity detector (TCD), and flame ionization detector (FID), respectively, and ultra-pure N$_2$ (>99.99%) was used as carrier gas. The calculation method of Faraday efficiency (FE) was as follows:

$$FE = \frac{n \times C \times v \times F}{Vm \times j} \times 100\%$$

where v is the gas flow rate of supplied CO$_2$, C is the concentration of the gaseous product, n is the number of electrons for producing a molecule of CO or H$_2$, F is the Faraday constant (96,485 C mol^{-1}), Vm is the molar volume of gas at 298 K, j is the recorded current (A).

4. Conclusions

In summary, we successfully prepared nitrogen-doped bismuth nanosheets (N-BiNSs) through a simple one-step activation and nitrogen-doping connective way. The N-Bi nanosheets exhibited very high activity, selectivity, and stability for formate production to BiNSs. The results showed that the selectivity of formate was 95.25% at −0.95 V (vs. RHE). Meanwhile, the N-BiNSs catalyst also showed excellent stability, and no apparent catalyst deactivation occurred after 20 h of electrolysis. Importantly, we also optimize the CO$_2$RR activity by flow-cell. The N-BiNSs catalyst possessed high durability at a current density of 300 mA cm^{-2} at a potential of -1.2 V (vs. RHE), and the average FE of formate was about 90.30%. The DFT simulation suggested that compared with other catalysts, N-BiNSs had the active site to adsorb CO$_2$ more easily. The free energy barrier for forming the critical intermediate *OCHO was reduced by nitrogen doping. This electrocatalyst with high catalytic activity, durability, and selectivity has great potential to improve the technical and economic feasibility of CO$_2$ to formate conversion. In addition, this study provides a reasonable catalyst design for CO_2RR studies by proposing a simple method for synthesizing nanostructured catalysts.

Supplementary Materials: The following supporting information can be downloaded at: https://www.mdpi.com/article/10.3390/ijms232214485/s1. References [40–59] are cited in Supplementary Materials.

Author Contributions: Conceptualization, J.P.; methodology, S.L., Y.K. and J.P.; formal analysis, S.L., Y.K. and C.M.; investigation, S.L., Y.K. and C.M.; resources, H.M. and J.P.; data curation, S.L., and Y.K.; writing—original draft preparation, S.L., C.M., Y.P., H.M. and J.P.; writing—review and editing, S.L., Y.P. and J.P.; visualization, S.L. and K.Y; project administration, J.P.; funding acquisition, Y.P. and J.P. All authors have read and agreed to the published version of the manuscript.

Funding: This work was funded by the support from the National Natural Science Foundation of China (No. 22262027), Ningxia leading scientific and technological innovation talents projects (No. KJT2018002), Ningxia Natural Science Foundation (No. 2022AAC03103, 2022AAC03113), and National First-rate Discipline Construction Project of Ningxia (NXYLXK2017A04). This paper was also supported by College Students' Innovative and Entrepreneurship Training Program of Ningxia University (No. S202210749054).

Institutional Review Board Statement: Not applicable.

Informed Consent Statement: Not applicable.

Data Availability Statement: All data in this study can be found in public data bases and Supplementary Materials, as described in the Material and Methods section (Section 3).

Conflicts of Interest: The authors declared that there was no competing financial interest.

References

1. Kumar, B.; Asadi, M.; Pisasale, D.; Sinha-Ray, S.; Rosen, B.A.; Haasch, R.; Abiade, J.; Yarin, A.L.; Salehi-Khojin, A. Renewable and metal-free carbon nanofibre catalysts for carbon dioxide reduction. *Nat. Commun.* **2013**, *4*, 2819. [CrossRef]
2. Feng, X.; Zou, H.; Zheng, R.; Wei, W.; Wang, R.; Zou, W.; Lim, G.; Hong, J.; Duan, L.; Chen, H. Bi_2O_3/BiO_2 Nanoheterojunction for Highly Efficient Electrocatalytic CO_2 Reduction to Formate. *Nano Lett.* **2022**, *22*, 1656–1664. [CrossRef] [PubMed]
3. Boutin, E.; Wang, M.; Lin, J.C.; Mesnage, M.; Mendoza, D.; Lassalle-Kaiser, B.; Hahn, C.; Jaramillo, T.F.; Robert, M. Aqueous Electrochemical Reduction of Carbon Dioxide and Carbon Monoxide into Methanol with Cobalt Phthalocyanine. *Angew. Chem. Int. Ed.* **2019**, *58*, 16172–16176. [CrossRef] [PubMed]
4. Guo, C.; Guo, Y.; Shi, Y.; Lan, X.; Wang, Y.; Yu, Y.; Zhang, B. Electrocatalytic Reduction of CO_2 to Ethanol at Close to Theoretical Potential via Engineering Abundant Electron-Donating Cu(delta+) Species. *Angew. Chem. Int. Ed.* **2022**, *134*, e202205909. [CrossRef]
5. Liu, B.; Yao, X.; Zhang, Z.; Li, C.; Zhang, J.; Wang, P.; Zhao, J.; Guo, Y.; Sun, J.; Zhao, C. Synthesis of Cu_2O Nanostructures with Tunable Crystal Facets for Electrochemical CO_2 Reduction to Alcohols. *ACS Appl. Mater. Interfaces* **2021**, *13*, 39165–39177. [CrossRef]
6. Gao, Y.; Wu, Q.; Liang, X.; Wang, Z.; Zheng, Z.; Wang, P.; Liu, Y.; Dai, Y.; Whangbo, H.M.; Huang, B. Cu_2O Nanoparticles with Both {100} and {111} Facets for Enhancing the Selectivity and Activity of CO_2 Electroreduction to Ethylene. *Adv. Sci.* **2020**, *7*, 1902820. [CrossRef]
7. Jung, H.; Lee, S.Y.; Lee, C.W.; Cho, M.K.; Won, D.H.; Kim, C.; Oh, H.S.; Min, B.K.; Hwang, Y.J. Electrochemical Fragmentation of Cu_2O Nanoparticles Enhancing Selective C-C Coupling from CO_2 Reduction Reaction. *J. Am. Chem. Soc.* **2019**, *141*, 4624–4633. [CrossRef]
8. Tan, X.; Yu, C.; Zhao, C.; Huang, H.; Yao, X.; Han, X.; Guo, W.; Cui, S.; Huang, H.; Qiu, J. Restructuring of Cu_2O to $Cu_2O@Cu$-Metal-Organic Frameworks for Selective Electrochemical Reduction of CO_2. *ACS Appl. Mater. Interfaces* **2019**, *11*, 9904–9910. [CrossRef]
9. Wang, Y.-H.; Jiang, W.-J.; Yao, W.; Liu, Z.-L.; Liu, Z.; Yang, Y.; Gao, L.-Z. Advances in electrochemical reduction of carbon dioxide to formate over bismuth-based catalysts. *Rare Met.* **2021**, *40*, 2327–2353. [CrossRef]
10. Pander, J.E.; Lum, J.W.J.; Yeo, B.S. The importance of morphology on the activity of lead cathodes for the reduction of carbon dioxide to formate. *J. Mater. Chem. A* **2019**, *7*, 4093–4101. [CrossRef]
11. Yang, Y.; Fu, J.J.; Tang, T.; Niu, S.; Zhang, L.B.; Zhang, J.N.; Hu, J.S. Regulating surface In–O in In@InO_x core-shell nanoparticles for boosting electrocatalytic CO_2 reduction to formate. *Chin. J. Catal.* **2022**, *43*, 1674–1679. [CrossRef]
12. Jiang, Z.; Wang, T.; Pei, J.; Shang, H.; Zhou, D.; Li, H.; Dong, J.; Wang, Y.; Cao, R.; Zhuang, Z.; et al. Discovery of main group single $Sb-N_4$ active sites for CO_2 electroreduction to formate with high efficiency. *Energy Environ. Sci.* **2020**, *13*, 2856–2863. [CrossRef]

13. Wen, G.; Lee, D.U.; Ren, B.; Hassan, F.M.; Jiang, G.; Cano, Z.P.; Gostick, J.; Croiset, E.; Bai, Z.; Yang, L.; et al. Orbital Interactions in Bi-Sn Bimetallic Electrocatalysts for Highly Selective Electrochemical CO_2 Reduction toward Formate Production. *Adv. Energy Mater.* **2018**, *8*, 1802427. [CrossRef]
14. Cheng, Y.; Hou, J.; Kang, P. Integrated Capture and Electroreduction of Flue Gas CO_2 to Formate Using Amine Functionalized SnOx Nanoparticles. *ACS Energy Lett.* **2021**, *6*, 3352–3358. [CrossRef]
15. Gao, S.; Jiao, X.C.; Sun, Z.T.; Zhang, W.H.; Sun, Y.F.; Wang, C.M.; Hu, Q.T.; Zu, X.L.; Yang, F.; Yang, S.Y.; et al. Ultrathin Co_3O_4 layers realizing optimized CO_2 electroreduction to formate. *Angew. Chem. Int. Ed.* **2016**, *55*, 698–702. [CrossRef] [PubMed]
16. Peng, L.; Wang, Y.; Wang, Y.; Xu, N.; Lou, W.; Liu, P.; Cai, D.; Huang, H.; Qiao, J. Separated growth of Bi-Cu bimetallic electrocatalysts on defective copper foam for highly converting CO_2 to formate with alkaline anion-exchange membrane beyond $KHCO_3$ electrolyte. *Appl. Catal. B Environ.* **2021**, *288*, 120003. [CrossRef]
17. Tao, Z.X.; Wu, Z.S.; Yuan, X.L.; Wu, Y.S.; Wang, H.L. Copper–gold interactions enhancing formate production from electrochemical CO_2 reduction. *ACS Catal.* **2019**, *9*, 10894–10898. [CrossRef]
18. Fan, J.; Zhao, X.; Mao, X.; Xu, J.; Han, N.; Yang, H.; Pan, B.; Li, Y.; Wang, L.; Li, Y. Large-Area Vertically Aligned Bismuthene Nanosheet Arrays from Galvanic Replacement Reaction for Efficient Electrochemical CO_2 Conversion. *Adv. Mater.* **2021**, *33*, e2100910. [CrossRef]
19. Xie, L.; Liu, X.; Huang, F.; Liang, J.; Liu, J.; Wang, T.; Yang, L.; Cao, R.; Li, Q. Regulating Pd-catalysis for electrocatalytic CO_2 reduction to formate via intermetallic PdBi nanosheets. *Chin. J. Catal.* **2022**, *43*, 1680–1686. [CrossRef]
20. Han, N.; Wang, Y.; Yang, H.; Deng, J.; Wu, J.; Li, Y.; Li, Y. Ultrathin bismuth nanosheets from in situ topotactic transformation for selective electrocatalytic CO_2 reduction to formate. *Nat. Commun.* **2018**, *9*, 1320. [CrossRef]
21. Wu, J.; Yu, X.; He, H.; Yang, C.; Xia, D.; Wang, L.; Huang, J.; Zhao, N.; Tang, F.; Deng, L.; et al. Bismuth-Nanosheet-Based Catalysts with a Reconstituted Bi0 Atom for Promoting the Electrocatalytic Reduction of CO_2 to Formate. *Ind. Eng. Chem. Res.* **2022**, *61*, 12383–12391. [CrossRef]
22. Wu, Z.; Wu, H.; Cai, W.; Wen, Z.; Jia, B.; Wang, L.; Jin, W.; Ma, T. Engineering Bismuth-Tin Interface in Bimetallic Aerogel with a 3D Porous Structure for Highly Selective Electrocatalytic CO_2 Reduction to HCOOH. *Angew. Chem. Int. Ed.* **2021**, *60*, 12554–12559. [CrossRef] [PubMed]
23. Zhang, W.; Hu, Y.; Ma, L.; Zhu, G.; Zhao, P.; Xue, X.; Chen, R.; Yang, S.; Ma, J.; Liu, J.; et al. Liquid-phase exfoliated ultrathin Bi nanosheets: Uncovering the origins of enhanced electrocatalytic CO_2 reduction on two-dimensional metal nanostructure. *Nano Energy* **2018**, *53*, 808–816. [CrossRef]
24. Yi, L.; Chen, J.; Shao, P.; Huang, J.; Peng, X.; Li, J.; Wang, G.; Zhang, C.; Wen, Z. Molten-Salt-Assisted Synthesis of Bismuth Nanosheets for Long-term Continuous Electrocatalytic Conversion of CO_2 to Formate. *Angew. Chem. Int. Ed.* **2020**, *59*, 20112–20119. [CrossRef] [PubMed]
25. Xing, Z.; Hu, X.; Feng, X. Tuning the microenvironment in gas-diffusion electrodes enables high-rate CO_2 electrolysis to formate. *ACS Energy Lett.* **2021**, *6*, 1694–1702. [CrossRef]
26. Fan, K.; Jia, Y.; Jia, Y.F.; Ji, Y.F.; Kuang, Y.P.; Zhu, B.C.; Liu, X.Y.; Yu, J.G. Curved surface boosts electrochemical CO_2 reduction to formate via bismuth nanotubes in a wide potential window. *ACS Catal.* **2019**, *10*, 358–364. [CrossRef]
27. Huang, J.; Guo, X.; Yang, J.; Wang, L. Electrodeposited Bi dendrites/2D black phosphorus nanosheets composite used for boosting formic acid production from CO_2 electroreduction. *J. CO_2 Util.* **2020**, *38*, 32–38. [CrossRef]
28. Li, W.; Bandosz, T.J. Analyzing the effect of nitrogen/sulfur groups' density ratio in porous carbons on the efficiency of CO_2 electrochemical reduction. *Appl. Surf. Sci.* **2021**, *569*, 151066. [CrossRef]
29. Zhao, M.; Gu, Y.; Gao, W.; Cui, P.; Tang, H.; Wei, X.; Zhu, H.; Li, G.; Yan, S.; Zhang, X.; et al. Atom vacancies induced electron-rich surface of ultrathin Bi nanosheet for efficient electrochemical CO_2 reduction. *Appl. Catal. B: Environ.* **2020**, *266*, 118625. [CrossRef]
30. Cheng, H.; Liu, S.; Zhang, J.; Zhou, T.; Zhang, N.; Zheng, X.S.; Chu, W.; Hu, Z.; Wu, C.; Xie, Y. Surface Nitrogen-Injection Engineering for High Formation Rate of CO_2 Reduction to Formate. *Nano Lett.* **2020**, *20*, 6097–6103. [CrossRef]
31. Chen, Z.; Mou, K.; Wang, X.; Liu, L. Nitrogen-Doped Graphene Quantum Dots Enhance the Activity of Bi_2O_3 Nanosheets for Electrochemical Reduction of CO_2 in a Wide Negative Potential Region. *Angew. Chem. Int. Ed.* **2018**, *57*, 12790–12794. [CrossRef] [PubMed]
32. Li, Q.; Wang, Z.; Zhang, M.; Hou, P.; Kang, P. Nitrogen doped tin oxide nanostructured catalysts for selective electrochemical reduction of carbon dioxide to formate. *J. Energy Chem.* **2017**, *26*, 825–829. [CrossRef]
33. Dong, W.; Zhang, N.; Li, S.; Min, S.; Peng, J.; Liu, W.; Zhan, D.; Bai, H. A Mn single atom catalyst with Mn–N_2O_2 sites integrated into carbon nanosheets for efficient electrocatalytic CO_2 reduction. *J. Mater. Chem. A* **2022**, *10*, 10892–10901. [CrossRef]
34. Guo, Y.; Yang, H.J.; Zhou, X.; Liu, K.L.; Zhang, C.; Zhou, Z.Y.; Wang, C.; Lin, W.B. Electrocatalytic reduction of CO_2 to CO with 100% faradaic efficiency by using pyrolyzed zeolitic imidazolate frameworks supported on carbon nanotube networks. *J. Mater. Chem. A* **2017**, *5*, 24867–24873. [CrossRef]
35. Varela, A.S.; Sahraie, N.R.; Steinberg, J.; Ju, W.; Oh, H.S.; Strasser, P. Metal-doped nitrogenated carbon as an efficient catalyst for direct CO_2 electroreduction to CO and hydrocarbons. *Angew. Chem. Int. Ed.* **2015**, *54*, 10758–10762. [CrossRef]
36. Kwon, I.S.; Debela, T.T.; Kwak, I.H.; Seo, H.W.; Park, K.; Kim, D.; Yoo, S.J.; Kim, J.-G.; Park, J.; Kang, H.S. Selective electrochemical reduction of carbon dioxide to formic acid using indium–zinc bimetallic nanocrystals. *J. Mater. Chem. A* **2019**, *7*, 22879–22883. [CrossRef]

37. Tian, W.L.; Zhang, J.; Feng, H.; Wen, H.; Sun, X.; Guan, X.; Zheng, D.C.; Liao, J.; Yan, M.L.; Yao, Y.D. Fuels, A hierarchical CoP@NiCo-LDH nanoarray as an efficient and flexible catalyst electrode for the alkaline oxygen evolution reaction. *Sustain. Energy Fuels* **2021**, *5*, 391–395. [CrossRef]
38. Yoo, J.S.; Christensen, R.; Vegge, T.; Norskov, J.K.; Studt, F. Theoretical Insight into the Trends that Guide the Electrochemical Reduction of Carbon Dioxide to Formic Acid. *ChemSusChem* **2016**, *9*, 358–363. [CrossRef]
39. Peng, H.; Ma, G.; Sun, K.; Mu, J.; Lei, Z. One-step preparation of ultrathin nitrogen-doped carbon nanosheets with ultrahigh pore volume for high-performance supercapacitors. *J. Mater. Chem. A* **2014**, *2*, 17297–17301. [CrossRef]
40. Kresse, G.; Furthmuller, J. Efficient iterative schemes for ab initio total-energy calculations using a plane-wave basis set. *Phys. Rev. B Condens Matter.* **1996**, *54*, 11169–11186. [CrossRef]
41. Kresse, G.; Hafner, J. Ab initio molecular-dynamics simulation of the liquid-metal–amorphous-semiconductor transition in germanium. *Phys. Rev. B Condens Matter.* **1994**, *49*, 14251–14269. [CrossRef] [PubMed]
42. Blochl, P.E. Projector augmented-wave method. *Phys. Rev. B Condens Matter.* **1994**, *50*, 17953–17979. [CrossRef] [PubMed]
43. Perdew, J.P.; Burke, K.; Ernzerhof, M. Generalized Gradient Approximation Made Simple. *Phys. Rev Lett.* **1996**, *77*, 3865–3868. [CrossRef] [PubMed]
44. Zhang, Y.; Yang, W. Comment on "Generalized Gradient Approximation Made Simple". *Phys. Rev Lett.* **1998**, *80*, 890. [CrossRef]
45. Hammer, B.; Hansen, L.B.; Nørskov, J.K. Improved adsorption energetics within density-functional theory using revised Perdew-Burke-Ernzerhof functionals. *Phys. Rev. B* **1999**, *59*, 7413–7421. [CrossRef]
46. Monkhorst, H.J.; Pack, J.D. Special points for Brillouin-zone integrations. *Phys. Rev. B* **1976**, *13*, 5188–5192. [CrossRef]
47. Nørskov, J.K.; Rossmeisl, J.; Logadottir, A.; Lindqvist, L.; Kitchin, J.R.; Bligaard, T.; Jónsson, H. Origin of the Overpotential for Oxygen Reduction at a Fuel-Cell Cathode. *J. Phys. Chem. B* **2004**, *108*, 17886–17892. [CrossRef]
48. Xing, Y.; Kong, X.; Guo, X.; Liu, Y.; Li, Q.; Zhang, Y.; Sheng, Y.; Yang, X.; Geng, Z.; Zeng, J. Bi@Sn Core-Shell Structure with Compressive Strain Boosts the Electroreduction of CO_2 into Formic Acid. *Adv Sci.* **2020**, *7*, 1902989. [CrossRef]
49. Li, F.; Xue, M.; Li, J.; Ma, X.; Chen, L.; Zhang, X.; MacFarlane, D.R.; Zhang, J. Unlocking the electrocatalytic activity of antimony for CO_2 reduction by two-dimensional engineering of the bulk material. *Angew. Chem. Int. Ed.* **2017**, *129*, 14910–14914. [CrossRef]
50. Watanabe, M.; Shibata, M.; Kato, A.; Azuma, M.; Sakata, T. Design of alloy electrocatalysts for CO_2 reduction: III. The selective and reversible reduction of on Cu alloy electrodes. *J. Electrochem. Soc.* **1991**, *138*, 3382. [CrossRef]
51. Sreekanth, N.; Nazrulla, M.A.; Vineesh, T.V.; Sailaja, K.; Phani, K.L. Metal-free boron-doped graphene for selective electroreduction of carbon dioxide to formic acid/formate. *Chem. Commun.* **2015**, *51*, 16061–16064. [CrossRef] [PubMed]
52. Won, D.H.; Choi, C.H.; Chung, J.; Chung, M.W.; Kim, E.H.; Woo, S.I. Rational design of a hierarchical tin dendrite electrode for efficient electrochemical reduction of CO_2. *ChemSusChem* **2015**, *8*, 3092–3098. [CrossRef] [PubMed]
53. Zheng, X.; De Luna, P.; García de Arquer, F.P.; Zhang, B.; Becknell, N.; Ross, M.B.; Li, Y.; Banis, M.N.; Li, Y.; Liu, M.; et al. Sulfur-modulated tin sites enable highly selective electrochemical reduction of CO_2 to formate. *Joule* **2017**, *1*, 794–805. [CrossRef]
54. He, S.; Ni, F.; Ji, Y.; Wang, L.; Wen, Y.; Bai, H.; Liu, G.; Zhang, Y.; Li, Y.; Zhang, B.; et al. The p-Orbital delocalization of main-group metals to boost CO_2 electroreduction. *Angew. Chem. Int. Ed.* **2018**, *130*, 16346–16351. [CrossRef]
55. Kumar, B.; Atla, V.; Brian, J.P.; Kumari, S.; Nguyen, T.Q.; Sunkara, M.; Spurgeon, J.M. Reduced SnO_2 porous nanowires with a high density of grain boundaries as catalysts for efficient electrochemical CO_2-into-HCOOH conversion. *Angew. Chem. Int. Ed.* **2017**, *56*, 3645–3649. [CrossRef]
56. Liang, C.; Kim, B.; Yang, S.; Yang Liu, Y.L.; Francisco Woellner, C.; Li, Z.; Vajtai, R.; Yang, W.; Wu, J.; Kenis, P.J.A.; et al. High efficiency electrochemical reduction of CO_2 beyond the two-electron transfer pathway on grain boundary rich ultra-small SnO_2 nanoparticles. *J. Mater. Chem. A* **2018**, *6*, 10313–10319. [CrossRef]
57. Lee, C.W.; Hong, J.S.; Yang, K.D.; Jin, K.; Lee, J.H.; Ahn, H.-Y.; Seo, H.; Sung, N.-E.; Nam, K.T. Selective Electrochemical Production of Formate from Carbon Dioxide with Bismuth-Based Catalysts in an Aqueous Electrolyte. *ACS Catal.* **2018**, *8*, 931–937. [CrossRef]
58. Zhang, Y.; Li, F.; Zhang, X.; Williams, T.; Easton, C.D.; Bond, A.M.; Zhang, J. Electrochemical reduction of CO_2 on defect-rich Bi derived from Bi_2S_3 with enhanced formate selectivity. *J. Mater. Chem. A* **2018**, *6*, 4714–4720. [CrossRef]
59. Zhao, Y.; Liang, J.; Wang, C.; Ma, J.; Wallace, G.G. Tunable and efficient tin modified nitrogen-doped carbon nanofibers for electrochemical reduction of aqueous carbon dioxide. *Adv. Energy Mater.* **2018**, *8*, 1702524. [CrossRef]

International Journal of
Molecular Sciences

Article

Selective Oxidation of Toluene to Benzaldehyde Using Co-ZIF Nano-Catalyst

Wei Long [1,2,*], Zhilong Chen [1], Yinfei Huang [3] and Xinping Kang [1]

1. College of Chemistry, Guangdong University of Petrochemical Technology, Maoming 525000, China
2. Guangdong Provincial Key Laboratory of Petrochemical Pollution Process and Control, Guangdong University of Petrochemical Technology, Maoming 525000, China
3. College of Electromechanical Engineering, Guangdong University of Petrochemical Technology, Maoming 525000, China
* Correspondence: longwei@gdupt.edu.cn; Tel.: +86-0668-2337639

Abstract: Nanometer-size Co-ZIF (zeolitic imidazolate frameworks) catalyst was prepared for selective oxidation of toluene to benzaldehyde under mild conditions. The typical characteristics of the metal-organic frameworks (MOFs) material were affirmed by the XRD, SEM, and TEM, the BET surface area of this catalyst was as high as 924.25 m^2/g, and the diameter of particles was near 200 nm from TEM results. The Co metal was coated with 2-methyl glyoxaline, and the crystalline planes were relatively stable. The reaction temperatures, oxygen pressure, mass amount of N-hydroxyphthalimide (NHPI), and reaction time were discussed. The Co-ZIF catalyst gave the best result of 92.30% toluene conversion and 91.31% selectivity to benzaldehyde under 0.12 MPa and 313 K. The addition of a certain amount of NHPI and the smooth oxidate capacity of the catalyst were important factors in the high yield of benzaldehyde. This nanometer-size catalyst showed superior performance for recycling use in the oxidation of toluene. Finally, a possible reaction mechanism was proposed. This new nanometer-size Co-ZIF catalyst will be applied well in the selective oxidation of toluene to benzaldehyde.

Keywords: selective oxidation; benzaldehyde; nanometer-size catalyst; MOFs material

1. Introduction

As the industry rapidly grows, volatile organic compounds (VOCs) are known as organic carbon-based chemicals that evaporate easily, causing serious human health and environmental pollution problems [1]. The relationship between many major environmental problems, such as global warming, photochemical ozone formation, stratospheric ozone depletion, and VOCs, is distinct and serious [2]. As significant air pollutants, VOCs are controlled at low concentrations when they are discharged from a chimney. However, the maximum emission concentration of VOCs is different in varied countries or regions [3]. Among the carbon-based VOCs, toluene, xylene, and benzene are found to be endocrine-disrupting chemicals, exerting severe toxicity to animals or humans [4]. They are the major source of very fine particulate matter or ozone that produces heavy, smoggy, unfriendly, harmful, and hazy weather [5]. Hence, it is essential to develop the technology level of VOCs transition.

Catalytic total oxidation has been considered the most appropriate method and economical route for VOCs removal, and many efforts have been made to design catalysts with suitable activity and selectivity [6–8]. Catalytic oxidation is an effective technology to completely convert VOCs into CO_2 and H_2O without any by-products at low temperatures [9]. The total oxidation of airborne VOCs into carbon dioxide and water at room temperature can be performed in the presence of a semiconductor catalyst [10]. However, the total oxidation of VOCs only can reduce pollution emissions and convert them directly into usable chemical material reuse wastes, which has great significance [11]. Selective oxidation products of VOCs are commonly used in various applications, including adhesives, paints, rubbers, leather tanning processes, and as chemical solvents to dissolve many organic substances [12]. Catalytic selective oxidation has been considered the most appropriate method for VOC removal, and many efforts have been made to design catalysts with suitable activity and selectivity [13–15].

The commercial catalysts for oxidation of VOCs can be classified into three categories: (1) supported noble metals [16,17]; (2) metal oxides or supported metals [18,19]; (3) mixtures of noble metals and metal oxides [20,21]. Noble metal catalysts generally show higher activity and selectivity to organic acid, transition metal oxides are one alternative to the noble metal-containing catalysts due to their resistance to halogens, low cost and high catalytic activity and selectivity to aldehyde [22]. As the high value chemicals and suitable chemical raw materials, aldehyde are very popular to chemical engineering researchers. Selective oxidation of VOCs is very suitable channel to reuse waste, but this process is difficult to unify considering the purification and separation of products [23]. As a typical example, selective oxidation of toluene to benzaldehyde have been received many attempts and reports from researchers in different areas of the world [24–27].

Many new catalysts have been invented and verified for this selective oxidation reaction, and many catalytic reaction mechanisms were proposed [28]. Xie et al. [29] prepared three-dimensionally ordered macroporous supported gold-palladium alloy catalyst (Au-Pd/3DOM Co_3O_4) which performed highly active and suitable stable for toluene oxidation, stronger noble metal-3DOM Co_3O_4 interaction was the key factor, but the catalyst cost is too high. New MnO_2 sample (MnO_2-P) exhibited the higher content of oxygen vacancy, which could enhance reducibility and mobility of lattice oxygen, so higher activity and best catalytic performance were obtained in selective oxidation reaction of toluene [30]. Yuan et al. [31] collected much experimental results with a wide range of conditions from low to high temperatures, normal atmospheric to high pressures, very lean to pyrolysis conditions and developed kinetic model of selective oxidation of toluene. Nevertheless, the ideal catalyst, which can show the properties as high product selectivity, suitable recycling use capacity, low cost, mild reaction conditions, and fast reaction speed has not been found. Hence, there is still space for the development of a catalyst that has higher activity and better selective to benzaldehyde in the oxidation of toluene.

Recently, MOF materials have shown a great advantage for CO_2 capture and separation [32], and Ni-MOF can exhibit great catalytic capacity in heterogeneous catalytic reactions such as CO_2 hydrogenation [33]. Usman et al. [34] proposed that MOF/g-C_3N_4 composites will be important photocatalysts for prospective applications in energy harvesting and controlling environmental pollution. Shafiq et al. [35] found the imidazole framework-95 (ZIF-95) MOF could be dispersed in polysulfone polymer to form mixed matrix membranes, so gas separation performance and selectivity of polysulfone membrane were enhanced greatly. Hence, Large surface area, tunable pore size and window, and tailored functionalities are the great advantages of MOFs, which can be used for the efficient catalyst.

In the present study, the selective catalytic oxidation process of toluene was shown by Scheme 1 [36]. Toluene can be oxidated to benzyl alcohol, and it is oxidated easily to benzaldehyde. The final oxidation product is benzoic acid. Recently, metal-organic frameworks (MOFs) materials have been used widely in the field of catalysis, and it shows obvious feature as high surface area, complex porous structure, and obvious cavity coordination [36]. As the stable catalyst, Co-ZIF is prepared and used in this oxidation process of toluene. Considering the control of selective oxidation of reactant, NHPI as the reaction initiator is added to the solution. The optimized reaction conditions are found, and better catalytic performance results are confirmed, so the advantages of this MOF catalyst are true and reliable.

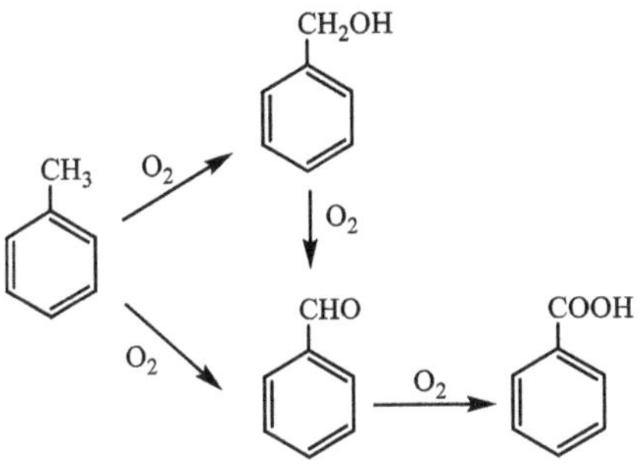

Scheme 1. The catalytic oxidation process of toluene [36].

2. Results

Characterization of Catalysts

FT-IR spectra of catalyst samples are shown in Figure 1, and there are some characteristic adsorption peaks of MOFs material in the curves. The obvious peak at 3407.5 and 3133.4 cm^{-1} are ascribed to the stretching vibration of the O-H bond of the crystal water molecule and the C-H bond of MOFs material, respectively. Many small peaks near 1467.3 cm^{-1} and 1141.6 cm^{-1} belong to the stretching vibration of the C-N bond in the MOFs material (Figure 1a). The acute peak at 1483.2 cm^{-1} belongs to the bending vibration of the C-N bond in the MOFs material. The distinct peaks near 1681.2 cm^{-1} are ascribed to the stretching vibration of the N-H bond in the 2-methyl glyoxaline molecule. Two overlap peaks at 755.0 cm^{-1} and 692.6 cm^{-1} are attributed to the bending vibration of the N-H bond.

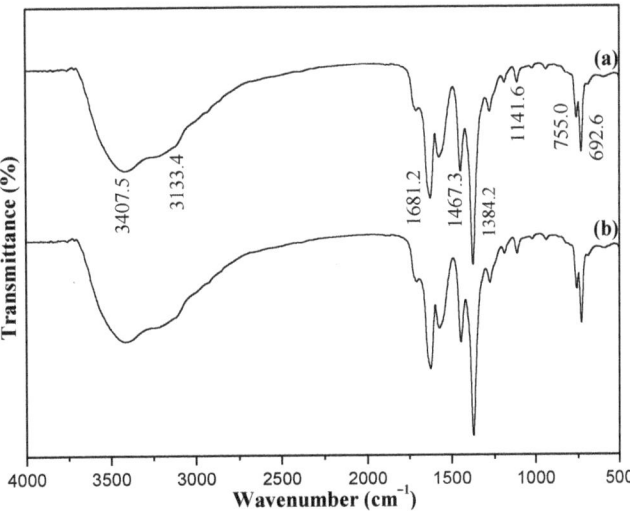

Figure 1. FT-IR spectrum of Co-ZIF catalyst. (a) fresh, (b) after reaction.

The characteristic adsorption peak at 1846.2 cm^{-1} is the important signal of the coordinate bond [37]. There are no peaks near 1846.2 cm^{-1}, indicating the coordinate bond of Co-N has been established successfully. Obviously, there are no apparent changes or weaknesses in the curve of the used catalyst (Figure 1b), so the Co-ZIF catalyst exhibits suitable stability.

The XRD spectra of the samples (Co-ZIF catalyst, fresh and after reaction) are shown in Figure 2. The characteristic diffraction peaks at 2θ = 10.4°, 12.7°, 13.3°, 14.8°, 16.5°, 18.0°, 22.1°, 24.5°, 25.5°, 26.7°, 29.5°, 30.6°, 31.6°, and 32.5° are ascribed to crystalline planes of

Co-ZIF material, which are different with crystalline planes of Co_3O_4 particles [38]. As the typical MOFs material, the crystallinity of the Co-ZIF catalyst was low, so the intensity of many characteristic diffraction peaks was low (Figure 2a). Two apparent peaks at $2\theta = 12.7°$ and $18.0°$ might are ascribed to miller indices of Co_3O_4 [211] and Co_3O_4 [111], respectively. This strange phenomenon might be elucidated because there was trace cobalt oxide (Co_3O_4) in the catalyst sample, but there are no other evident diffraction peaks of Co_3O_4 particles, so it is not pure cobalt oxide, and the MOFs constituent is affirmed.

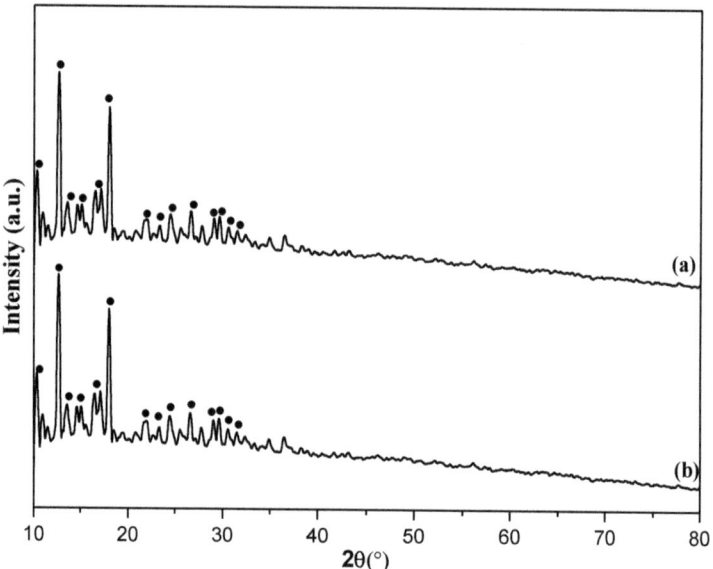

Figure 2. X-ray diffraction (XRD) patterns of Co-ZIF catalyst. (a) fresh, (b) after reaction.

We can see similar XRD signals for the used catalyst sample in Figure 2b, so the crystalline planes are relatively stable, and the stability of this MOF material maybe is suitable. These strange phenomena in the XRD spectra might be due to the forceful coordinate effect in the MOFs material. The average particle size of Co-ZIF species was approx. 18.4 nm, which was calculated from the XRD data based on the Scherrer equation.

TG curves of the Co-ZIF catalyst are shown in Figure 3, and there are two obvious weightlessness stages in the curve. The first weightlessness rate is 20.67% since the temperature reaches 120 °C due to the loss of crystal water in catalyst powder. The reason might be ascribed to the drying temperature being low in the preparation processing. The second weightlessness rate is 55.08% since the temperature reaches 380 °C due to the deformation or decomposition of MOFs material. This weightlessness occurs from 300 °C to 380 °C, so the stability of the Co-ZIF catalyst is poor, and it only is used in the reaction under 300 °C. The curve shows very flat from 400 °C to 700 °C might indicate that the microstructures of the catalyst have collapsed, and cobaltous oxide and carbide is the major constituent. These results are consistent with the typical characteristics of MOFs materials, and the reaction temperature should be controlled below 300 °C in the oxidizing reaction of toluene.

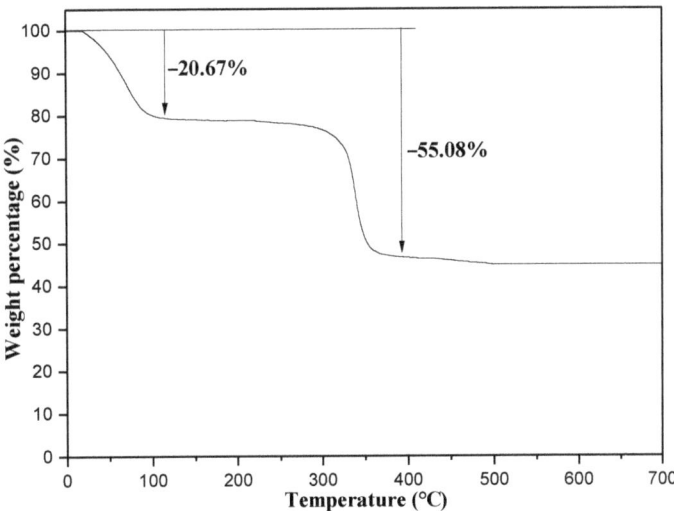

Figure 3. TG curves of Co-ZIF catalyst.

The high specific surface area and lots of homogeneous pores are the typical features of MOFs materials, so the BET characterization can be used for the textural properties of this catalyst. N_2 adsorption-desorption isotherms of the Co-ZIF catalyst are displayed in Figure 4, and the pore size distributions of the Co-ZIF catalyst are shown in Figure S1.

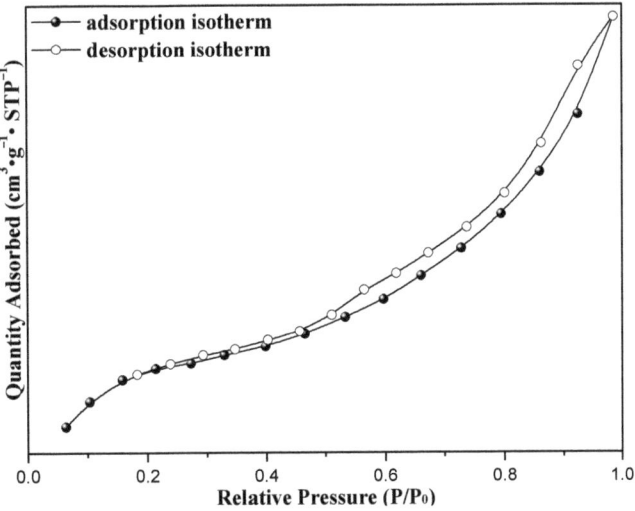

Figure 4. N_2 adsorption-desorption isotherms of the Co-ZIF catalyst.

Obviously, the N_2 isothermal adsorption-desorption curve is of type II according to the Brunauer–Deming–Deming–Teller classification, which means the strong interaction between adsorbate and surface. The hugeous multi-layer adsorption exists when the vapor pressure reaches saturation. The hystersis loop is of type H3 according to the IUPAC classification, which means the catalyst has the feature of irregular pores. The initial isotherm curvatures of the prepared catalyst sample are assigned to micro- and mesopores, while the ascending curvature of the plateau in the high relative pressure range (P/P_0 = 0.85–1.00) is inherent to macropores [39]. This strange phenomenon might be ascribed to the slit holes by cumulus of MOF flaky structures, and adsorption saturation does not occur at higher pressure.

The BET surface area of this catalyst is 924.25 $m^2 \cdot g^{-1}$, the pore volume data are 0.429 $cm^3 \cdot g^{-1}$, and there is a little solvent blocking the pore in the preparation process, and it does not affect the reaction. The pore diameter distribution is relatively concentrated in Figure S1, and the homogeneous pores of this MOF catalyst are affirmed, and this structure might be beneficial for the reaction. However, a wide pore size distribution ranging from ~2.0 nm to ~22.0 nm in Figure S1 further confirms the microporous-dominated structure of this catalyst. The obtained BJH adsorption average pore width of this catalyst is 4.0 nm. In contrast, the wider range of pore size distribution, BJH average pore width, and higher adsorbed volume indicate partial mesopores in this catalyst [40]. The presence of simultaneous micro-, meso-, and macropores in this catalyst indicate the formation of the hierarchical porous structure. Such a combination of the assessable porous structure is advantageous for capturing reactant molecules through the micropores, while the meso- and macropores play an effective role in developing contact probability between the reactant molecule and active centers of the catalyst.

The SEM images of the Co-ZIF catalyst are displayed in Figure 5. The microparticle shape of the catalyst is clear and has apparent coral characteristics in Figure 5a. After further magnification, the particles of the MOFs catalyst are piled, and there are many pores on the surface. The diameter of major particles is less than 1 µm, and many dark channels might be the overlap of particles. This result is consistent with the BET results, and the typical characteristics of the MOFs material are confirmed.

Figure 5. SEM images of the Co-ZIF catalyst. (**a**) fresh catalyst at magnification of 2000; (**b**) fresh catalyst at magnification of 10,000.

The TEM images of the Co-ZIF catalyst are displayed in Figure 6, and there are many circular dark balls shown in Figure 6a. There is a certain gap among the ball's site, and the arrangement model is not uniform. In addition, we can measure the particle size in Figure 6b and the diameter near 200 nm, so the nano-scale catalyst has been confirmed. There are many rough convex objects and dark ditches on the surface of particles, so the Co metal might be coated with 2-methyl glyoxaline tightly.

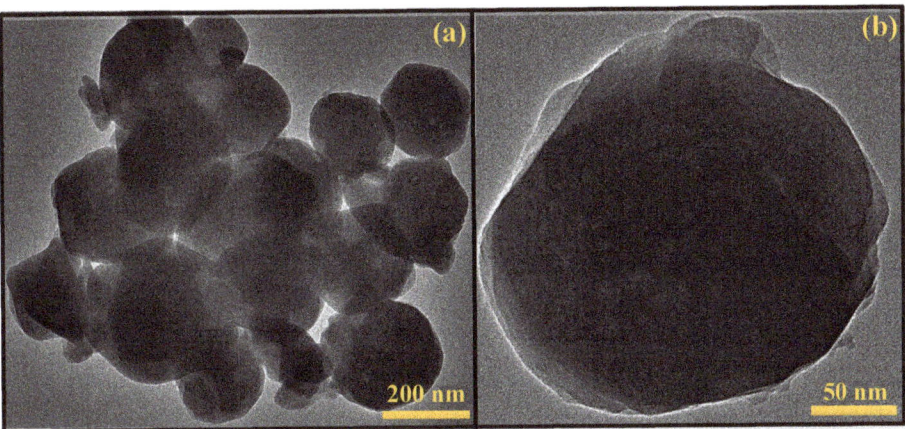

Figure 6. TEM images of the Co-ZIF catalyst. (a) fresh catalyst at magnification of 100,000; (b) fresh catalyst at magnification of 400,000.

The EDS spectra of the Co-ZIF catalyst and elemental mapping are shown in Figure 7. The elemental peaks correspond to C, N, Co, and Au. The major contribution of the C, N, and Co EDS peaks suggests the well surface element distribution. The obvious peaks of Au in Figure 7 are not evitable due to the conventional preparation method used. The uniform distribution of the major elements as C, N, and Co can reflect the successful combination of MOFs materials.

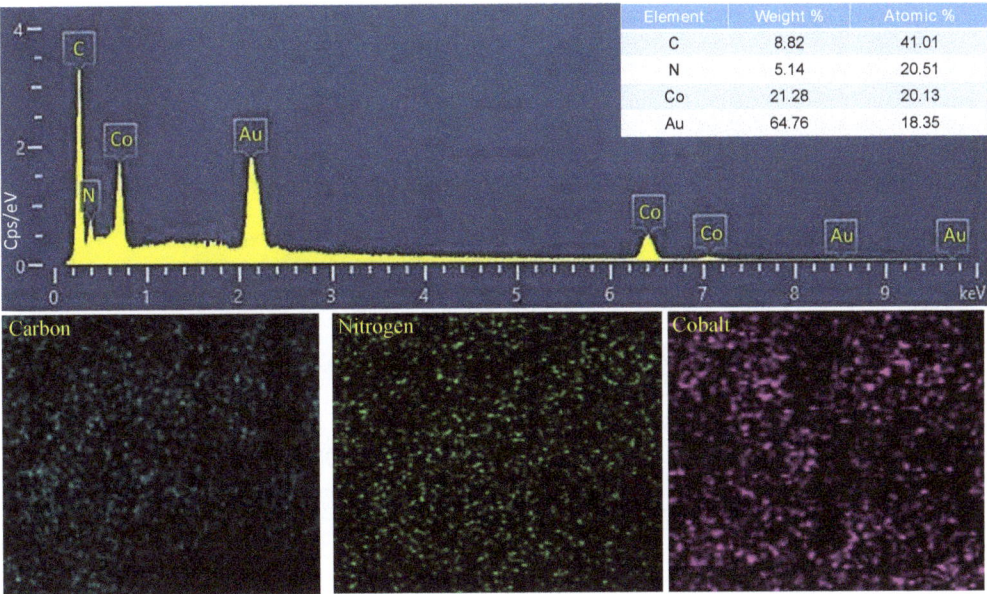

Figure 7. EDS spectra of Co-ZIF catalyst and elemental mapping.

3. Discussion

3.1. Catalytic Performance

The reaction temperature is the important factor for the selective oxidation of toluene, and the effects of the reaction temperature are shown in Figure 8. It is clearly found that reaction temperature has a great effect on this catalytic performance, and the selectivity of products is different at different temperatures. The conversion of toluene increases distinctly from 60.72% to 92.30% since the reaction temperature rises from 303 K to 318 K, and the selectivity to benzaldehyde increases from 75.64% to 91.31%. However, this increasing trend slows down since the reaction temperature exceeds 318 K, and the selectivity to benzaldehyde decreases clearly. The selectivity to benzoic acid increases rapidly, so the higher reaction temperature may promote the transformation of benzaldehyde to benzoic acid. The optimum temperature is 313 K, and the maximum yield of benzaldehyde reached 84.28%.

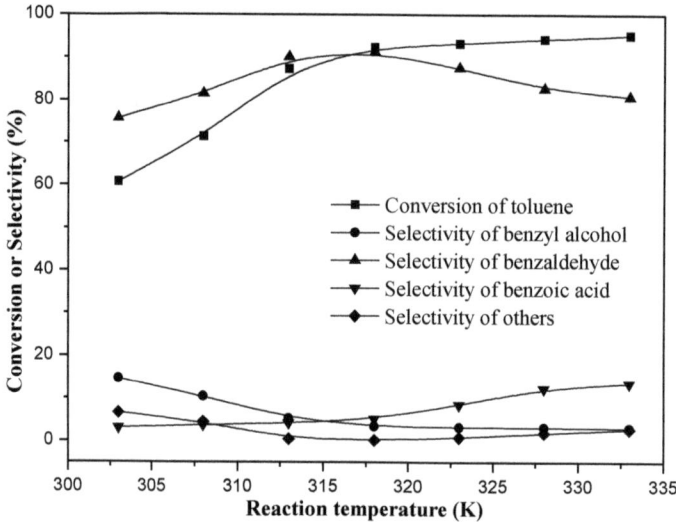

Figure 8. Effect of reaction temperature. Reaction conditions 0.20 g Co-ZIF, 0.500 mmol toluene, 20 mL HFIP, 0.004 g NHPI, 0.12 MPa O_2 and 240 min.

The effects of O_2 pressure are listed in Table 1, and the effect of the O_2 pressure is not as obvious as the reaction temperature. The conversion of toluene and the selectivity to benzoic acid increase with the increment of O_2 pressure, but the selectivity to benzaldehyde reach the maximum value (91.31%) when the O_2 pressure is 0.12 MPa. The selectivity to benzyl alcohol is significantly reduced distinctly from 8.14% to 2.67%, so the higher O_2 pressure may benefit the deep oxidation of toluene. The selectivity to others reaches the lowest value when the O_2 pressure is 0.12 MPa, so the inevitable side reaction maybe can not be neglected in this reaction system.

Table 1. Effects of the O_2 pressure.

O_2 Pressure (MPa)	Conversion (%)	Selectivity (%)			
		Benzyl Alcohol	Benzaldehyde	Benzoic Acid	Others
0.06	79.35	8.14	82.17	2.17	7.52
0.08	86.32	6.42	86.58	3.24	3.76
0.10	90.11	4.13	89.32	4.39	2.16
0.12	92.30	3.42	91.31	5.13	0.14
0.14	92.56	2.98	89.14	6.57	1.31
0.16	92.88	2.67	86.34	8.46	2.53

Reaction conditions 0.20 g Co-ZIF, 0.500 mmol toluene, 20 mL HFIP, 0.004 g NHPI, 313 K and 240 min.

The effects of toluene concentrations are listed in Table 2. We can identify that the conversion capacity of the catalyst is limited. When the toluene concentration was 25.0 mmol/L, the conversion of toluene was not as high as 100%. The conversion of toluene cut down with the increment of toluene concentrations, so the appropriate initialized toluene concentrations were an important factor in developing the yield of major products. The selectivity of benzyl alcohol increased clearly with the increment of the initialized toluene concentrations, which indicated that the oxidation reaction was insufficient. The maximum selectivity of benzaldehyde was 91.31% when the initialized toluene concentration was 25.0 mmol/L, and the corresponding yield of benzaldehyde was as high as 84.28%.

Table 2. Effects of the initialized toluene molar.

Toluene Concentration (mmol/L)	Conversion (%)	Selectivity (%)			
		Benzyl Alcohol	Benzaldehyde	Benzoic Acid	Others
6.25	100.00	1.03	72.69	22.46	3.82
12.5	100.00	2.37	83.79	12.37	1.47
25.0	92.30	3.42	91.31	5.13	0.14
37.5	73.48	12.49	80.49	4.16	2.86
50.0	52.87	23.17	70.34	3.38	3.11

Reaction conditions 0.20 g Co-ZIF, 20 mL HFIP, 0.004 g NHPI, 0.12 MPa O_2, 313 K and 240 min.

The effects of the mass of the catalyst are listed in Table S1. The conversion of toluene increased clearly with the increment of the catalyst. The selectivity of benzaldehyde reached maximum since the mass of the catalyst was 0.20 g. Too much catalyst can develop the selectivity of benzoic acid, so the content of benzaldehyde was obviously reduced. Hence, the suitable mass of the catalyst in this reaction was 0.20 g.

The effects of the mass of NHPI are shown in Figure 9. It is obvious that the addition of NHPI can develop the catalytic activity of the catalyst, which is proved by the variation tendency of selectivity of benzoic acid. The conversion of toluene is as low as 6.74%, and benzoic acid is the major product when the NHPI is not added to this reaction system. The catalytic activity of the catalyst was low without the NHPI. As the role of initiator, the addition of NHPI is necessary. The suitable mass of NHPI in this reaction was 0.004 g, and the selectivity of benzaldehyde reached the maximum.

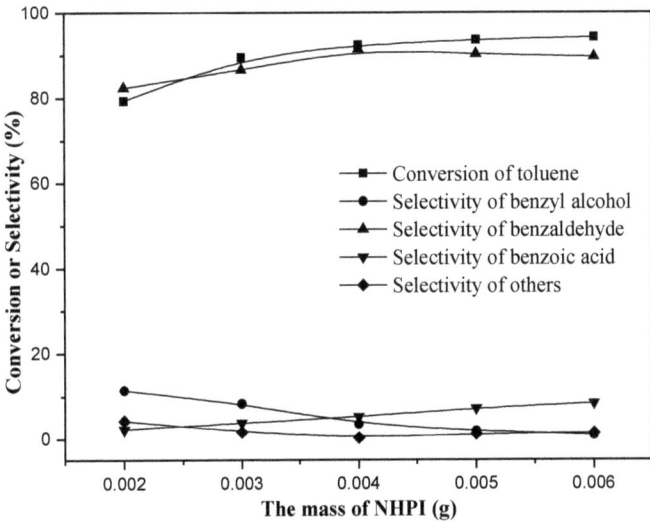

Figure 9. Effect of the mass amount of NHPI. Reaction conditions 0.20 g Co-ZIF, 0.500 mmol toluene, 20 mL HFIP, 0.12 MPa O_2, 313 K and 240 min.

Although NHPI and toluene can be oxidized by nanocatalysts, the further oxidation of benzyl alcohol and benzaldehyde maybe occur at the same time. Too much NHPI also is not suitable in Figure 9. The selectivity of benzoic acid increases rapidly, and the selectivity of benzyl alcohol decreases distinctly with the increment of NHPI mass. This strange phenomenon might be ascribed to the activity of an intermediate compound that was affected by NHPI.

The effects of reaction time are displayed in Figure 10. It is obvious that the conversion of toluene increases rapidly with the prolonged reaction time. The conversion of toluene can reach 90% since the reaction time is 240 min, but the variation tendency of the conversion of toluene is not obvious since the reaction time is greater than 240 min. So the catalytic reaction reached a dynamic equilibrium since the reaction time was 240 min.

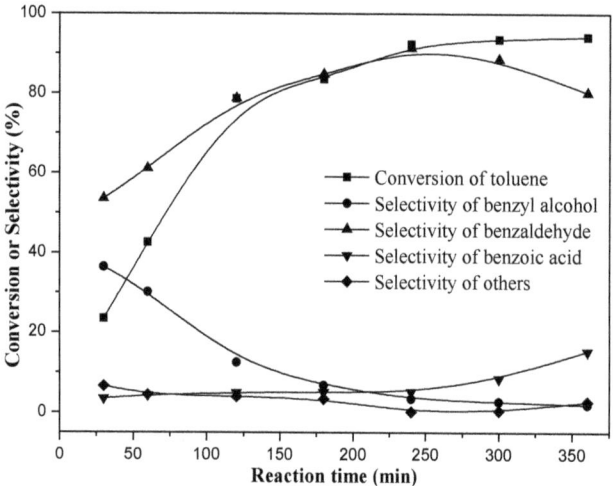

Figure 10. Effect of the reaction time. Reaction conditions 0.20 g Co-ZIF, 0.500 mmol toluene, 20 mL HFIP, 0.004 g NHPI, 0.12 MPa O_2 and 313 K.

For the selectivity of products, the reaction time might be an important factor. We saw the selectivity of benzaldehyde cut down since the reaction time was more than 240 min. The selectivity to benzoic acid increased as the reaction time went on. We also found that the selectivity of benzyl alcohol cut down clearly with the increment of reaction time. Toluene was oxidized to benzyl alcohol, and the benzyl alcohol was oxidized to benzaldehyde or benzoic acid. Hence, proper oxidation is beneficial to the formation of benzaldehyde and the suitable reaction time for selective oxidation of toluene over a nanometer-size Co-ZIF catalyst is 240 min under 313 K and 0.12 MPa.

Many nano catalysts containing Mn or Co were prepared for the oxidation of toluene, and permanganate was used for the oxidation of toluene as early as 1975 [41]. The ability of permanganate to abstract a hydrogen atom is rationalized on the basis of the strong O-H bond formed on H· addition to permanganate, which is a key step of C-H bond oxidations. A comparison of the oxidation of toluene catalyzed by various catalysts is listed in Table 3. Different Mn or Co metal catalysts are better for the oxidation of toluene, and the major products are different. A complete oxidation process also can be performed to generate carbon dioxide, which is significant for the removal of toluene. The catalytic activity and advantage of the Co-ZIF catalyst are clear in Table 3, and this oxidation by the Co-ZIF catalyst belonged to selective oxidation. Benzaldehyde was the major product, and the yield (84.3%) was higher than other catalysts.

Table 3. Comparison of oxidation of toluene catalyzed by various catalysts.

Catalysts	w (%)	Major Product	S_i (%)	Years	Ref.
Co(OAc)$_2$·4H$_2$O	91.0	Benzaldehyde	90.0	2002	[42]
Mn$_{0.67}$-Cu$_{0.33}$	99.0	Carbon dioxide	-	2004	[43]
Cobalt tetraphenylporphyrin	8.9	Benzaldehyde and benzyl alcohol	60.0	2005	[44]
Fe$_2$O$_3$-Mn$_2$O$_3$	100.0	Carbon monoxide and carbon dioxide	-	2009	[45]
CoO$_x$/SiO$_2$	91.2	Benzaldehyde	68.8	2019	[46]
ZIF-67-24	87.9	Benzaldehyde	66.9	2020	[47]
Au/γ-Al$_2$O$_3$	99.0	Carbon dioxide	-	2022	[48]
Co-ZIF	92.3	Benzaldehyde	91.3	in this work	

The results of the reusability of the Co-ZIF catalyst are shown in Figure 11. Typically, the catalysts are separated by filtration, washed and dried in vacuum, and then used in the next run. NHPI is the important initiator in this reaction system, which is added in every new cycle of the process (0.004 g). The catalytic performance decreases a little until the fourth recycle. Nevertheless, the toluene conversion decreases to 86.53%, and the selectivity to benzaldehyde decreases to 87.46% for the fifth time. Hence, the catalytic activity of Co-ZIF is very stable, and it is suitable for recycling use in the oxidation of toluene.

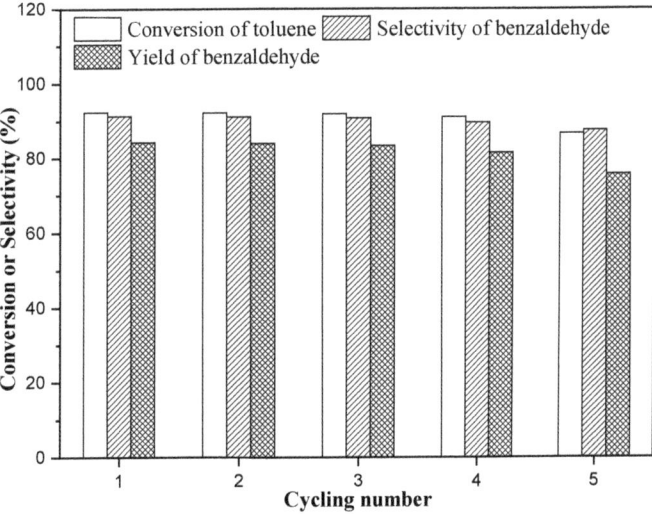

Figure 11. The recycling of Co-ZIF catalyst. Reaction conditions 0.20 g Co-ZIF, 0.500 mmol toluene, 20 mL HFIP, 0.004 g NHPI, 0.12 MPa O$_2$, 313 K and 240 min.

Additionally, the effects of solvent over the Co-ZIF catalyst are shown in Supplementary Table S2. It can be seen that the polar solvent is helpful for the selective oxidation of toluene, and the high polar solvent of HFIP gives the best catalytic performance of 92.30% conversion and 91.31% selectivity to benzaldehyde. As the strong polar solvent, HFIP is beneficial to the oxidation of the C-H bond, and the dehydrogenation reaction is easier to proceed by the action of free radicals. So benzyl alcohol is easier to produce. Another important factor is that the hydrogen bond between benzaldehyde and HFIP can inhibit the further oxidation of benzaldehyde, so the yield of benzaldehyde can be improved clearly.

3.2. Probable Catalytic Mechanism

Based on the results from this work and the literature [38,49], a possible reaction path for the oxidation of toluene over the Co-ZIF catalyst was proposed and shown in Figure 12. The free radical reaction mechanism maybe is a popular reasonable explanation. PINO can be generated by the deprotonation of NHPI. It also is related to the participation of a catalyst. $Co^{III}L_n$ represents the elementary unit of the Co-ZIF catalyst, and it is oxidized to the strong oxidation active intermediate as $L_nCo^{III}OOCo^{III}L_n$ by oxygen [47]. PINO can be generated by the deproton reaction under this strong oxidation active intermediate. The dehydrogenation of toluene is the first step of the main chain reaction, so the attractive proton capacity of PINO is the decisive influence in these steps, which indicates that NHPI can be restored. Another way is the oxidation of $L_nCo^{III}OH$, which can transform NHPI to PINO.

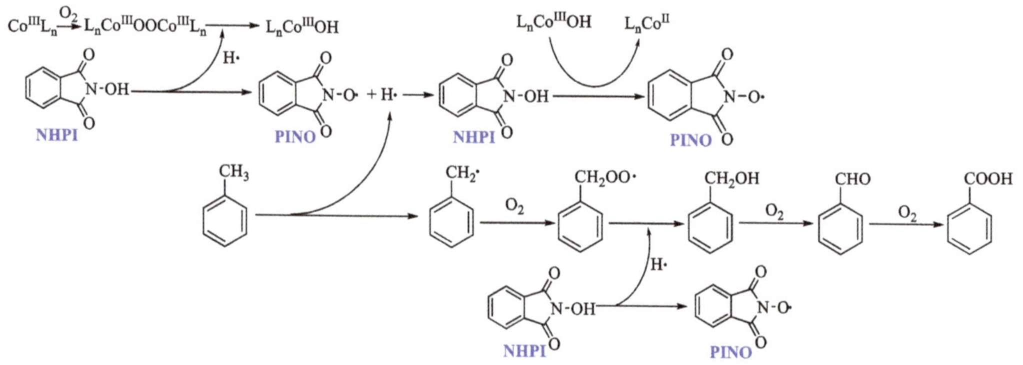

Figure 12. The proposed mechanism of selective oxidation of toluene by Co-ZIF catalyst.

Toluene radical can be generated from toluene by the strong proton attraction of PINO, and it is oxidized by oxygen into the peroxide intermediate (benzyl peroxy radical). Then, benzyl alcohol can be formed by the transformation of benzyl peroxy radical with a free proton from NHPI [50]. In the step of chain termination, benzyl alcohol is oxidized to benzaldehyde and further oxidized to benzoic acid. The intermediate products of toluene radical, PINO and benzyl peroxy radical were testified by GC-MS. Considering the boiling point of PINO is much higher than the others, vacuum distillation or rectification can be used for the further separation of products and the PINO radical. Hence, the yield of benzaldehyde is affected by the mass amount of NHPI, but the catalytic activity of NHPI is affected by the Co-ZIF catalysts.

4. Materials and Methods

4.1. Materials

$Co(NO_3)_2 \cdot 6H_2O$, 2-methyl glyoxaline, dimethylformamide (DMF), toluene, benzyl alcohol, benzaldehyde were purchase from Shanghai Macklin Biochemical Co., Ltd. (Shanghai, China) N-hydroxyl phthalimide (NHPI), benzoic acid, 1,1,1,3,3,3-hexafluoroisopropanol (HFIP), ethanol, and methanol were analytical grade and purchased from Sinopharm Chemical Reagent Co., Ltd (Tianjin, China). As an internal standard in the meteorological chromatographic analysis procedure, acetophenone were chromatographically pure and purchased from Shanghai Macklin Biochemical Co., Ltd (Shanghai, China). O_2 (99.99%) was provided by Maoming Deyuan Gas Company (Maoming, China).

4.2. Catalyst Preparation

The hydrothermal synthesis method [51] was chosen for the preparation of catalysts. Firstly, 1.478 g of 2-methyl glyoxaline was dispersed in 20 mL of DMF solution with stirred continuously for 1 h. Then, a certain amount of $Co(NO_3)_2 \cdot 6H_2O$ crystal was added to this mixture solution and stirred continuously for 2 h. All liquid reactants were transferred into the crystallization autoclave and stayed in the cabinet dryer at 140 °C for 24 h. After being cooled to room temperature, the product powder was washed with distilled water and methanol several times. The purple crystalline powder was dried in air overnight at 80 °C and labeled as Co-ZIF.

4.3. Catalyst Characterization

Fourier transform infrared (FT-IR) spectra of the catalysts were recorded on a PerkinElmer spectrum One (C) spectrometer in the wave number range of 500–4000 cm^{-1}. TG/DTG curves of the catalyst were recorded by Netzsch 210C thermogravimetric using air as purge gas (40 mL/min) over a temperature range of 35–800 °C with a heating rate of 10 °C/min. Powder X-ray diffraction (XRD) patterns of catalysts were recorded by the Bruker D/max 2515TC diffractometer device with Cu Kα radiation (λ = 1.535 Å). The scan range was 5–90° with with a scanning rate of 2°/min, and the tube voltage and current were 45 kV and 35 mA, respectively. The specific surface area, pore volume, and pore size distribution of the catalysts were analyzed by the nitrogen adsorption-desorption on a Quantachrome NOVA-2301D automated gas sorption system. Specific surface areas were calculated by Brunauer–Emmett–Teller (BET), and pore size distributions were calculated by Barrett-Joyner-Halenda (BJH) methods. The morphologies of the catalysts were observed by scanning electron microscope (SEM, JEOL 6510 ED) device. The microstructure of the catalyst was observed by transmission electron microscopy (TEM, TecnaiG350 SE) device, the working voltage at less than 200 kV, and the catalyst powder was deposited on a copper grid.

4.4. Catalytic Test Method

The catalytic oxidation process of toluene was performed in a 50 mL Teflon-lined stainless steel autoclave with a magnetic stirrer at 600 rpm. Typically, 20 mL HFIP, 0.500 mmol toluene, and 0.04 g NHPI were mixed into the quartz lining, 0.20 g catalyst powder was added into the mixture solution, and the quartz lining was transferred into the autoclave. The reactor was sealed and purged with O_2 to exclude air three times, and then it was pressurized to 0.12 MPa with O_2 under continuous stirring after the designed temperature was reached.

After a certain reaction time, the catalysts were separated by filtration carefully, and the gas phase products and the liquid phase products were analyzed by GC-MS. The gas phase products include oxygen, carbon dioxide, carbon monoxide, benzene, etc. The liquid phase products and contents of the reactants were analyzed by gas chromatography (Agilent Technologies, 8960B) equipped with a DB-7501 capillary column (diameter 0.60 mm, length 28 m) and a flame ionization detector (FID) using acetophenone as the internal standard substance (sample was diluted by ethanol). The major product was benzaldehyde, and by-products were benzyl alcohol, benzoic acid, benzyl benzoate, biphenyl, etc.

Considering the gas phase products are rare and the liquid phase products are important, all the concerned products contain an aromatic nucleus. We can ignore the carbon proportion. The toluene conversion (w) and product selectivity (S_i) were calculated by the following equations [47].

$$w\ (\%) = \frac{Mol(toluene)_{in} - Mol(toluene)_{out}}{Mol(toluene)_{in}} \times 100 \qquad (1)$$

$$S_i(\%) = \frac{Mol_i}{Mol_{reacted\cdot toluene}} \times 100 \qquad (2)$$

5. Conclusions

In summary, nanometer-size Co-ZIF catalyst was prepared and characterized [37,38]. The catalyst presents a suitable catalytic performance in the selective oxidation of toluene to benzaldehyde under mild conditions. The typical characteristics of the MOFs material were affirmed by the SEM and TEM. The diameter of particles is near 200 nm, and the BET surface area of this catalyst is as high as 924.25 m^2/g. The Co metal is coated with 2-methyl glyoxaline, and the crystalline planes are relatively stable. The addition of a certain amount of NHPI and the smooth oxidate capacity of the catalyst is in favor of a high yield of benzaldehyde. Under the optimized reaction conditions, the best catalytic performance of 92.30% conversion of toluene and the selectivity to benzaldehyde is up to

91.31%. This nanometer-size catalyst showed superior performance for recycling use in the selective oxidation of toluene. The possible reaction path is proposed based on GC-MS and the works of other researchers.

Supplementary Materials: The following supporting information can be downloaded at: https://www.mdpi.com/article/10.3390/ijms232112881/s1.

Author Contributions: Conceptualization, W.L. and X.K.; methodology, W.L. and Z.C.; formal analysis and investigation, W.L., Z.C. and Y.H.; resources, W.L. and X.K.; data curation, W.L.; writing—original draft preparation, W.L. and Z.C.; writing—review and editing, W.L.; project administration and funding acquisition, W.L. All authors have read and agreed to the published version of the manuscript.

Funding: This work was supported by Open Fund of Guangdong Provincial Key Laboratory of Petrochemical Pollution Process and Control, Guangdong University of Petrochemical Technology (No.2018B030322017), Advanced Talents of Guangdong University of Petrochemical Technology (No.2018rc50), and Science and Technology Plan Project of Maoming City (No.2019395, No.2021624).

Institutional Review Board Statement: Not applicable.

Informed Consent Statement: Not applicable.

Data Availability Statement: Data are contained within the article or Supplementary Materials. The data presented in this study are available in Supplementary Materials.

Conflicts of Interest: The authors declared there are no conflict of interest with respect to the research, publication of this article, and authorship.

References

1. Qi, Y.; Shen, L.; Zhang, J.; Yao, J.; Lu, R.; Miyakoshi, T. Species and release characteristics of VOCs in furniture coating process. *Environ. Pollut.* **2019**, *245*, 810–819. [CrossRef] [PubMed]
2. Lerner, J.C.; Sanchez, E.Y.; Sambeth, J.E.; Porta, A.A. Characterization and health risk assessment of VOCs in occupational environments in buenos airs, argentina. *Atmos. Environ.* **2012**, *55*, 440–447. [CrossRef]
3. Carriero, G.; Neri, L.; Famulari, D.; Di Lonardo, S.; Piscitelli, D.; Manco, A.; Baraldi, R. Composition and emission of VOC from biogas produced by illegally managed waste landfills in giugliano (Campania, Italy) and potential impact on the local population. *Sci. Total Environ.* **2018**, *640–641*, 377–386. [CrossRef] [PubMed]
4. Wang, S.; Bai, P.; Wei, Y.; Liu, W.; Ren, X.; Bai, J.; Yu, J. Three-Dimensional-Printed core-shell structured MFI-type zeolite monoliths for volatile organic compound capture under humid conditions. *ACS Appl. Mater. Interfaces* **2019**, *11*, 38955–38963. [CrossRef]
5. Schiavon, M.; Scapinello, M.; Tosi, P.; Ragazzi, M.; Torretta, V.; Rada, E.C. Potentical of non-thermal plasmas for helping the biodegradation of volatile organic compounds(VOCs) released by waste management plants. *J. Clean. Prod.* **2015**, *104*, 211–219. [CrossRef]
6. Zhang, Z.X.; Jiang, Z.; Shangguan, W.F. Low-temperature catalysis for VOCs removal in technology and application: A state of the art review. *Catal. Today* **2016**, *264*, 270–278. [CrossRef]
7. Nguyen, H.P.; Park, M.J.; Kim, S.B.; Kim, H.J.; Baik, L.J.; Jo, Y.M. Effective dielectric barrier discharge reactor operation for decomposition of volatile organic compounds. *J. Clean. Prod.* **2018**, *198*, 1232–1238. [CrossRef]
8. Tidahy, H.L.; Hosseni, M.; Siffert, S.; Cousin, R.; Lamonier, J.F.; Aboukaïs, A.; Leclercq, G. Nanostructured macro-mesoporous zirconia impregnated by noble metal for catalytic total oxidation of toluene. *Catal. Today* **2008**, *137*, 335–339. [CrossRef]
9. Lin, T.; Yu, L.; Sun, M.; Cheng, G.; Lan, B.; Fu, Z. Mesoporous α-MnO_2 microspheres with high specific surface area: Controlled synthesis and catalytic activities. *Chem. Eng. J.* **2016**, *286*, 114–121. [CrossRef]
10. Maira, A.J.; Yeung, K.L.; Soria, J.; Coronado, J.M.; Belver, C.; Lee, C.Y.; Augugliaro, V. Gas-phase photo-oxidation of toluene using nanometer-size TiO_2 catalysts. *Appl. Catal. B Environ.* **2001**, *29*, 327–336. [CrossRef]
11. Liotta, L.F. Catalytic oxidation of volatile organic compounds on supported noble metals. *Appl. Catal. B.* **2010**, *100*, 403–412. [CrossRef]
12. Fu, J.L.; Dong, N.; Ye, Q.; Cheng, S.Y.; Kang, T.F.; Dai, H.X. Enhanced performance of the OMS-2 catalyst by Ag Loading for the oxidation of benzene, toluene, and formaldehyde. *New J. Chem.* **2018**, *42*, 18117–18127. [CrossRef]
13. Bertinchamps, F.; Gregoire, C.; Gaineaux, E.M. Systematic investigation of supported transition metal oxide based formulations for the catalytic oxidative elimination of (chloro)-aromatics. Part II. Influence of the nature and addition protocol of secondary phases to VO_x/TiO_2. *Appl. Catal. B.* **2006**, *66*, 10–22. [CrossRef]
14. Dos Santos, A.A.; Lima, K.M.; Figueiredo, R.T.; Egues, S.M.d.S.; Ramos, A.L.D. Toluene deep oxidation over nble metals, Copper and Vanadium Oxides. *Catal. Lett.* **2007**, *114*, 59–63. [CrossRef]
15. Gaur, V.; Sharma, A.; Verma, N. Catalytic oxidation of toluene and m-xylene by activated carbon fiber impregnated with transition metals. *Carbon* **2005**, *43*, 3041–3053. [CrossRef]

16. Wu, J.C.S.; Chang, T.Y. VOC deep oxidation over Pt catalysts using hydrophobic supports. *Catal. Today* **1998**, *44*, 111–118.
17. Chen, C.; Wu, Q.; Chen, F.; Zhang, L.; Pan, S.; Bian, C.; Xiao, F.S. Aluminium-rich beta zeolite-supported platinum nanoparticles for the low-temperature catalytic removal of toluene. *J. Mater. Chem. A* **2015**, *3*, 5556–5562. [CrossRef]
18. Kim, S.C.; Shim, W.G. Influence of physicochemical treatments on iron-based spent catalyst for catalytic oxidation of toluene. *J. Hazard. Mater.* **2007**, *154*, 310–316. [CrossRef]
19. Zou, X.; Rui, Z.; Ji, H. Core-Shell NiO@PdO nanoparticles supported on alumina as an advanced catalyst for methane oxidation. *ACS Catal.* **2017**, *7*, 1615–1625. [CrossRef]
20. Feijen-Jeurissen, M.M.R.; Jorna, J.J.; Nieuwenhuys, B.E.; Sinquin, G.; Petit, C.; Hindermann, J.P. Mechanism of catalytic destruction of 1,2-dichloroethane and trichloroethylene over g-Al_2O_3 and g-Al_2O_3 supported chromium and palladium catalysts. *Catal. Today* **1999**, *54*, 65–79. [CrossRef]
21. Yang, H.; Deng, J.; Liu, Y.; Xie, S.; Wu, Z.; Dai, H. Preparation and catalytic performance of Ag, Au, Pd or Pt nanoparticles supported on 3dom CeO_2-Al_2O_3 for toluene oxidation. *J. Mol. Catal. A Chem.* **2016**, *414*, 9–18. [CrossRef]
22. He, C.; Cheng, J.; Zhang, X.; Douthwaite, M.; Pattisson, S.; Hao, Z.P. Recent advances in the catalytic oxidation of volatile organic compounds: A review based on pollutant sorts and sources. *Chem. Rev.* **2019**, *119*, 4471–4568. [CrossRef] [PubMed]
23. Xia, Y.; Xia, L.; Liu, Y.; Yang, T.; Deng, J.; Dai, H. Concurrent catlytic removal of typical volatile organic compound mixtures over Au-Pd/α-MnO_2 nanotubes. *J. Environ. Sci.* **2018**, *64*, 276–288. [CrossRef] [PubMed]
24. Li, X.; Wang, L.; Xia, Q.; Liu, Z.; Li, Z. Catalytic oxidation of toluene over copper and manganese based catalysts: Effect of water vapor. *Catal. Commun.* **2011**, *14*, 15–19. [CrossRef]
25. Yang, H.; Deng, J.; Liu, Y.; Xie, S.; Xu, P.; Dai, H. Pt/Co_3O_4/3DOM Al_2O_3: Highly effective catalysts for toluene combustion. *Chin. J. Catal.* **2016**, *37*, 934–946. [CrossRef]
26. Chen, C.; Zhu, J.; Chen, F.; Meng, X.; Zheng, X.; Gao, X.; Xiao, F.S. Enhanced performance in catalytic combustion of toluene over mesoporous beta zeolite-supported platinum catalyst. *Appl. Catal. B Environ.* **2013**, *140–141*, 199–205. [CrossRef]
27. Zhao, S.; Hu, F.; Li, J. Hierarchical core-shell Al_2O_3@Pd-CoAlO microspheres for low-temperature toluene combustion. *ACS Catal.* **2016**, *6*, 3433–3441. [CrossRef]
28. Deng, J.G.; Ultra, H.X. Low loading of silver nanoparticles on Mn_2O_3 nanowires derived with molten salts: A high-efficiency catalyst for the oxidative removal of toluene. *Environ. Sci. Technol.* **2015**, *49*, 11089–11095. [CrossRef]
29. Xie, S.H.; Deng, J.G.; Zang, S.M.; Yang, H.G.; Guo, G.S.; Arandiyan, H.; Dai, H.X. Au-Pd/3DOM Co_3O_4: Highly active and stable nanocatalysts for toluene oxidation. *J. Catal.* **2015**, *322*, 38–48. [CrossRef]
30. Dong, C.; Wang, H.; Ren, Y.W.; Qu, Z.P. Layer MnO_2 with oxygen vacancy for improved toluene oxidation activity. *Surf. Interfaces* **2021**, *22*, 100897–100905. [CrossRef]
31. Yuan, W.H.; Li, Y.Y.; Dagaut, P.; Yang, J.Z.; Qi, F. Investigation on the pyrolysis and oxidation of toluene over a wide range conditions. II. A comprehensive kinetic modeling study. *Combust. Flame* **2015**, *162*, 22–40. [CrossRef]
32. Usman, M.; Iqbal, N.; Noor, T.; Zaman, N.; Asghar, A.; Abdelnaby, M.M.; Galadima, A.; Helal, A. Advanced strategies in metal-organic-frameworks for CO_2 capture and separation. *Chem. Rec.* **2022**, *22*, e202100230–e202100258. [CrossRef] [PubMed]
33. Helal, A.; Shah, S.S.; Usman, M.; Khan, M.Y.; Aziz, M.A.; Rahman, M.M. Potential applications of Nickel-based metal-organic-frameworks and their derivatives. *Chem. Rec.* **2022**, *22*, e202200055. [CrossRef] [PubMed]
34. Usman, M.; Zeb, Z.; Ullah, H.; Suliman, M.H.; Humayun, M.; Ullash, L.; Shah, S.N.Z.; Ahmed, U.; Saeed, M. A review of metal-organic frameworks/graphitic carbon nitride composites for solar-driven green H_2 production, CO_2 reduction, and water purification. *J. Environ. Chem. Eng.* **2022**, *10*, 107548–107574. [CrossRef]
35. Shafiq, S.; Al-Maythalony, B.A.; Usman, M.; Ba-Shammakh, M.S.; Al-Shammari, A.A. ZIF-95 as a filler for enhanced gas separation performance of polysulfone membrane. *RSC Adv.* **2021**, *11*, 34319–34328. [CrossRef]
36. Zhang, Q.; Jiang, Y.W.; Gao, J.H.; Fu, M.L.; Zou, S.B.; Li, Y.X.; Ye, D.Q. Interfaces in MOF-derived CeO_2-MnO_x composites as high-activity catalysts for toluene oxidation: Monolayer dispersion threshold. *Catalysts* **2020**, *10*, 681. [CrossRef]
37. Zhou, K.; Mousavi, B.; Luo, Z.; Phatanasri, S.; Chaemchuen, S.; Verpoort, F. Characterization and properties of Zn/Co zeolitic imidazolate frameworks vs. ZIF-8 and ZIF-67. *J. Mater. Chem. A* **2017**, *5*, 952–957. [CrossRef]
38. Wei, G.; Zhou, Z.; Zhao, X.; Zhang, W.; An, C. Ultrathin metal-organic framework nanosheet-derived ultrathin Co_3O_4 nanomeshes with robust oxygen-evolving performance and asymmetric supercapacitors. *ACS Appl. Mater. Inter.* **2018**, *10*, 23721–23730. [CrossRef]
39. Shah, S.S.; Qasem, M.A.A.; Berni, R.; Casino, C.D.; Cai, G.; Contal, S.; Ahmad, I.; Siddiqui, K.S.; Gatti, E.; Predieri, S.; et al. Physico-chemical properties and toxicological effects on plant and algal models of carbon nanosheets from a nettle fibreclone. *Sci. Rep.* **2021**, *11*, 6945–6960. [CrossRef]
40. Shah, S.S.; Cevik, E.; Aziz, M.A.; Qahtan, T.F.; Bozkurt, A.; Yamani, Z.H. Jute sticks derived and commercially available activated carbons for symmetric supercapacitors with bio-electrolyte: A comparative study. *Synth. Met.* **2021**, *277*, 116765–116780. [CrossRef]
41. Gardner, K.A.; Mayer, J.M. Understanding C-H bond oxidations:$H\cdot$ and H^- transfer in the oxidation of toluene by permanganate. *Science* **1995**, *269*, 1849–1851. [CrossRef]
42. Grasselli, R.K. Fundamental principles of selective heterogeneous oxidation catalysis. *Top. Catal.* **2002**, *21*, 79–88. [CrossRef]
43. Li, W.B.; Chu, W.B.; Zhuang, M.; Hua, J. Catalytic oxidation of toluene on Mn-containing mixed oxides prepared in reverse microemulsions. *Catal. Today* **2004**, *93–95*, 205–209. [CrossRef]

44. Guo, C.C.; Liu, Q.; Wang, X.T.; Hu, H.Y. Selective liquid phase oxidation of toluene with air. *Appl. Catal. A: Gen.* **2005**, *282*, 55–59. [CrossRef]
45. Florea, M.; Alifanti, M.; Parvulescu, V.I.; Mihaila-Tarabasanu, D.; Diamandescu, L.; Feder, M.; Negrila, C.; Frunza, L. Total oxidation of toluene on ferrite-type catalysts. *Catal. Today* **2009**, *141*, 361–366. [CrossRef]
46. Shi, G.; Xu, S.; Bao, Y.; Xu, j.; Liang, Y. Selective aerobic oxidation of toluene to benzaldehyde on immobilized CoO_x on SiO_2 catalyst in the presence of N-hydroxyph thalimide and hexafluoropropan-2-ol. *Catal. Commun.* **2019**, *123*, 73–78. [CrossRef]
47. Xiao, Y.P.; Song, B.C.; Chen, Y.J.; Cheng, L.H.; Ren, Q.G. ZIF-67 with precursor concentration-dependence morphology for aerobic oxidation of toluene. *J. Organomet. Chem.* **2020**, *930*, 121597–121602. [CrossRef]
48. Wu, Z.L.; Zhu, D.D.; Chen, Z.Z.; Yao, S.L.; Li, J.; Gao, E.; Wang, W. Enhanced energy efficiency and reduced nanoparticle emission on plasma catalytic oxidation of toluene using $Au/\gamma-Al_2O_3$ nanocatalyst. *Chem. Eng. J.* **2022**, *427*, 130983–130992. [CrossRef]
49. Gaster, E.; Kozuch, S.; Pappo, D. Selective aerobic oxidation of methylarenes to benzaldehydes catalyzed by N-hydroxyphthalimide and cobalt(II) acetate in hexafluoropropanol. *Angew. Chem. Int. Ed.* **2017**, *56*, 5912–5915. [CrossRef]
50. Wang, L.; Zhang, Y.; Du, R.; Yuan, H.; Wang, Y.; Yao, J.; Li, H. Selective one-step aerobic oxidation of cyclohexane to e-caprolactone mediated by N-hydroxyphthalimide (NHPI). *Chem. Cat Chem.* **2019**, *11*, 2260–2264.
51. Zakzeski, J.; Dczak, A.; Bruijnincx, P.C.A. Catalytic oxidation of aromatic oxygenates by the heterogeneous catalyst Co-ZIF-9. *Appl. Catal. A Gen.* **2011**, *394*, 79–85. [CrossRef]

Article

Dendritic Mesoporous Silica Nanoparticle Supported PtSn Catalysts for Propane Dehydrogenation

Ning Zhang [1], Yiou Shan [1], Jiaxin Song [2], Xiaoqiang Fan [1,*], Lian Kong [1], Xia Xiao [1], Zean Xie [1] and Zhen Zhao [1,2,*]

[1] Institute of Catalysis for Energy and Environment, Shenyang Normal University, Shenyang 110034, China
[2] State Key Laboratory of Heavy Oil Processing, China University of Petroleum, Beijing 102249, China
* Correspondence: fanxiaoqiang@synu.edu.cn (X.F.); zhenzhao@cup.edu.cn (Z.Z.); Tel.: +86-24-86578737 (X.F.)

Abstract: PtSn catalysts were synthesized by incipient-wetness impregnation using a dendritic mesoporous silica nanoparticle support. The catalysts were characterized by XRD, N_2 adsorption–desorption, TEM, XPS and Raman, and their catalytic performance for propane dehydrogenation was tested. The influences of Pt/Sn ratios were investigated. Changing the Pt/Sn ratios influences the interaction between Pt and Sn. The catalyst with a Pt/Sn ratio of 1:2 possesses the highest interaction between Pt and Sn. The best catalytic performance was obtained for the Pt_1Sn_2/DMSN catalyst with an initial propane conversion of 34.9%. The good catalytic performance of this catalyst is ascribed to the small nanoparticle size of PtSn and the favorable chemical state and dispersion degree of Pt and Sn species.

Keywords: propane dehydrogenation; dendrimer-like silica nanoparticle support; Pt/Sn ratios

Citation: Zhang, N.; Shan, Y.; Song, J.; Fan, X.; Kong, L.; Xiao, X.; Xie, Z.; Zhao, Z. Dendritic Mesoporous Silica Nanoparticle Supported PtSn Catalysts for Propane Dehydrogenation. Int. J. Mol. Sci. 2022, 23, 12724. https://doi.org/10.3390/ijms232112724

Academic Editors: Junjiang Zhu and Luísa Margarida Martins

Received: 8 August 2022
Accepted: 18 October 2022
Published: 22 October 2022

Publisher's Note: MDPI stays neutral with regard to jurisdictional claims in published maps and institutional affiliations.

Copyright: © 2022 by the authors. Licensee MDPI, Basel, Switzerland. This article is an open access article distributed under the terms and conditions of the Creative Commons Attribution (CC BY) license (https://creativecommons.org/licenses/by/4.0/).

1. Introduction

Propene is an important raw material that is widely used in the production of polypropene, propene oxide, acrylonitrile, etc. Currently, it is mainly obtained by steam cracking of naphtha, and the fluid catalytic cracking process. Due to the restriction of the oil price and storage and the growing demand for propene, some alternative methods, including the methanol-to-olefins (MTO) reaction and the dehydrogenation of propane (PDH) process, have been used to increase its production [1]. Among them, PDH is advantageous because of the price gap between propene and propane and the abundant resources of shale gas and natural gas. The PDH process is endothermic. In order to obtain a desirable yield of propene, the PDH process operates at a high temperature, above 450 °C. The high temperature is favorable for coke formation and hydrocarbon cracking, which increase the deactivation of catalysts. Therefore, the development of PDH catalysts with high catalytic activity and stability is key. Two main types of catalysts are used for PDH: noble metal-based and non-noble metal-based catalysts. Cr-based catalysts, as the most prominent example of non-noble metal-based catalysts with high catalytic activity for PDH, have been commercialized. Other non-noble metal-based catalysts including Zn [2], Mo [3], V [1,4], Co [5], Ga [6,7], Fe [8,9], Cu [10], etc. [11–13] also show catalytic activity for PDH. As the extensively investigated precious metal catalyst for PDH, Pt-based catalysts are also used in commercial processes. However, Pt-based catalysts still face the challenge of deactivation due to coke deposition and Pt nanoparticle sintering. In order to improve the stability of Pt-based catalysts, many strategies have been adopted, including adding metal promoters. Sn is the most effective promoter for the Pt catalyst in PDH [14–18]. The role of Sn is mainly explained in terms of geometric effects and electronic effects [19]. It can assist in the separation of large Pt ensembles into small clusters, which can suppress structure-sensitive side reaction. The electron transfer between Sn and Pt atoms can change the Pt electron density, thereby affecting the adsorption–desorption of reactants and products, promoting the selectivity and stability. Sexton et al. [20] found that Sn(II) is a surface modifier of γ-Al$_2$O$_3$ and the change of electronic interactions between Pt and Sn(II)-γ-Al$_2$O$_3$ changes the reactivity. Motagamwala et al. [21] reported a silica-supported Pt$_1$Sn$_1$ nanoparticle catalyst with excellent stability during PDH. Xiong et al. [22] found that the formation of small PtSn clusters allows the catalyst to achieve high propene selectivity due to the facile desorption of propene. Wang et al. [23] described the structural evolution of Pt–Sn bimetallic nanoparticles for PDH and found the recovery of Pt-Sn alloy by Sn segregation important for sustaining the high catalytic performance of PDH. Liu et al. [24] also investigated the evolution of Pt and Sn species during high-temperature treatments. Gao et al. [25] reported that Pt clusters that were partially modified by reduced Sn exhibited higher activity than PtSn alloy counterparts for PDH. Although Sn as the promoter has been widely studied, the nature of the PtSn active sites still needs further research. Moreover, Zn [26,27], Cu [28], Fe [29,30], Co [31], Ga [32], In [33,34] and Mn [35,36] are also used as promoters in PDH. The other strategy is to improve metal–support interactions so as to increase the catalytic performance. Many supports have been used, including Al$_2$O$_3$, SiO$_2$ and zeolites [37–41]. Compared with Al$_2$O$_3$ and zeolites, SiO$_2$ shows chemical inertness and is suitable for use as a support to study the relationship between active components and catalytic performance.

Dendritic mesoporous silica nanoparticles (DMSNs) are attractive materials as support due to their unique feature. They show high surface area, which mainly originates from their dendrimer-like morphology, not from their mesoporous channels as with SBA-15 or MCM-41 [42]. This unique feature shows the usefulness compared to conventional ordered mesoporous silica for some catalytic reactions [43]. Moreover, the monodispersed nanoparticles with a short diffusion length as the support might promote the diffusion of reactants and products and influence the catalytic behavior. It has been reported that support with suitable physical morphology is of fundamental importance for PDH. Compared with SBA-15 or MCM-41, the diffusion distance in the pores is shortened on DMSN support (radius of the nanoparticle about 100 nm for DMSN and usually several μm for SBA-15 or MCM-41) [44]. This makes the PtSn nanoparticles in the pores easily accessible for reactants and makes the products easy to diffuse to avoid further reaction. Compared with the zeolites of micropores, the PtSn nanoparticles are usually loaded on the external surface area rather than in the microporous area of zeolites, which will decrease the confined effect of the pores. Therefore, in this paper, DMSNs are prepared as the support and used in a PDH reaction. Catalysts with different ratios of Pt and Sn were prepared by the impregnation method and were characterized by means of various techniques to investigate the dispersion and state of Pt and Sn species on different ratios of Pt and Sn. The catalytic performances were tested with PDH to investigate the influence of a PtSn active component on the catalytic performance.

2. Results and Discussion

2.1. Analysis of Physico-Chemical Properties

Nitrogen sorption measurements were conducted to characterize the pore parameters of the catalysts with different ratios of Pt and Sn. The corresponding nitrogen adsorption–desorption isotherms and pore size distributions are shown in Figure 1A,B, respectively. All the catalysts showed typical type IV adsorption–desorption isotherms with H$_4$-like hysteresis loops at values of P/P$_0$ ranging from 0.40 to 1.00, which indicated the presence of mesopores. The similarity of the catalysts demonstrated that the loading of metals did not cause damage to the porous structure of the support. Figure 1B shows the pore size calculated according to the BJH method. The pore sizes of these samples through Pt/DMSN to Pt$_1$Sn$_4$/DMSN catalyst were found to be consistent with the value of 3.1 or 3.2 nm. This may be caused by the low metal loading amounts or the high dispersion of the metals.

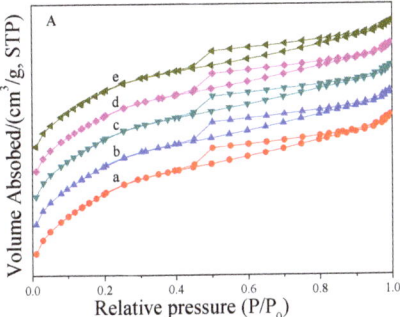 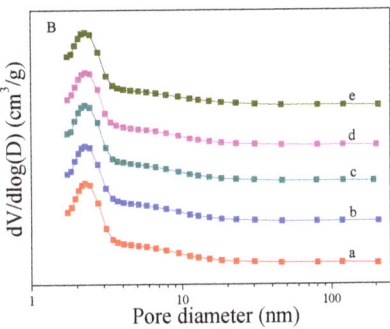

Figure 1. (**A**) N$_2$ adsorption–desorption isotherms and (**B**) pore size distributions of the calcined PtSn/DMSN catalysts with different ratios of Pt and Sn: (a) Pt/DMSN, (b) Pt$_1$Sn$_1$/DMSN, (c) Pt$_1$Sn$_2$/DMSN, (d) Pt$_1$Sn$_3$/DMSN, (e) Pt$_1$Sn$_4$/DMSN.

According to the adsorption–desorption isotherms, the textural properties of these samples were calculated by specific algorithms, and the results are shown in Table 1. The Pt/DMSN sample showed a surface area and total pore volume of 1091.8 m^2/g and 0.79 cm^3/g, respectively. The high surface area could be attributed to the fine features in dendritic spheres in the form of wrinkles or lamellar structures [45], and the pore volume was due to the plentiful pore construction in the DMSN support. With the increasing of Sn loadings, the BET surface area of the catalysts decreased from 1091.8 m^2/g on the Pt/DMSN catalyst to 1005.5 m^2/g on the Pt$_1$Sn$_4$/DMSN catalyst. The decrease in surface area with the increasing of Sn loadings indicated that the metals entered into the mesopores, which hindered the nitrogen molecule diffusion into the pore, leading to a decrease in the BET surface area. It also was demonstrated by the decrease in total pore volume and mesoporous volume, which decreased from 0.79 cm^3/g to 0.71 cm^3/g and 0.55 cm^3/g to 0.49 cm^3/g, respectively. The gradual reduction in the surface area, total pore volume and mesoporous volume from the Pt/DMSN catalyst to the Pt$_1$Sn$_4$/DMSN catalyst may have been due to the increasing of the total loading amounts of the metals. Moreover, the pore sizes of these catalysts were nearly unchanged, which implies the formation of nanoparticles in the mesopores.

Table 1. Textural properties of the calcined PtSn/DMSN catalysts with different ratios of Pt/Sn.

Samples	S_{BET} [a] (m^2·g^{-1})	V_t [b] (cm^3·g^{-1})	V_{mes} [c] (cm^3·g^{-1})	d_{BJH} [d] (nm)
Pt/DMSN	1091.8	0.79	0.55	3.2
Pt$_1$Sn$_1$/DMSN	1061.9	0.75	0.53	3.1
Pt$_1$Sn$_2$/DMSN	1042.1	0.74	0.52	3.2
Pt$_1$Sn$_3$/DMSN	1018.6	0.72	0.51	3.1
Pt$_1$Sn$_4$/DMSN	1005.5	0.71	0.49	3.2

[a] Calculated by the BET method. [b] The total pore volume was obtained at a relative pressure of 0.98. [c] The mesoporous volume was calculated using the BJH method. [d] Mesopore diameter was calculated using the BJH method.

To investigate the influence of the ratios of Pt and Sn on the crystalline structure of Pt and Sn, XRD characterization was carried out; the results are shown in Figure 2. It can be seen that the catalysts with different ratios of Pt and Sn showed similar diffraction peaks in the range of 10° to 90°, and the intensity of these peaks was considerable. The broad diffraction peak at around 23.5° was a typical reflection of amorphous silica. The weak peaks at 2θ of 39.8°, 46.4°, 67.7° and 81.4° were assigned to the (111), (200), (220) and (311) reflections of cubic Pt, respectively [46]. The similar intensity of these peaks indicated the similar crystallinity of Pt, which indicated that the ratio of Pt and Sn (the amounts of Sn) had a negligible influence on the size of Pt. Moreover, there were no apparent diffraction peaks attributed to SnO_x in the XRD patterns, indicating high dispersion of SnO_x or the formation of amorphous SnO_x phases.

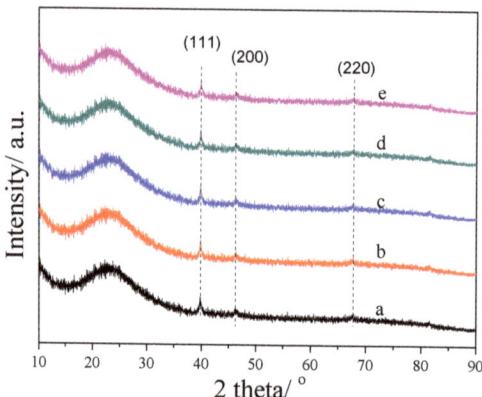

Figure 2. XRD patterns of calcined PtSn/DMSN catalysts with different ratios of Pt and Sn: (a) Pt/DMSN, (b) Pt_1Sn_1/DMSN, (c) Pt_1Sn_2/DMSN, (d) Pt_1Sn_3/DMSN, (e) Pt_1Sn_4/DMSN.

Raman spectra of fresh catalysts with different Pt and Sn ratios are presented in Figure 3. It can be seen that the spectra of PtSn/DMSN samples showed typical vibrations of SiO_2 in the range of 100–1200 cm^{-1} [47]. For the Pt/DMSN catalyst, there was no diffraction peaks attributed to Pt species, indicating the Pt species existing in the form of the metallic state. For the catalysts with high ratios of Pt and Sn (Pt/Sn < 1/4), the Raman spectra for the PtSn/DMSN catalysts showed no obvious changes, which may have been caused by the low metal loading amounts. However, for Pt_1Sn_4/DMSN, a slight redshift of about 12 cm^{-1} was noticed for the band at 980 cm^{-1}, which suggests the incorporation of Sn in the framework of silica. This redshift could be due to an appreciable decrease in the bond strength and/or bond angle caused by the formation of oxygen–metal–oxygen (O–Sn–O) bridges in the silica framework [48]. Li et al. [49] reported that the band at about 960 cm^{-1} was associated with Si–O–Si linkages next to M–O–Si bonds and identified contributions attributed to transition-metal ions bonded to the framework. It can be seen from the Raman spectra that Sn incorporated into the framework of silica resulted in a strong interaction between Sn and the support. The strong interaction between Sn and the support can favor the interaction between Pt and the support so as to increase the stability of the PtSn/DMSN catalyst during the PDH reaction.

The chemical state of Pt and Sn species on the catalysts were examined by XPS. The XPS spectra of Pt4f and Sn3$d_{5/2}$ are presented in Figure 4. As shown in Figure 4A, the Pt4f spectra of all the PtSn/DMSN catalysts could be deconvoluted into two peaks that could be assigned to metallic Pt (Pt°) [50], which indicated that the Pt species existed in a metallic state on the PtSn/DMSN catalysts. This was also in agreement with the XRD results. Moreover, the binding energy for Pt/DMSN was 71.0 eV, while the binding energy for PtSn/DMSN was 71.2 eV. The slight shift to high binding energy in PtSn/DMSN catalysts indicated the electronic effects between Pt and Sn [51]. Figure 4B shows the Sn3$d_{5/2}$ XPS spectra of calcined samples. There were three peaks at 485.8, 486.6 and 487.3 eV after deconvolution, corresponding to the Sn°, Sn^{2+} and Sn^{4+} species, respectively [52]. This indicated that the Sn species on the catalysts existed both in a metallic state and in the form of SnO_x. Figure 4C shows the Sn3$d_{5/2}$ XPS spectra of reduced samples. It can be seen that the Sn species still existed both in a metallic state and in the form of SnO_x after reduction treatment. Furthermore, the fraction of metallic Sn species increased compared with the calcined samples. According to the XPS results, it can be seen that the state of Pt species on PtSn/DMSN catalysts with different ratios of Pt and Sn was a metallic state, while the state of Sn species was both an oxidation state and a metallic state.

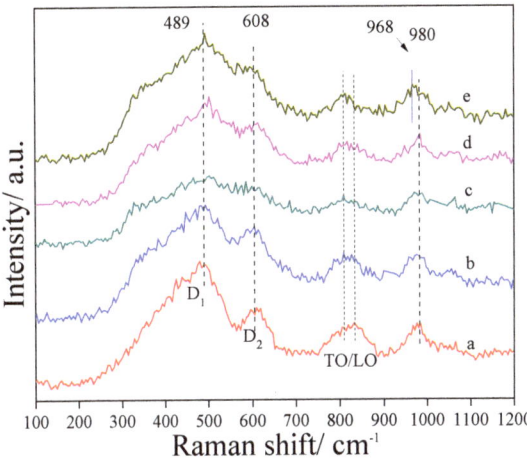

Figure 3. Raman spectra of calcined PtSn/DMSN catalysts with different ratios of Pt and Sn: (a) Pt/DMSN, (b) Pt_1Sn_1/DMSN, (c) Pt_1Sn_2/DMSN, (d) Pt_1Sn_3/DMSN, (e) Pt_1Sn_4/DMSN.

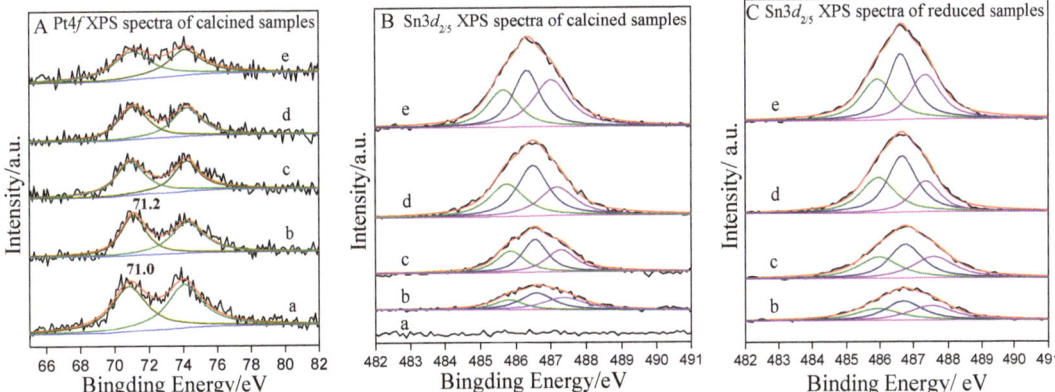

Figure 4. Pt4f (**A**) and Sn3$d_{5/2}$ (**B,C**) XPS spectra of calcined and reduced PtSn/DMSN catalysts with different ratios of Pt and Sn: (a) Pt/DMSN, (b) Pt_1Sn_1/DMSN, (c) Pt_1Sn_2/DMSN, (d) Pt_1Sn_3/DMSN, (e) Pt_1Sn_4/DMSN.

To investigate the morphology of PtSn/DMSN and to provide evidence for the distribution of Pt and Sn, TEM analyses were carried out. Figure 5A1 shows the TEM images of the Pt_1Sn_2/DMSN catalyst. It can be seen that the morphology of the DMSN was porous spheres with a dendritic pore structure arranged in three dimensions to form spheres. The unique hierarchical pore structures gave such DMSN material potential as a catalyst support. There were some nanoparticles that appeared in the DMSN support, which may be assigned to PtSn nanoparticles. The EDS elemental mapping in Figure 5A2 shows that Si, O and Sn species were uniformly dispersed on the DMSN support. In order to investigate the interaction between metal and the support, the high loading PtSn/DMSN catalyst was prepared and characterized by TEM. Figure 5B1 shows the TEM images of Pt_3Sn_9/DMSN with the loading amounts of 3 wt% of Pt and 9 wt% of Sn. It can be seen that the sizes of the nanoparticles were larger than those of Figure 5A1 because of the increasing of Pt and Sn loading amounts. In order to investigate the dispersion of Pt and Sn, EDS elemental mapping was carried out. The EDS elemental mapping in Figure 5B3 of Pt apparently showed a hot spot, indicating the aggregation of Pt. It can be seen from

Figure 5B3 that the Pt elemental mapping showed corresponding dispersion with the STEM image, while Sn elemental mapping showed more uniform dispersion with no apparent hot spot shown corresponding to the nanoparticles in the STEM images. It was indicated that the dispersions of Pt and Sn were different. Sn species possessed strong interactions with the support so as to show highly dispersion, which was demonstrated by the Raman results that some of Sn species were incorporated into the framework of the silica. Pt species, by contrast, possessed weak interactions with the support so as to show lower dispersion than Sn species. Furthermore, the strong interaction of Sn and the support may have been favorable to the interaction of Pt and the support so as to increase the dispersion of Pt. Figure 5B2 is Figure 5B1 under several minutes of high voltage exposure. It can be seen that a large amount of small nanoparticles are shown (the red circle). This indicates that there were both highly dispersed small nanoparticles and nanoparticles (yellow circle) in the Pt_3Sn_9/DMSN sample. Therefore, there were both Pt and Sn element dispersions on the DMSN support except the hot spot. Furthermore, the uniform dispersal of Pt and Sn in Figure 5A2 indicates the high dispersion of both Pt and Sn, which may have been caused by the strong interactions of Sn and the support.

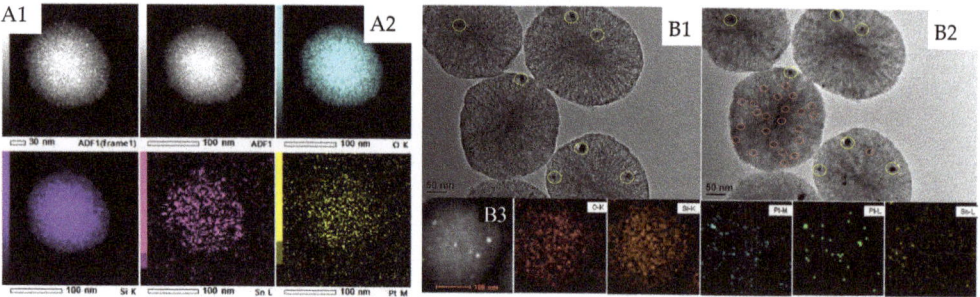

Figure 5. TEM images and EDS elemental mapping of calcined (**A1,A2**) Pt_1Sn_2/DMSN and (**B1–B3**) Pt_3Sn_9/DMSN catalyst.

2.2. Catalytic Performance of Propane Dehydrogenation

The PtSn/DMSN catalysts were subjected to the propane dehydrogenation reaction. Figure 6 shows the catalytic performances of PtSn/DMSN catalysts with different ratios of Pt and Sn. As shown in Figure 6A, the initial conversion of propane over the Pt/DMSN catalyst was 15.6%, indicating that Pt works to activate propane. Compared with Pt/DMSN, the catalysts loading both Pt and Sn showed higher initial propane conversion. The initial propane conversion was 27.3% for the Pt_1Sn_1/DMSN catalyst, and it increased to 34.9% for the Pt_1Sn_2/DMSN catalyst. It decreased to 32.2% for the Pt_1Sn_3/DMSN catalyst. With a further increase in the amount of Sn, the initial propane conversion decreased to 27.6% for the Pt_1Sn_4/DMSN catalyst. The lowest initial propane conversion and fast deactivation rate (k_d = 0.21) were obtained for the Pt_1Sn_1/DMSN catalyst. As shown in Figure 6A, the Pt_1Sn_2/DMSN catalyst showed the highest propane conversion, even after 6 h of reaction. Compared with high initial propane conversions, high stabilities are more favorable for non-oxidative dehydrogenation, and high selectivity is essential for high stability. Figure 6B shows the propene selectivity for the PtSn/DMSN catalysts. Of these PtSn/DMSN catalysts, the lowest propene selectivity was obtained for the Pt/DMSN catalyst. Compared with Pt/DMSN, all PtSn/DMSN catalysts showed high propene selectivity. This indicates that the Sn promoter could increase the propene selectivity. As shown in Figure 6B, the selectivity of propene was below 90% for the Pt_1Sn_1/DMSN catalyst, while the selectivity of propene was higher than 95% for the Pt_1Sn_2/DMSN, Pt_1Sn_3/DMSN and Pt_1Sn_4/DMSN catalysts. This indicates that the low Sn loading catalyst showed lower propene selectivity compared with the high Sn loading catalyst. Moreover, the selectivity decreased from the

Pt_1Sn_2/DMSN catalyst to the Pt_1Sn_4/DMSN catalyst. One reason for this tendency may be the excess Sn results in the side reaction.

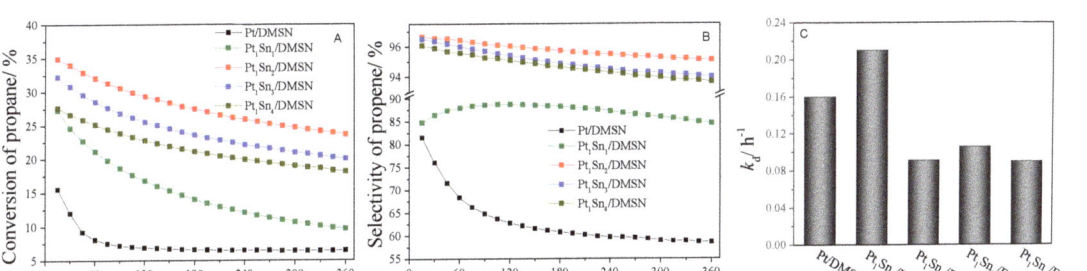

Figure 6. Propane conversion (**A**) and propene selectivity (**B**) during 6 h PDH over PtSn/DMSN catalysts; (**C**) the deactivation rate constant (k_d) of PtSn/DMSN catalysts.

In order to analyze the effect of the Pt/Sn ratio on the PDH catalytic performance, the relationship of Pt/Sn ratios over deactivation rate constant k_d was investigated and can be seen in Figure 6C. It can also be seen that k_d was 0.21 for the Pt_1Sn_1/DMSN catalyst, indicating fast deactivation. When the ratio of Sn to Pt increased to 2, k_d apparently decreased. The catalyst with a Pt/Sn ratio of 1/2 showed a k_d of 0.09, indicating high stability among these catalysts. Compared with the Pt_1Sn_1/DMSN catalyst, the catalysts with high loading amounts of Sn showed high stability.

Table 2 lists the catalytic data of $PtSn/SiO_2$ catalysts used in the PDH reaction. It can be seen that the reaction conditions were different over the $PtSn/SiO_2$ catalysts. Because the propane dehydrogenation reaction is an endothermic reversible reaction, the reaction temperature and feed composition resulted in different equilibrium conversions of propane. Table 2 lists the equilibrium conversion over the different conditions. It can be seen that the initial propane conversion was lower than the equilibrium conversion, and the initial propane conversion was about 85% of the equilibrium conversion in this work, indicating high activity of the Pt_1Sn_2/DMSN catalyst. On the other hand, the k_d of the Pt_1Sn_2/DMSN catalyst seemed high because of the feed composition of pure propane.

Table 2. Catalytic data of $PtSn/SiO_2$ catalysts used in the PDH reaction.

Catalyst	Reaction Temperature (°C)	WHSV (h^{-1})	Feed Composition	Equilibrium Conversion (%)	Initial Conversion (%)	Initial Selectivity (%)	k_d (h^{-1})	Ref
Pt/Sn-MFI	600	3.2	Pure C_3H_8	49.3	42	95	0.012	[17]
Pt_1Sn_1/SiO_2	580	4.7	C_3H_8:He = 4:21	67	66.5	~99	0.008	[21]
Pt-Sn/SBA-15	600	16.5	C_3H_8:H_2:N_2 = 14:14:72	60.8	~40	~92	~0.09	[23]
K-PtSn@MFI	650	29.5	C_3H_8:N_2 = 5:16	83.5	~55	~99	~0.006	[24]
4SnPt/SiO_2	500	17.7	C_3H_8:H_2:N_2 = 2:3:95	45.7	~11	99	0.004	[53]
Pt-Sn/xAlSBA-15	600	11.8	C_3H_8:N_2 = 3:10	52.7	25.5	~97	0.22	[54]
Pt_1Sn_2/DMSN	590	2.4	Pure C_3H_8	40.8	34.9	96.7	0.09	This work

2.3. Characterization Results of the Spent Catalysts

To provide evidence for the distribution of Pt and Sn species before and after the PDH reaction, HRTEM analyses were carried out on the spent Pt_1Sn_2/DMSN catalyst. Figure 7 shows the HRTEM images of the EDS elemental mapping of the spent Pt_1Sn_2/DMSN catalyst. Compared with Figure 5A1, the morphological structure of the spent Pt_1Sn_2/DMSN catalyst was similar to that of the calcined sample after a 6 h PDH reaction, indicating the high stability of the DMSN support. No significant aggregation occurred on the PtSn nanoparticles compared with the calcined sample.

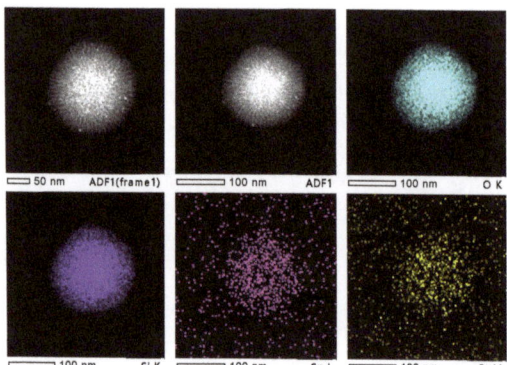

Figure 7. HRTEM images and EDS elemental mapping of spent Pt_1Sn_2/DMSN catalyst.

In order to investigate the influence of coke deposition on the PDH catalytic performance, the amount of coke on the spent catalysts was investigated by TG characterization. As shown in Figure 8, the Pt_1Sn_1/DMSN catalyst had a mass loss of 4.7%, which was the highest among the PtSn/DMSN catalysts, indicating the largest amount of coke deposition. This is because of the occurrence of side reactions that demonstrated the low propene selectivity of this sample. The Pt_1Sn_2/DMSN catalyst showed higher mass loss than that of Pt_1Sn_3/DMSN and Pt_1Sn_4/DMSN catalysts. This indicates that both the dehydrogenation reaction and the coke deposition were enhanced compared to Pt_1Sn_3/DMSN and Pt_1Sn_4/DMSN catalysts. Furthermore, the Pt_1Sn_3/DMSN catalyst showed the lowest mass loss, indicating the lowest amount of coke deposition on the sample.

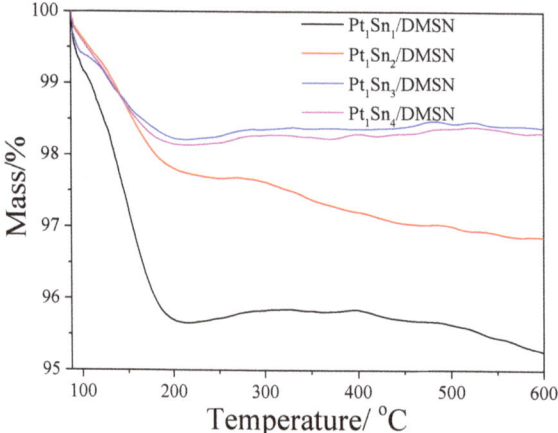

Figure 8. TG profiles of the spent PtSn/DMSN catalysts with different ratios of Pt and Sn.

Pt-based catalysts have been widely used in the PDH process. As the most effective promoter, Sn is usually combined with Pt in PDH. The promoting effect of Sn has been understood both geometrically and electronically. Therefore, the ratio of Pt/Sn will show some influence on the geometric and electronic effects, as confirmed by the DFT calculations [55]. The catalyst with a Pt/Sn ratio of 2 showed the best catalytic performance with propane initial conversion of 34.9%, and the deactivation rate constant on the high Sn loading PtSn/DMSN catalysts was lower. The coke analysis shown in Figure 8 indicates the inhibition by Sn of coke formation.

3. Materials and Methods

3.1. Catalyst Preparation

DMSN support was prepared by the emulsion method as described previously [30]. PtSn/DMSN catalysts were prepared by the incipient-wetness impregnation method. $SnCl_4 \cdot 5H_2O$ and $H_2PtCl_6 \cdot 6H_2O$ precursors were dissolved in deionized water to form a solution, and powder DMSN was impregnated in the solution. After that, the mixture was sonicated for 30 min and dried at 25 °C for 24 h. After being completely dried, the catalysts were calcined at 500 °C for 4 h and reduced with H_2 at 500 °C for 4 h. The composition of the catalyst with different ratios of Pt/Sn were 1 wt% Pt and 0, 1, 2, 3 and 4 wt% Sn. For the sake of brevity, these catalysts were named as Pt/DMSN, Pt_1Sn_1/DMSN, Pt_1Sn_2/DMSN, Pt_1Sn_3/DMSN and Pt_1Sn_4/DMSN.

3.2. Characterization

Nitrogen adsorption/desorption isotherms at -196 °C were recorded using a Micromeritics TriStar II 3020 porosimetry analyzer. The samples were degassed at 300 °C for 8 h prior to the measurements. Wide-angle XRD patterns were obtained by a powder X-ray diffractometer (Shimadzu XRD 6000, Kyoto, Japan) using Cu Kα ($\lambda = 0.15406$ nm) radiation with a nickel filter operating at 40 kV and 40 mA in the 2θ range of 10–90 ° at a scanning rate of 4°/min. Transmission electron microscopy (TEM) images were taken on a JEOL JEM 2100 electron microscope equipped with a field emission source at an acceleration voltage of 200 kV. The TEM samples were sonicated and well suspended in ethanol. Drops of the suspension were applied, and after drying the fine particles were well dispersed on a copper grid coated with carbon. X-ray photoelectron spectra (XPS) were recorded on a Perkin-Elmer PHI-1600 ESCA spectrometer using an Mg Kα (hv = 1253.6 eV, 1 eV = 1.603×10^{-19} J) X-ray source. The binding energies were calibrated using the C1s peak of contaminant carbon (BE = 284.6 eV) as an internal standard. Raman spectra were performed on a Renishaw inVia Reflex Raman spectrometer with a 325 nm laser at room temperature under ambient conditions. The amount of coke deposited was determined by a NETZSCH STA 449 F5 thermogravimetric (TG) analyzer. The samples were exposed to 20% O_2/N_2 flowing and oxidized from ambient temperature to 600 °C at a rate of 10 °C min^{-1}.

3.3. Catalytic Activity Test

The PDH reaction was carried out in a conventional quartz tubular micro-reactor. Then, 0.2 g of the catalyst was placed in the center of the reactor and reduced under 10% H_2/Ar at 500 °C for 4 h. Subsequently, the pure C_3H_8 with a weight hourly space velocity (WHSV) of 2.4 h^{-1} was fed to the reactor, and the reaction temperature was raised to and maintained at 590 °C. The reaction products were analyzed by using an online Agilent-7890B gas chromatograph equipped with an HP-Al$_2$O$_3$ capillary column for the separation of CH_4, C_2H_4, C_2H_6, C_3H_6 and C_3H_8. The conversion of propane and the selectivity for propene were defined as follows:

$$X_{C_3H_8} = \frac{CH_4 + 2C_2H_4 + 2C_2H_6 + 3C_3H_6}{CH_4 + 2C_2H_4 + 2C_2H_6 + 3C_3H_6 + 3C_3H_8} \times 100\%$$

$$S_{C_3H_6} = \frac{3C_3H_6}{CH_4 + 2C_2H_4 + 2C_2H_6 + 3C_3H_6} \times 100\%$$

The deactivation rate constant k_d was calculated as follows:

$$k_d = \frac{\ln\left[\frac{1-X_{final}}{X_{final}}\right] - \ln\left[\frac{1-X_{initial}}{X_{initial}}\right]}{t}$$

where $X_{initial}$ and X_{final} represent the propane conversion at the initial and final stages of an experiment, respectively, and t (h) represents the duration of the PDH reaction.

4. Conclusions

PtSn/DMSN catalysts were synthesized by using DMSN support. The influence of the Pt/Sn ratios was investigated. The addition of Sn could enhance the catalytic performance of PDH, and the Pt/Sn ratios affected coke deposition. The addition of Sn modified the dispersion and electronic properties of Pt. The Pt_1Sn_2/DMSN catalyst with a Pt/Sn ratio of 1:2 showed the highest catalytic performance with initial propane conversion of 34.9%. The good catalytic performance of the Pt_1Sn_2/DMSN catalyst was ascribed to the small nanoparticle size of PtSn and the favorable chemical state and dispersion degree of Pt and Sn species.

Author Contributions: Conceptualization, X.F. and Z.Z.; methodology, N.Z. and Y.S.; software, N.Z. and J.S.; validation, J.S.; formal analysis, N.Z. and X.F.; investigation, Y.S.; data curation, X.F. and N.Z.; writing—original draft preparation, X.F.; writing—review and editing, X.F.; visualization, L.K., X.X. and Z.X.; supervision, Z.Z.; project administration, Z.Z.; funding acquisition, X.F. and Z.Z. All authors have read and agreed to the published version of the manuscript.

Funding: This research was funded by the National Natural Science Foundation of China, grant numbers 22172101, 91845201 and 92145301, and the Scientific Research Project of Education Office of Liaoning Province (LQN202009).

Institutional Review Board Statement: Not applicable.

Informed Consent Statement: Not applicable.

Data Availability Statement: Data are available upon reasonable request from the corresponding authors.

Acknowledgments: We acknowledge the support of the University level innovation team of Shenyang Normal University and the Program for Excellent Talents in Shenyang Normal University.

Conflicts of Interest: The authors declare no conflict of interest.

References

1. Bai, P.; Ma, Z.; Li, T.; Tian, Y.; Zhang, Z.; Zhong, Z.; Xing, W.; Wu, P.; Liu, X.; Yan, Z. Relationship between surface chemistry and catalytic performance of mesoporous γ-Al_2O_3 supported VO_X catalyst in catalytic dehydrogenation of propane. *ACS Appl. Mater. Interfaces* **2016**, *8*, 25979–25990. [CrossRef] [PubMed]
2. Zhao, D.; Guo, K.; Han, S.; Doronkin, D.E.; Lund, H.; Li, J.; Grunwaldt, J.-D.; Zhao, Z.; Xu, C.; Jiang, G.; et al. Controlling reaction-induced loss of active sites in ZnO_x/silicalite-1 for durable nonoxidative propane dehydrogenation. *ACS Catal.* **2022**, *12*, 4608–4617. [CrossRef]
3. Frank, B.; Cotter, T.P.; Schuster, M.E.; Schlögl, R.; Trunschke, A. Carbon dynamics on the molybdenum carbide surface during catalytic propane dehydrogenation. *Chem. Eur. J.* **2013**, *19*, 16938–16945. [CrossRef] [PubMed]
4. Zhang, T.; Guo, X.; Song, C.; Liu, Y.; Zhao, Z. Fabrication of isolated VO_x sites on alumina for highly active and stable non-oxidative dehydrogenation. *J. Phys. Chem. C* **2021**, *125*, 19229–19237. [CrossRef]
5. Bian, Z.; Dewangan, N.; Wang, Z.; Pati, S.; Xi, S.; Borgna, A.; Kus, H.; Kawi, S. Mesoporous-silica-stabilized cobalt(II) oxide nanoclusters for propane dehydrogenation. *ACS Appl. Nano Mater.* **2021**, *4*, 1112–1125. [CrossRef]
6. Searles, K.; Siddiqi, G.; Safonova, O.V.; Copéret, C. Silica-supported isolated gallium sites as highly active, selective and stable propane dehydrogenation catalysts. *Chem. Sci.* **2017**, *8*, 2661–2666. [CrossRef] [PubMed]
7. Castro-Fernández, P.; Mance, D.; Liu, C.; Abdala, P.M.; Willinger, E.; Rossinelli, A.A.; Serykh, A.I.; Pidko, E.A.; Copéret, C.; Fedorov, A.; et al. Bulk and surface transformations of Ga_2O_3 nanoparticle catalysts for propane dehydrogenation induced by a H_2 treatment. *J. Catal.* **2022**, *408*, 155–164. [CrossRef]
8. Yun, J.H.; Lobo, R.F. Catalytic dehydrogenation of propane over iron-silicate zeolites. *J. Catal.* **2014**, *312*, 263–270. [CrossRef]
9. Sun, Y.; Wu, Y.; Shan, H.; Wang, G.; Li, C. Studies on the promoting effect of sulfate species in catalytic dehydrogenation of propane over Fe_2O_3/Al_2O_3 catalysts. *Catal. Sci. Technol.* **2015**, *5*, 1290–1298. [CrossRef]
10. Schäferhans, J.; Gómez-Quero, S.; Andreeva, D.V.; Rothenberg, G. Novel and effective copper–aluminum propane dehydrogenation catalysts. *Chem. Eur. J.* **2011**, *17*, 12254–12256. [CrossRef]
11. Zhang, B.; Song, M.; Liu, H.; Li, G.; Liu, S.; Wang, L.; Zhang, X.; Liu, G. Role of Ni species in ZnO supported on Silicalite-1 for efficient propane dehydrogenation. *Chin. J. Chem. Eng.* **2022**, *43*, 240–247. [CrossRef]
12. Sharma, L.; Baltrus, J.; Rangarajan, S.; Baltrusaitis, J. Elucidating the underlying surface chemistry of Sn/Al_2O_3 catalysts during the propane dehydrogenation in the presence of H_2S co-feed. *Appl. Surf. Sci.* **2022**, *573*, 151205. [CrossRef]
13. Wang, G.; Zhang, H.; Wang, H.; Zhu, Q.; Li, C.; Shan, H. The role of metallic Sn species in catalytic dehydrogenation of propane: Active component rather than only promoter. *J. Catal.* **2016**, *344*, 606–608. [CrossRef]

14. Zhao, S.; Xu, B.; Yu, L.; Fan, Y. Honeycomb-shaped PtSnNa/γ-Al$_2$O$_3$/cordierite monolithic catalyst with improved stability and selectivity for propane dehydrogenation. *Chin. Chem. Lett.* **2018**, *29*, 884–886. [CrossRef]
15. Zhao, S.; Xu, B.; Yu, L.; Fan, Y. Catalytic dehydrogenation of propane to propylene over highly active PtSnNa/γ-Al$_2$O$_3$ catalyst. *Chin. Chem. Lett.* **2018**, *29*, 475–478. [CrossRef]
16. Xie, J.; Jiang, H.; Qian, Y.; Wang, H.; An, N.; Chen, S.; Dai, Y.; Guo, S. Fine tuning the morphology of spinel as ultra-stable catalyst support in propane dehydrogenation. *Adv. Mater. Interfaces* **2021**, *8*, 2101325. [CrossRef]
17. Lezcano-González, I.; Cong, P.; Campbell, E.; Panchal, M.; Agote-Arán, M.; Celorrio, V.; He, Q.; Oord, R.; Weckhuysen, B.M.; Beale, A.M. Structure-activity relationships in highly active platinum-tin MFI-type zeolite catalysts for propane dehydrogenation. *ChemCatChem* **2022**, *14*, e202101828. [CrossRef]
18. Wang, X.; Hu, H.; Zhang, N.; Song, J.; Fan, X.; Zhao, Z.; Kong, L.; Xiao, X.; Xie, Z. One-Pot synthesis of MgAlO support for PtSn catalysts over propane dehydrogenation. *ChemistrySelect* **2022**, *7*, e202104367. [CrossRef]
19. Llorca, J.; Homs, N.; León, J.; Sales, J.; Fierro, J.L.G.; Ramirez de la Piscina, P. Supported Pt-Sn catalysts highly selective for isobutane dehydrogenation: Preparation, characterization and catalytic behavior. *Appl. Catal. A Gen.* **1999**, *189*, 77–86. [CrossRef]
20. Sexton, B.A.; Hughes, A.E.; Foger, K. An X-ray photoelectron spectroscopy and reaction study of Pt Sn catalysts. *J. Catal.* **1984**, *88*, 466–477. [CrossRef]
21. Motagamwala, A.H.; Almallahi, R.; Wortman, J.; Igenegbai, V.O.; Linic, S. Stable and selective catalysts for propane dehydrogenation operating at thermodynamic limit. *Science* **2021**, *373*, 217–222. [CrossRef] [PubMed]
22. Xiong, H.; Lin, S.; Goetze, J.; Pletcher, P.; Guo, H.; Kovarik, L.; Artyushkova, K.; Weckhuysen, B.M.; Datye, A.K. Thermally stable and regenerable platinum–tin clusters for propane dehydrogenation prepared by atom trapping on ceria. *Angew. Chem. Int. Ed.* **2017**, *56*, 8986–8991. [CrossRef] [PubMed]
23. Wang, J.; Chang, X.; Chen, S.; Sun, G.; Zhou, X.; Vovk, E.; Yang, Y.; Deng, W.; Zhao, Z.-J.; Mu, R.; et al. On the role of Sn segregation of Pt-Sn catalysts for propane dehydrogenation. *ACS Catal.* **2021**, *11*, 4401–4410. [CrossRef]
24. Liu, L.; Lopez-Haro, M.; Lopes, C.W.; Meira, D.M.; Concepcion, P.; Calvino, J.J.; Corma, A. Atomic-level understanding on the evolution behavior of subnanometric Pt and Sn species during high-temperature treatments for generation of dense PtSn clusters in zeolites. *J. Catal.* **2020**, *391*, 11–24. [CrossRef]
25. Gao, X.; Xu, W.; Li, X.; Cen, J.; Xu, Y.; Lin, L.; Yao, S. Non-oxidative dehydrogenation of propane to propene over Pt-Sn/Al$_2$O$_3$ catalysts: Identification of the nature of active site. *Chem. Eng. J.* **2022**, *443*, 136393. [CrossRef]
26. Sun, Q.; Wang, N.; Fan, Q.; Zeng, L.; Mayoral, A.; Miao, S.; Yang, R.; Jiang, Z.; Zhou, W.; Zhang, J.; et al. Subnanometer bimetallic platinum–zinc clusters in zeolites for propane dehydrogenation. *Angew. Chem. Int. Ed.* **2020**, *59*, 2–11. [CrossRef]
27. Wang, Y.; Hu, Z.P.; Lv, X.; Chen, L.; Yuan, Z.Y. Ultrasmall PtZn bimetallic nanoclusters encapsulated in silicalite-1 zeolite with superior performance for propane dehydrogenation. *J. Catal.* **2020**, *385*, 61–69. [CrossRef]
28. Sun, G.; Zhao, Z.J.; Mu, R.; Zha, S.; Li, L.; Chen, S.; Zang, K.; Luo, J.; Li, Z.; Purdy, S.C.; et al. Breaking the scaling relationship via thermally stable Pt/Cu single atom alloys for catalytic dehydrogenation. *Nat. Commun.* **2018**, *9*, 4454. [CrossRef]
29. Cai, W.; Mu, R.; Zha, S.; Sun, G.; Chen, S.; Zhao, Z.J.; Li, H.; Tian, H.; Tang, Y.; Tao, F.; et al. Subsurface catalysis-mediated selectivity of dehydrogenation reaction. *Sci. Adv.* **2018**, *4*, 5418–5425. [CrossRef]
30. Liu, D.; Hu, H.; Yang, Y.; Cui, J.; Fan, X.; Zhao, Z.; Kong, L.; Xiao, X.; Xie, Z. Restructuring effects of Pt and Fe in Pt/Fe-DMSN catalysts and their enhancement of propane dehydrogenation. *Catal. Today* **2022**, *402*, 161–171. [CrossRef]
31. Cesar, L.G.; Yang, C.; Lu, Z.; Ren, Y.; Zhang, G.; Miller, J.T. Identification of a Pt$_3$Co surface intermetallic alloy in Pt–Co propane dehydrogenation catalysts. *ACS. Catal.* **2019**, *9*, 5231–5244. [CrossRef]
32. Searles, K.; Chan, K.W.; Mendes Burak, J.A.; Zemlyanov, D.; Safonova, O.; Copéret, C. Highly productive propane dehydrogenation catalyst using silica-supported Ga–Pt nanoparticles generated from single-sites. *J. Am. Chem. Soc.* **2018**, *140*, 11674–11679. [CrossRef] [PubMed]
33. Sun, P.; Siddiqi, G.; Vining, W.C.; Chi, M.; Bell, A.T. Novel Pt/Mg (In)(Al) O catalysts for ethane and propane dehydrogenation. *J. Catal.* **2011**, *282*, 165–174. [CrossRef]
34. Srisakwattana, T.; Watmanee, S.; Wannakao, S.; Saiyasombat, C.; Praserthdam, P.; Panpranot, J. Comparative incorporation of Sn and In in Mg(Al)O for the enhanced stability of Pt/MgAl(X)O catalysts in propane dehydrogenation. *Appl. Catal. A Gen.* **2021**, *615*, 118053. [CrossRef]
35. Rochlitz, L.; Pessemesse, Q.; Fischer, J.W.A.; Klose, D.; Clark, A.H.; Plodinec, M.; Jeschke, G.; Payard, P.-A.; Copéret, C. A robust and efficient propane dehydrogenation catalyst from unexpectedly segregated Pt$_2$Mn nanoparticles. *J. Am. Chem. Soc.* **2022**, *144*, 13384–13393. [CrossRef] [PubMed]
36. Fan, X.; Liu, D.; Sun, X.; Yu, X.; Li, D.; Yang, Y.; Liu, H.; Diao, J.; Xie, Z.; Kong, L.; et al. Mn-doping induced changes in Pt dispersion and Pt$_x$Mn$_y$ alloying extent on Pt/Mn-DMSN catalyst with enhanced propane dehydrogenation stability. *J. Catal.* **2020**, *389*, 450–460. [CrossRef]
37. Tong, Q.; Zhao, S.; Liu, Y.; Xu, B.; Yu, L.; Fan, Y. Design and synthesis of the honeycomb PtSnNa/ZSM-5 monolithic catalyst for propane dehydrogenation. *Appl. Organomet. Chem.* **2019**, *34*, e5380. [CrossRef]
38. Deng, L.; Miura, H.; Shishido, T.; Hosokawa, S.; Teramura, K.; Tanaka, T. Strong metal-support interaction between Pt and SiO$_2$ following high-temperature reduction: A catalytic interface for propane dehydrogenation. *Chem. Commun.* **2017**, *53*, 6937–6940. [CrossRef]

39. Shi, L.; Deng, G.M.; Li, W.C.; Miao, S.; Wang, Q.N.; Zhang, W.P.; Lu, A.H. Al_2O_3 Nanosheets rich in pentacoordinate Al^{3+} ions stabilize Pt-Sn clusters for propane dehydrogenation. *Angew. Chem. Int. Ed.* **2015**, *54*, 13994–13998. [CrossRef]
40. Zhou, H.; Gong, J.; Xu, B.; Deng, S.; Ding, Y.; Yu, L.; Fan, Y. PtSnNa/SUZ-4: An efficient catalyst for propane dehydrogenation. *Chin. J. Catal.* **2017**, *38*, 529–536. [CrossRef]
41. Smoliło-Utrata, M.; Tarach, K.A.; Samson, K.; Gackowski, M.; Madej, E.; Korecki, J.; Mordarski, G.; Śliwa, M.; Jarczewski, S.; Podobiński, J.; et al. Modulation of ODH propane selectivity by zeolite support desilication: Vanadium species anchored to Al-rich shell as crucial active sites. *Int. J. Mol. Sci.* **2022**, *23*, 5584. [CrossRef]
42. Singh, B.; Polshettiwar, V. Design of CO_2 sorbents using functionalized fibrous nanosilica (KCC-1): Insights into the effect of the silica morphology (KCC-1 vs. MCM-41). *J. Mater. Chem. A* **2016**, *4*, 7005–7019. [CrossRef]
43. Fihri, A.; Bouhrara, M.; Patil, U.; Cha, D.; Saih, Y.; Polshettiwar, V. Fibrous nano-silica supported ruthenium (KCC-1/Ru): A sustainable catalyst for the hydrogenolysis of alkanes with good catalytic activity and lifetime. *ACS. Catal.* **2012**, *2*, 1425–1431. [CrossRef]
44. Fan, X.; Li, J.; Zhao, Z.; Wei, Y.; Liu, J.; Duan, A.; Jiang, G. Dehydrogenation of propane over PtSnAl/SBA-15 catalysts: Al addition effect and coke formation analysis. *Catal. Sci. Technol.* **2015**, *5*, 339–350. [CrossRef]
45. Febriyanti, E.; Suendo, V.; Mukti, R.R.; Prasetyo, A.; Arifin, A.F.; Akbar, M.A.; Triwahyono, S.; Marsih, I.N. Further insight into the definite morphology and formation mechanism of mesoporous silica KCC-1. *Langmuir* **2016**, *32*, 5802–5811. [CrossRef]
46. Deng, L.; Miura, H.; Shishido, T.; Hosokawa, S.; Teramura, K.; Tanaka, T. Dehydrogenation of propane over silica-supported platinum–tin catalysts prepared by direct reduction: Effects of tin/platinum ratio and reduction temperature. *ChemCatChem* **2014**, *6*, 2680–2691. [CrossRef]
47. Fan, X.; Li, J.; Zhao, Z.; Wei, Y.; Liu, J.; Duan, A.; Jiang, G. Dehydrogenation of propane over PtSn/SBA-15 catalysts: Effect of the amount of metal loading and state. *RSC. Adv.* **2015**, *5*, 28305–28315. [CrossRef]
48. Wachs, I.E. Raman and IR studies of surface metal oxide species on oxide supports: Supported metal oxide catalysts. *Catal. Today* **1996**, *27*, 437–455. [CrossRef]
49. Li, C.; Xiong, G.; Liu, J.; Ying, P.; Xin, Q.; Feng, Z. Identifying framework titanium in TS-1 zeolite by UV resonance Raman spectroscopy. *J. Phys. Chem.* **2001**, *105*, 2993–2997. [CrossRef]
50. Yu, X.; Li, J.; Wei, Y.; Zhao, Z.; Liu, J.; Jin, B.; Duan, A.; Jiang, G. Three-dimensionally ordered macroporous $Mn_xCe_{1-x}O_\delta$ and $Pt/Mn_{0.5}Ce_{0.5}O_\delta$ catalysts: Synthesis and catalytic performance for soot oxidation. *Ind. Eng. Chem. Res.* **2014**, *53*, 9653–9664. [CrossRef]
51. Gong, N.; Zhao, Z. Efficient supported Pt-Sn catalyst on carambola-like alumina for direct dehydrogenation of propane to propene. *Mol. Catal.* **2019**, *477*, 110543. [CrossRef]
52. Li, B.; Xu, Z.; Jing, F.; Luo, S.; Chu, W. Facile one-pot synthesized ordered mesoporous Mg-SBA-15 supported PtSn catalysts for propane dehydrogenation. *Appl. Catal. A Gen.* **2017**, *533*, 17–27. [CrossRef]
53. Ye, C.; Mao, P.; Wang, Y.; Zhang, N.; Wang, D.; Jiao, M.; Miller, J.T. Surface hexagonal Pt_1Sn_1 intermetallic on Pt nanoparticles for selective propane dehydrogenation. *ACS Appl. Mater. Interfaces* **2020**, *12*, 25903–25909. [CrossRef] [PubMed]
54. Vu, B.K.; Shin, E.W.; Ahn, I.Y.; Ha, J.M.; Suh, D.J.; Kim, W.I.; Koh, H.L.; Choi, Y.G.; Lee, S.B. The effect of tin–support interaction on catalytic stability over Pt–Sn/xAl–SBA-15 catalysts for propane dehydrogenation. *Catal. Lett.* **2012**, *142*, 838–844. [CrossRef]
55. Yang, M.L.; Zhu, Y.A.; Zhou, X.G.; Sui, Z.J.; Chen, D. First-principles calculations of propane dehydrogenation over PtSn catalysts. *ACS Catal.* **2012**, *2*, 1247–1258. [CrossRef]

Article

Solid Fe Resources Separated from Rolling Oil Sludge for CO Oxidation

Wei Gao [†], Sai Tang [†], Ting Wu, Jianhong Wu, Kai Cheng * and Minggui Xia *

Hubei Key Laboratory of Biomass Fibers and Eco-Dyeing & Finishing, School of Chemistry and Chemical Engineering, Wuhan Textile University, Wuhan 430200, China
* Correspondence: chengkai@wtu.edu.cn (K.C.); xiaminggui@wtu.edu.cn (M.X.)
† These authors contributed equally to this work.

Abstract: The efficient recycling of valuable resources from rolling oil sludge (ROS) to gain new uses remains a formidable challenge. In this study, we reported the recycling of solid Fe resources from ROS by a catalytic hydrogenation technique and its catalytic performance for CO oxidation. The solid Fe resources, after calcination in air (Fe_2O_3-H), exhibited comparable activity to those prepared by the calcinations of ferric nitrate (Fe_2O_3-C), suggesting that the solid resources have excellent recycling value when used as raw materials for CO oxidation catalyst preparation. Further studies to improve the catalytic performance by supporting the materials on high surface area 13X zeolite and by pretreating the materials with CO atmosphere, showed that the CO pretreatment greatly improved the CO oxidation activity and the best activity was achieved on the 20 wt.%Fe_2O_3-H/13X sample with complete CO conversion at 250 °C. CO pretreatment could produce more oxygen vacancies, facilitating O_2 activation, and thus accelerate the CO oxidation reaction rate. The excellent reducibility and sufficient O_2 adsorption amount were also favorable for its performance. The recycling of solid Fe resources from ROS is quite promising for CO oxidation applications.

Keywords: ROS; solid Fe resources; recycling; CO oxidation

1. Introduction

Rolling oil is extensively used for refrigeration and lubrication during the cool rolling process of stainless steel, but it will deteriorate after long-term use and transform into rolling oil sludge (ROS) by mixing with iron powder. ROS is one type of refractory hazardous waste [1–3]. With the rapid development of the steel industry, the enormous usage of rolling oil produces a tremendous increase of ROS, causing serious threats to both the ecological environment and human health [3–5]. Conventional treatment technologies to dispose of ROS include incineration, distillation, brick or briquette making, solvent extraction, etc. [6–8], but encounter more problems in practical use, such as high operation cost, complex process, serious waste pollution, insufficient use of calorific value, etc. [8,9]. Hydrofining technology is one of the most common technologies for the recycling of spent lubrication oil [10–12]. It possesses many advantages, such as a large treatment capacity, high recovery rate, good product quality, etc., despite its huge initial input costs. For ROS treatment, the distillation process is generally employed to cut components prior to the hydrofining process, and solid wastes are not recycled effectively [13–15]. Moreover, the solid wastes are also hard to separate from ROS even after the hydrofining treatment due to the addition of solid catalysts, which could mix with the solid wastes and be hardly separated. The Fe element does not have full electron fill in d-shell with a body-centered cubic lattice structure, thereby exhibiting a certain hydrogenation capability [16,17]. Hence, the development of a simple technology that can effectively recycle the high-value solid resources from ROS and thereby find the potential application of these solid resources is of great importance.

Catalytic CO oxidation is a well-known reaction in heterogeneous catalysis. It can not only act as a prototypical reaction to understand the reaction mechanism but also receive wide applications in practice, such as CO abatement in closed systems (e.g., submarines, aircraft, spacecraft, etc.), automotive exhaust catalytic purification, gas purification in gas masks, removal of CO impurity in feed gas, etc. [18–20]. It has been reported that the pretreatment of catalysts with the reactant is of great importance and can significantly affect the catalytic performance for CO oxidation. For example, the pretreatment of Pd-supported catalysts with H_2 or CO showed excellent catalytic performance due to the formation of new active sites [21] and the improved ability for O_2 activation [22]. Fe-based catalysts are extensively studied as promising alternatives to the commercial noble metal-based three-way catalysts, such as Fe-Ce [23], Fe-Co [24], Fe-Ni [25], etc. Tian et al. [26] synthesized Cu-Fe-Co ternary oxide thin film supported on copper grid mesh for CO oxidation, showing excellent catalytic activity, which was owed to the synergistic effects of chemisorbed oxygen species, electrical resistivity, easy mass transferability, etc. Moreover, 13X zeolite is usually employed as catalyst support due to its high specific surface area, easy availability, and low cost. It has been widely applied in various reactions, including catalytic removal of ethylbenzene, cyclohexane, and hexadecane [27,28]. Therefore, it is quite interesting to develop Fe-supported 13X zeolite catalysts for CO oxidation.

In this work, ROS collected from a food-grade stainless steel facility in Wuhan Iron and Steel Co. (WISCO) was disposed of by the hydrogenation technique. The solid Fe resources can act as catalysts for the hydrogenation of organics in the ROS and be automatically separated from the oil without adding extra hydrogenation catalysts. To explore the application of the recycled solid Fe resources from the ROS, they were utilized as raw materials to prepare catalysts by supporting them on 13X zeolite. The catalytic performance of such prepared catalysts was evaluated for CO oxidation reaction at a stoichiometric feed and the effect of CO pretreatment on the catalytic activity was investigated.

2. Results and Discussion

2.1. Physical Properties and Morphology of Fe Based Catalysts

Figure 1 shows the images of the ROS and the separated oil and solid phase after the catalytic hydrogenation process. The ROS was seriously emulsified with high viscosity and was difficult to dry naturally, owing to the poor volatile ability of the oily sludge components. Meanwhile, it emitted funky odors. After catalytic hydrogenation, the product was readily filtered to achieve complete oil/solid separation. The oil color changed from black to brown, and the viscosity became quite low with the capability of free flow. The solid was quite loose and easily crushed into powder. It was then used to prepare Fe-based catalysts for CO oxidation.

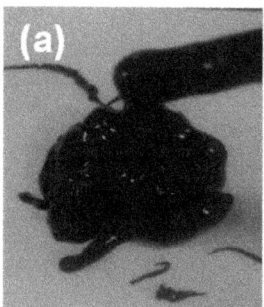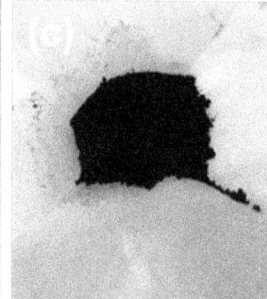

Figure 1. Images of the (**a**) rolling oily sludge, (**b**) separate oil phase and (**c**) solid phase after the catalytic hydrogenation process.

HRTEM/TEM images in Figure 2 show that the Fe_2O_3-H sample (the solid Fe resource after air calcination) was made of nanoparticles in the size range of 30~40 nm. The interplanar distances were measured to be 0.27 and 0.25 nm, which were assigned to the (104) and (111) planes of Fe_2O_3, respectively. To detect the crystalline phases of the Fe-based catalysts, XRD measurements were performed and the results are presented in Figure 3a. By screening the XRD database, the peaks located at 24.1, 33.2, 35.6, 40.8, and 49.6° are assigned to those of the α-Fe_2O_3 phase (PDF #73-2234), [29,30], which is consistent with the TEM/HRTEM results. For the 13X zeolite-supported catalysts, the catalysts displayed the characteristic diffraction peaks ascribed to 13X zeolite besides α-Fe_2O_3 phase. Meanwhile, the intensity of diffraction peaks of 13X zeolite weakened in accordance with that reported elsewhere [27]. Figure 3b depicts the XRD patterns of samples pretreated with CO. For the 13X zeolite-supported samples (20% Fe_2O_3-C/13X@CO and 20% Fe_2O_3-H/13X@CO), the new diffraction peaks at 30.3, 37.4, 43.4, 53.9, and 57.5° were attributed to Fe_3O_4 (PDF#88-0315) indicated that the Fe_2O_3 in these two samples was reduced to Fe_3O_4 by CO. While for the unsupported samples (Fe_2O_3-C@CO and Fe_2O_3-H@CO), the peaks ascribed to Fe_2O_3 co-existed with those ascribed to Fe_3O_4. These results showed that a portion of Fe^{3+} in the fresh samples was converted into Fe^{2+} after exposure to 1% CO/N_2 at 450 °C.

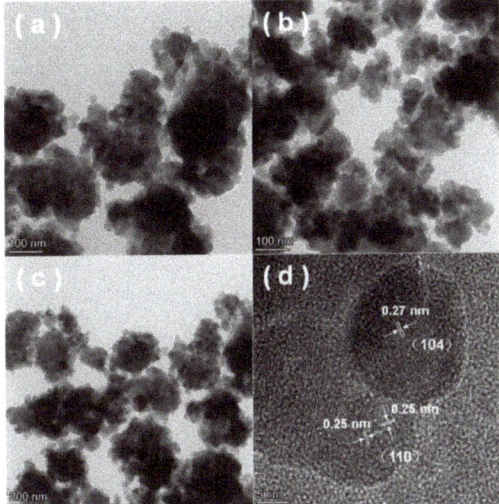

Figure 2. (**a**–**c**) TEM and (**d**) HRTEM images of the Fe_2O_3-H.

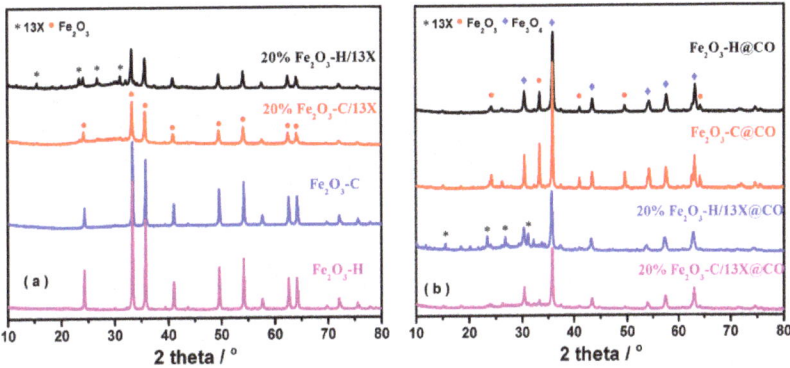

Figure 3. X-ray diffraction patterns on the samples (**a**) without and (**b**) with CO pretreatment.

2.2. Redox and Surface Properties

To investigate the redox properties of the catalysts, H_2-TPR experiments were carried out. As shown in Figure 4a, the reduction behavior of iron species was divided into three stages for the Fe_2O_3-H sample: Fe_2O_3 to Fe_3O_4 in the temperature range of 300–450 °C, and Fe_3O_4 to FeO, then to Fe^0 in the high temperature range of 450–800 °C [31–33]. The introduction of 13X zeolite substantially changed the character of the reduction curves. Only one reduction peak at approximately 533 °C was observed for the 20% Fe_2O_3-H/13X sample. The reduction of Fe_2O_3-H/13X was almost completed at 700 °C, which was much lower than that of the Fe_2O_3-H sample, indicating that the presence of 13X zeolite facilitated the reduction of iron species. In contrast, the reduction was quite complex for 20% Fe_2O_3-C/13X. It started off at a similar temperature to that of the 20% Fe_2O_3-H/13X, but the profile then became flat until a major peak appeared at 792 °C. The poor reducibility of 20% Fe_2O_3-C/13X could account for the low activity as discussed below. The CO pretreatment affected the redox ability of the catalysts, and the corresponding results are shown in Figure 4b. Fe species of Fe_2O_3-H@CO were readily reduced as the reduction process was completed at lower temperatures compared to Fe_2O_3-H. Notably, CO treatment led to the increment of the reduction peak area of the Fe_2O_3-H and 20% Fe_2O_3-H/13X samples compared to their respective fresh samples, but the opposite trend was observed on the Fe_2O_3-C-based catalysts. The results indicated that CO pretreatment facilitated the generation of reducibly Fe active species on the Fe_2O_3-H-based samples while inhibiting it on the Fe_2O_3-C-based samples. Moreover, the reduction peak of 20% Fe_2O_3-C/13X after CO pretreatment remained at around 790 °C, while that of 20% Fe_2O_3-H/13X shifted from 533 to 515 °C after CO exposure, showing that the redox ability was improved on the 20% Fe_2O_3-H/13X@CO.

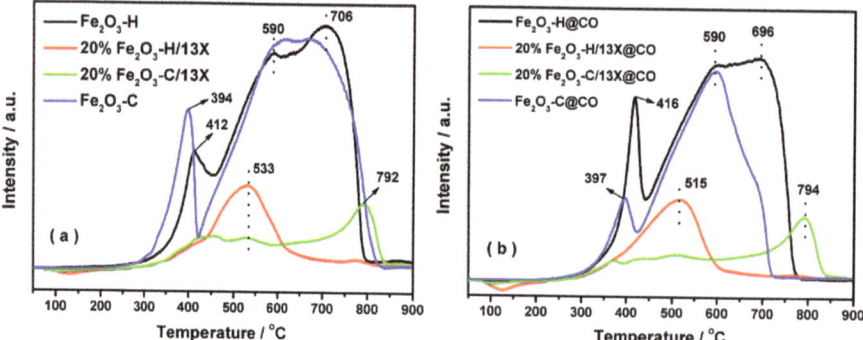

Figure 4. H_2-TPR profiles on the samples (**a**) without and (**b**) with the exposure to CO.

To investigate the O_2 adsorption behavior of samples, O_2-TPD experiments of fresh samples were conducted, and the results are shown in Figure 5a. The desorption peaks below 300 °C and above 600 °C are generally attributed to chemisorbed oxygen species and bulk lattice oxygen, respectively [34,35]. Both Fe_2O_3-C and Fe_2O_3-H showed a weak ability for O_2 adsorption, while the introduction of 13X zeolite profoundly enhanced the O_2 adsorption ability, which was a crucial factor influencing the activity of CO oxidation. Compared to 20% Fe_2O_3-C/13X, 20% Fe_2O_3-H/13X exhibited a bigger O_2 desorption peak, showing its superior adsorption capability. The effect of CO pretreatment on O_2 adsorption ability was investigated, and the results are shown in Figure 5b. Lower adsorption temperatures were observed on all the samples after CO pretreatment. Similarly, the 13X zeolite-supported Fe_2O_3 samples showed a larger O_2 adsorption amount than pure Fe_2O_3 samples after CO pretreatment.

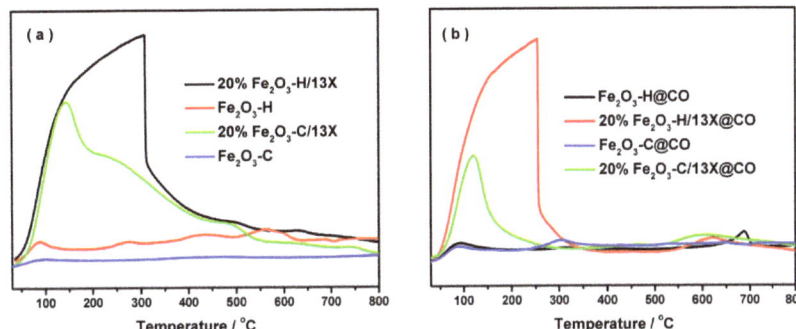

Figure 5. O_2-TPD profiles on the samples (**a**) without and (**b**) with the exposure to CO.

An XPS measurement was employed to investigate the chemical state of the samples. XPS spectra of Fe 2p of samples before and after CO pretreatment are presented in Figure 6. The peaks at 710.97 and 724.45 eV are assigned to Fe $2p_{3/2}$ and Fe $2p_{1/2}$, respectively, and the satellite peak at 719.16 eV is attributed to Fe_2O_3 [36,37]. No peaks assigned to Fe^0 and Fe^{2+} were detected, as expected. The binding energies of Fe 2p shifted to lower values at 0.5 eV when Fe_2O_3-H and Fe_2O_3-C were supported on 13X zeolite. This indicated that the introduction of 13X zeolite affected the chemical environments of Fe^{3+} in the samples. Three peaks at 709.6–709.9 eV, 710.2–711.7 eV, and 712.9–713.7 eV appeared after CO pretreatment, which can be assigned to octahedral Fe(II) species, octahedral Fe(III) species, and tetrahedral Fe(III) species [38]. This suggests that the iron in the samples existed both in Fe^{2+} and Fe^{3+} species. Furthermore, a shakeup satellite attributed to the Fe $2p_{3/2}$ peak appeared, which was ~9 eV higher than that of the main Fe $2p_{3/2}$ peak. Table 1 lists the oxidation states of Fe species in the samples after CO pretreatment. The high Fe^{2+} / (Fe^{2+} + Fe^{3+}) ratio caused the imperfect structure of iron oxide and resulted in the formation of oxygen vacancies that facilitated CO oxidation. The lowest Fe^{2+} content was found in 20% Fe_2O_3-C/13X@CO, and Fe_2O_3-H@CO exhibited a higher Fe^{2+} / (Fe^{2+} + Fe^{3+}) ratio than Fe_2O_3-C@CO. The Fe^{2+} / (Fe^{2+} + Fe^{3+}) ratio increased from 0.296 for Fe_2O_3-H@CO to 0.346 for 20% Fe_2O_3-H/13X@CO, which can be attributed to the interaction between iron and 13X zeolite.

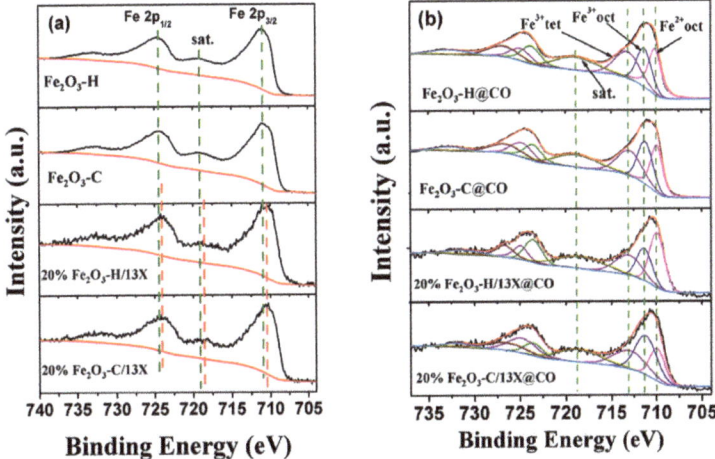

Figure 6. Fe 2p XPS spectra of Fe_2O_3-H, Fe_2O_3-C, 20% Fe_2O_3-H/13X, and 20% Fe_2O_3-C/13X (**a**) before and (**b**) after CO pretreatment. Magenta, navy and purple lines represented octahedral Fe(II) species, octahedral Fe(III) species, and tetrahedral Fe(III) species, respectively.

Table 1. Peak-fitting quantitative results of Fe 2p and O 1s spectra of different samples.

Sample	Peak Location (eV)			$Fe^{2+}/(Fe^{2+} + Fe^{3+})$ [a]	O_a/O [b]	O_b/O [c]
	Fe^{2+}oct	Fe^{3+}oct	Fe^{3+}tet			
Fe_2O_3-H	-	-	-	-	66.4%	33.6%
Fe_2O_3-C	-	-	-	-	71.2%	28.8%
20% Fe_2O_3-H/13X	-	-	-	-	18.0%	82.0%
20% Fe_2O_3-C/13X	-	-	-	-	19.4%	80.6%
Fe_2O_3-H@CO	710.06	711.24	713.12	29.6%	61.2%	38.8%
Fe_2O_3-C@CO	709.92	711.15	712.92	27.1%	68.2%	31.8%
20% Fe_2O_3-H/13X@CO	709.90	711.30	713.06	34.6%	11.5%	88.5%
20% Fe_2O_3-C/13X@CO	709.9	711.15	712.98	20.1%	17.5%	82.5%

[a] $Fe^{2+}/(Fe^{2+} + Fe^{3+}) = S_{Fe^{2+}}/S_{(Fe^{2+} + Fe^{3+})}$. [b] $O_a/O = O_a/S_{(Oa+Ob)}$. [c] $O_b/O = O_a/S_{(Oa+Ob)}$.

Figure 7 shows the XPS spectra of O 1s of samples before and after CO pretreatment, which could be fitted into three peaks assigned to chemisorbed oxygen $O_β$ species (O_2^{2-}, ~531.3 eV; and O_2^-, ~533.0 eV) and lattice oxygen $O_α$ species (O^{2-}, ~529.8 eV), respectively. Table 1 lists the $O_α$ and $O_β$ ratio calculated from the fitted peak areas ascribed to $O_α$ and $O_β$ species. The $O_β$ ratio increased from 28.8 to 33.6% on Fe_2O_3-H catalysts when compared to Fe_2O_3-C catalysts. As lattice oxygen $O_α$ species were closely related to iron oxides in the samples, the introduction of 13X zeolite changed the major feature of the profile, showing that $O_β$ ratio was much higher than $O_α$ ratio due to low Fe contents. CO pretreatment also markedly promoted the generation of chemisorbed oxygen $O_β$ species (O_2^{2-}, ~531.3 eV; and O_2^-, ~533.0 eV), e.g., the $O_β$ ratio increased from 82.0% to 88.5% after the 20% Fe_2O_3-H/13X was pretreated with CO. The results were consistent with the Fe 2p XPS spectra, i.e., more Fe species with low valence states were favorable for the generation of more oxygen vacancies. $O_β$ species are widely recognized as more reactive than $O_α$ species because of their higher mobility.

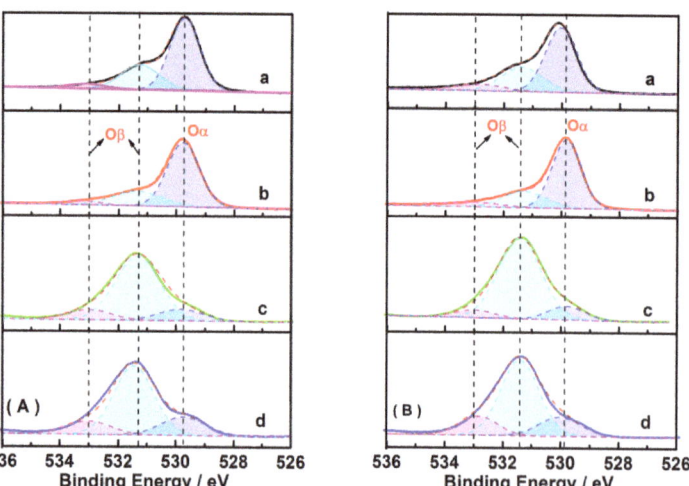

Figure 7. O 1s XPS spectra of the following samples (**A**) without and (**B**) with CO pretreatment: (a) Fe_2O_3-H, (b) Fe_2O_3-C, (c) 20% Fe_2O_3-H/13X, and (d) 20% Fe_2O_3-C/13X. Blue peak area represented lattice oxygen $O_α$ species, while green and pink peak area, chemisorbed oxygen $O_β$ species.

2.3. CO Oxidation Application

2.3.1. Effect of Calcination Temperatures, Different Synthesis Methods, and Pretreatment Conditions

The phase of Fe_2O_3-H, determined by calcination temperatures, showed a significant influence on CO oxidation performance. Figure S1 presents XRD patterns of Fe_2O_3-H under different calcination temperatures. Several main characteristic diffraction peaks at 30.2, 35.6, 43.3, 53.7, and 57.3° appeared on the samples under 400 and 500 °C calcination, which are assigned to metastable maghemite phase γ-Fe_2O_3 (PDF#39-1346). Further increasing the calcination temperatures to 600 and 700 °C led to the phase transformation to the more stable hematite phase α-Fe_2O_3 (PDF#33-0664), which displayed main characteristic diffraction peaks at 24.1, 33.2, 35.6, 40.8, and 49.6°. The average crystalline size, estimated by the Scherrer equation, slightly increased from 25.1 to 27.9 nm when increasing the calcination temperatures from 400 to 600 °C, while it markedly increased to 35.8 nm when further elevating the calcination temperature to 700 °C. This suggests that the crystalline size would significantly grow large at calcination temperatures above 600 °C. Figure S2 exhibited CO oxidation activity over Fe_2O_3-H after calcined at 400–700 °C. The CO oxidation performance increased in the following sequence of 400 < 500 < 700 < 600 °C, suggesting that the optimal calcination temperature was 600 °C.

The activities of the original solid Fe resources, Fe_2O_3-H and Fe_2O_3-C, were then compared, and the corresponding results are shown in Figure 8. The solid Fe resources showed the lowest CO conversion in the whole temperature window compared with the samples after calcination, which may be due to residual oil or other impurities in the solid Fe resources that inhibited its catalytic activity. Fe_2O_3-H and Fe_2O_3-C showed similar activity with T10 at 200 °C and T100 at 300 °C, demonstrating that the performance of Fe_2O_3-H, obtained from the ROS, was comparable to that of the synthetic Fe_2O_3 in the lab, and hence the excellent recycling value of the solid Fe resources in CO oxidation application. CO pretreatment could change the chemical state of active species and catalyst structure, thus further affecting CO oxidation activity. Figure 8b shows CO oxidation performance over Fe_2O_3-H, Fe_2O_3-H@CO, and Fe_2O_3-C@CO. It was found that the activity of Fe_2O_3-H and Fe_2O_3-C can be significantly improved after CO pretreatment, and the Fe_2O_3-H@CO exhibited slightly higher activity than the Fe_2O_3-C@CO. This indicates that CO pretreatment can evidently boost the activity of Fe_2O_3 for CO oxidation. The best activity was obtained from Fe_2O_3-H@CO, which showed 54% and 100% CO conversions at 200 and 250 °C, respectively. According to XRD and XPS results, a portion of Fe^{3+} was reduced to low valence Fe^{2+} species during the CO pretreatment, resulting in the formation of more oxygen vacancies that facilitated O_2 adsorption and activation. Enhanced O_2 adsorption and activation would accelerate CO oxidation rates on the Fe_2O_3-H and Fe_2O_3-C after CO pretreatment.

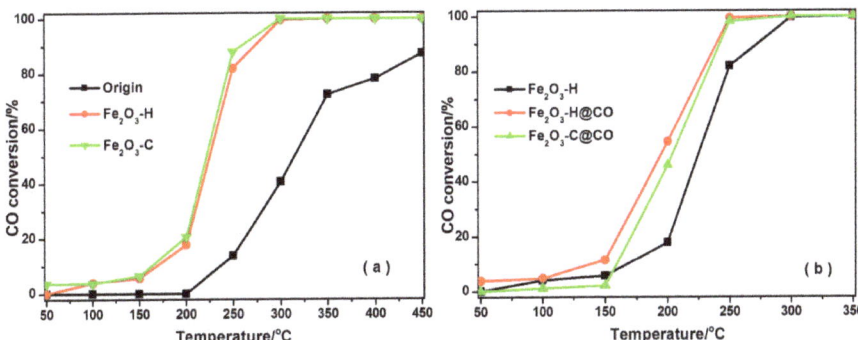

Figure 8. CO oxidation activities on (**a**) the original Fe resources, Fe_2O_3-H and Fe_2O_3-C; and (**b**) the Fe_2O_3-H, Fe_2O_3-H@CO and Fe_2O_3-C@CO.

2.3.2. CO Oxidation Activity on 13X Zeolite-Supported Catalysts

Pure nanometer oxides readily agglomerate and grow when exposed to heat. High surface area supports are typically employed to disperse those nanometer oxide particles to enhance their durability and thermal stability. On this basis, we prepared Fe_2O_3-H supported on the 13X zeolite, 5A, FCC and γ-Al_2O_3 catalysts, and their XRD patterns are shown in Figure S3. All the samples displayed the characteristic peaks ascribed to the corresponding supports, indicating that each support was successfully introduced. Their CO oxidation activities were further compared, and the results are shown in Figure S4. The CO oxidation activity increased in the following order of Fe_2O_3-H/FCC < Fe_2O_3-H/5A < Fe_2O_3-H/γ-Al_2O_3 < Fe_2O_3-H/13X. To further clarify the effect of pure supports on the performance, we tested their catalytic activities. All the supports showed quite poor activities in the whole temperature window. The results suggested that the activity came from Fe_2O_3 and the interactions between Fe_2O_3 and the support, and that the optimal catalyst support was 13X zeolite.

After selecting 13X zeolite to be a support, the effect of different Fe_2O_3-H loadings on the activity was investigated. Figure 9a shows that the CO oxidation activity increased with the increase of Fe_2O_3-H loading, and the pure Fe_2O_3-H exhibited the best performance. After CO pretreatment, the activity of samples for CO oxidation was markedly enhanced, and the best activity was observed on 20% Fe_2O_3-H/13X@CO in Figure 9b. In particular, the 20% Fe_2O_3-H/13X@CO showed 71% CO conversion at 200 °C, compared with 5% CO conversion on the 20% Fe_2O_3-H/13X at the same temperature, revealing the importance of CO pretreatment in improving the activity of catalysts. CO pretreatment was also introduced to the 20% Fe_2O_3-C/13X for comparison. The CO conversion of 20% Fe_2O_3-C/13X@CO was 15% at 200 °C, which is much lower than that of 20% Fe_2O_3-H/13X@CO. This clearly demonstrated the value of recycling the solid Fe resources from the ROS. The 20% Fe_2O_3-H/13X@CO exhibited the highest CO oxidation activity due to better redox ability and superior O_2 adsorption and activation ability after CO treatment. It had T100 (the catalyst temperature required to reach 100% CO conversion) at 250 °C, which was obviously improved compared with these Fe_2O_3-based catalysts reported in the literature, as listed in Table 2. The stability of 20% Fe_2O_3-H/13X@CO for CO oxidation was also investigated by evaluating the activities with time-on-stream tests. Figure S5 showed that minor CO conversion loss was observed in 30 h, which further proved its great promise for practical applications.

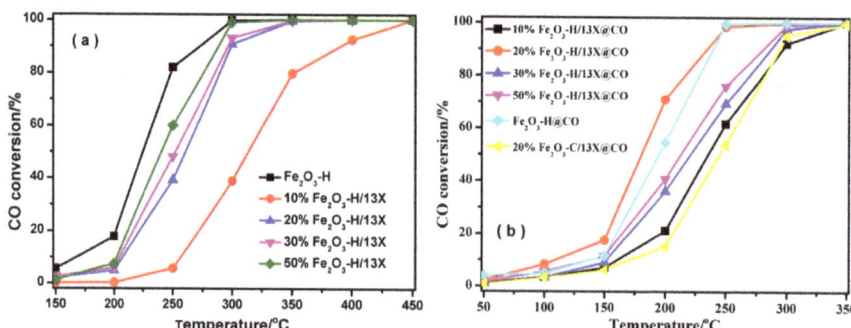

Figure 9. CO conversion as a function of reaction temperature on (**a**) the x% Fe_2O_3-H/13X; and (**b**) the x% Fe_2O_3-H/13X@CO and 20% Fe_2O_3-C/13X @CO.

Based on the above results, the catalytic mechanism of CO oxidation on the Fe_2O_3-H/13X@CO was then discussed. It is generally believed that iron oxide catalysts follow the redox mechanism in CO oxidation. For supported Fe catalysts, CO molecular first adsorbs on the surface active sites, and is subsequently oxidized by lattice oxygen in iron oxides. Several medium CO_3^{2-} or HCO_3^{-} species are formed, which are further converted into

CO_2. Correspondingly, reduced active sites in the former step are, in turn, oxidized to the initial state by adsorbed O_2 to become available for further reaction cycles. Combined with characterization and activity results, Fe_2O_3-H/13X@CO showed the best CO oxidation activity, possibly due to the following reasons: Fe species with low valence states can generate more oxygen vacancies, involving the reaction. Furthermore, the 20% Fe_2O_3-H/13X@CO surface provided sufficient chemical chemisorbed oxygen species for reduced active sites' regeneration and exhibited excellent reducibility for CO molecules' oxidation, which also contributed to its best CO oxidation performance.

Table 2. Summary of reported T100 on Fe based catalysts in CO oxidation.

NO	Catalysts	T100/°C	Ref.
1	Fe_2O_3/13X	250	This work
2	Ce-Fe	275	[23]
3	Fe_2O_3/Al_2O_3	300	[39]
4	Fe_2O_3/Al_2O_3	278	[40]
5	Fe_2O_3 rod	370	[41]
6	Fe_2O_3	288	[42]
7	Fe_2O_3/TiO_2	260	[43]
8	Fe_2O_3/Al_2O_3	300	[44]
9	Fe_2O_3	350	[45]
10	Fe/Al-pillared bentonite	>400	[46]

3. Methods and Materials

3.1. Materials and Reagents

ROS was collected from a food-grade stainless steel facility in WISCO. The facilities used in experiments include a heating mantle, a thermostatic water bath, an electrically-heated drying cabinet, an oven, and a large capacity centrifuge. Raw materials, such as iron(III) nitrate nonahydrate ($Fe(NO_3)_3 \cdot 9H_2O$, 99%), ethyl acetate, ammonium hydroxide ($NH_3 \cdot H_2O$, 25 wt.%), and 13X zeolite, purchased from Sinopharm Group Chemical Reagent Beijing Co. Ltd., Beijing, China are analytically pure and can be used without further purification.

3.2. Catalytic Hydrogenation Reaction

To recycle the solid Fe resources from ROS, catalytic hydrogenation of the ROS was carried out on a 500 mL high-pressure reactor. Typically, 250 g of ROS is added to the reactor. Prior to reaction, the reactor was flushed with nitrogen twice to remove the air residue. Afterwards, H_2 was filled into the reactor up to 6 MPa at room temperature, and the temperature was raised at a ramp rate of 5 °C/min with a stirring speed of 300 r/min. The reaction was carried out for 4 h at 320 °C to guarantee completion of the hydrogenation process. After cooling to room temperature, the mixture was filtered and the solid Fe resources were finally obtained.

3.3. Catalyst Preparation

The solid Fe resources were washed repeatedly with ethyl acetate to remove oil and other impurities, then dried at 100 °C overnight. The powder was calcined at 400–700 °C for 5.5 h, and the resultant product was denoted as Fe_2O_3-H. The supported catalysts were synthesized by a mechanical milling method through mixing the solid Fe resources and 13X zeolite. Specifically, the Fe_2O_3-H was mixed with 13X zeolite at the desired mass ratio and milled in a rock grinder for 10 min. The Fe_2O_3 loading was set at 10~50 wt.% and the catalysts were denoted as x% Fe_2O_3-H/13X. The obtained samples were further crushed into 40~60 mesh before use. For comparison, a sample denoted as Fe_2O_3-C was obtained through direct calcination of $Fe(NO_3)_3 \cdot 9H_2O$ at 600 °C for 5.5 h, which was then supported on 13X zeolite (denoted as x%Fe_2O_3-C/13X) with the mechanical milling method described above. To avoid confusion, for catalysts pretreated with CO before the reaction, they were denoted as catalyst@CO, e.g., Fe_2O_3-H@CO means the fresh Fe_2O_3-H was pretreated with 1%CO at 450 °C for 1 h. The scheme of the Fe_2O_3-H/13X@CO preparation process was depicted in Figure 10. Moreover, 5A zeolite, fluid catalytic cracking (FCC) spent catalysts, and γ-Al_2O_3 carriers were chosen to investigate the effect of different supports on CO oxidation activity. The preparation method was the same as Fe_2O_3-H/13X except for replacing 13X with 5A, FCC or γ-Al_2O_3 carriers and Fe_2O_3 loading content was fixed at 20%. The catalysts were denoted as Fe_2O_3-H/5A, Fe_2O_3-H/FCC and Fe_2O_3-H/γ-Al_2O_3.

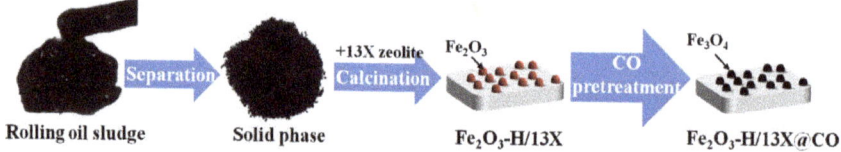

Figure 10. Schematic illustration of the Fe_2O_3-H/13X@CO preparation process.

3.4. Catalyst Characterization

X-ray Diffraction (XRD) measurements were performed on a Bruker D8 diffractometer equipped with Cu Kα radiation, with 2θ range between 10° and 80° and a scanning rate of 5°/min. N_2 sorption isotherms of the catalysts were recorded with a JWGB Sci & Tech Ltd. JW-BK112 analyzer at −196 °C. The samples were degassed at 350 °C for 12 h prior to measurements. A Transmission Electron Microscope (TEM) was conducted on a Tecnai G^2 F20 S-TWIN to observe surface morphology and measure particle size. X-ray photoelectron spectroscopy (XPS) measurements were performed on a Thermo Escalab 250 Xi spectrometer (Al Kα radiation) to analyze surface chemical composition and chemical states. Binding energies were calibrated using the C1s peak of adventitious carbon at 284.6 eV. A temperature programmed reduction with hydrogen (H_2-TPR) was performed on a Chemisorb 2720 TPx chemisorption analyzer. A 100 mg sample was pretreated by Ar (25 mL/min) at 350 °C for 30 min and then cooled to room temperature in Ar flow. Afterwards, the TPR spectrum was then collected by raising the temperature at a heating rate of 10 °C/min in 10% H_2/Ar (50 mL/min). Temperature-programmed desorption of oxygen (O_2-TPD) was performed on the same apparatus. The first step was to heat up the catalyst to 350 °C in flowing N_2 to desorb gases adsorbed on the catalyst during catalyst preparation. A 50 mg sample was subjected to pure O_2 gas at room temperature for 1 h. Afterwards, the gas was switched to pure N_2 to remove the physically adsorbed O_2 molecules. The TPD spectrum was then collected by raising the temperature at a heating rate of 10 °C/min.

3.5. CO Oxidation Performance Evaluation

CO oxidation performance was evaluated in a home-made fixed bed quartz reactor operating at atmospheric pressure. Typically, 3 g of catalyst was packed in the reactor and supported by a quartz wool plug. To activate the fresh catalysts, the catalysts were heated up to 450 °C for 30 min in flowing N_2 (100 mL/min). All activities were tested under steady state conditions with a temperature range of 50~450 °C and a temperature increment of 50 °C. The feed gas composition was 0.5 mol.% CO and 0.25 mol.% O_2 (N_2 balance) with a stoichiometric feed condition (λ = 1.0). The total flow rate was 100 mL/min and the gaseous hourly space velocity, 2000 h^{-1}. CO and CO_2 concentrations in the effluent were quantified by an on-line gas chromatograph (GC-9560, Zhongke Huifen) equipped with a flame ionization detector (FID). Before entering FID, CO and CO_2 were fully converted to CH_4 by a Ni catalyst maintained at 380 °C. The CO conversion was calculated using the following equation:

$$X_{CO} (\%) = \frac{C_{CO,in} - C_{CO,out}}{C_{CO,in}} \times 100$$

where $C_{CO,in}$ and $C_{CO_2,out}$ are the inlet and outlet concentrations of CO, respectively.

4. Conclusions

The solid Fe resources, recycled from ROS by the catalytic hydrogenation process, were used as raw materials to prepare Fe-based catalysts for CO oxidation. The XRD and HRTEM results confirmed that the Fe_2O_3-H, prepared by calcinating the solid Fe resources in air, mainly consists of the Fe_2O_3 phase. TEM results showed that the particle morphology is spherical in nature and the particle size was around 30–40 nm. The CO oxidation activity of Fe_2O_3-H was comparable to that of the Fe_2O_3-C synthesized in the lab. In particular, the activity of Fe_2O_3-H for CO oxidation can be dramatically enhanced after mixing with 13X zeolite and pre-treating in a CO atmosphere, with the best activity obtained from 20% Fe_2O_3-H/13X@CO, which showed 71% CO conversion at 200 °C and 100% conversion at 250 °C. This potentiates the great value of recycling the solid Fe resources from ROS. The excellent activity of 20% Fe_2O_3-H/13X@CO was attributed to the generation of low-valence Fe species, enhanced reducibility, and improved O_2 adsorption ability. This work provides a new vista to develop promising alternatives to noble metal-based three-way catalysts via recycling the solid phase of waste ROS.

Supplementary Materials: The following supporting information can be downloaded at: https://www.mdpi.com/article/10.3390/ijms232012134/s1.

Author Contributions: Conceptualization, K.C. and M.X.; methodology, K.C. and M.X.; software, W.G. and S.T., validation, W.G. and S.T.; formal analysis, K.C. and J.W.; investigation, W.G., S.T. and T.W.; resources, K.C.; data curation, J.W.; writing—original draft preparation, S.T., K.C. and M.X. All authors have read and agreed to the published version of the manuscript.

Funding: This work was supported by NSFC (22002111), Opening Project of Hubei Key Laboratory of Biomass Fibers and Eco-Dyeing & Finishing (STRZ202106), Young Talent Project of Hubei Provincial Department of Education (214096) and the Science and Technology Department of Hubei Province (2021CFA034).

Conflicts of Interest: The authors declare no conflict of interest.

References

1. Zhu, S.; Li, T.; Wu, Y.; Chen, Y.; Su, T.; Ri, K.; Huo, Y. Effective purification of cold-rolling sludge as iron concentrate powder via a coupled hydrothermal and calcination route: From laboratory-scale to pilot-scale. *J. Clean. Prod.* **2020**, *276*, 124274. [CrossRef]
2. Liu, B.; Zhang, S.-g.; Steenari, B.-M.; Ekberg, C. Synthesis and properties of $SrFe_{12}O_{19}$ obtained by solid waste recycling of oily cold rolling mill sludge. *Int. J. Miner. Metall. Mater.* **2019**, *26*, 642–648. [CrossRef]
3. Liang, Q.; Han, D.; Cao, Z.; Du, J. Studies on kinetic and reaction mechanism of oil rolling sludge under a wide temperature range. *Energy Sources Part A* **2021**, 1–13. [CrossRef]
4. Qin, L.; Han, J.; He, X.; Zhan, Y.; Yu, F. Recovery of energy and iron from oily sludge pyrolysis in a fluidized bed reactor. *J. Environ. Manag.* **2015**, *154*, 177–182. [CrossRef]
5. Williams, D. Turning waste oil into profit. *J. Can. Pet. Technol.* **2009**, *48*, 7–8.
6. Wang, Z.; Gong, Z.; Wang, Z.; Li, X.; Chu, Z. Application and development of pyrolysis technology in petroleum oily sludge treatment. *Environ. Eng. Res.* **2020**, *26*, 190460. [CrossRef]
7. Chen, H.S.; Zhang, Q.M.; Yang, Z.J.; Liu, Y.S. Research on Treatment of Oily Sludge from the Tank Bottom by Ball Milling Combined with Ozone-Catalyzed Oxidation. *ACS Omega* **2020**, *5*, 12259–12269. [CrossRef]
8. Chang, W.; Zhao, D.; Lian, J.; Han, S.; Xue, Y. Regeneration of waste rolling oil via inorganic flocculation-adsorption process. *Pet. Sci. Technol.* **2019**, *37*, 837–844. [CrossRef]
9. Zhou, J.; Zhao, Z.; Xue, Y.; Chang, W.; Liu, P.; Han, S.; Lin, H. Regeneration of used rolling oil for sustainable use. *Pet. Sci. Technol.* **2017**, *35*, 1642–1647. [CrossRef]
10. Zhao, J.; Yuan, C.; Yao, X. A simple method to evaluate solvent for lubricating oil by solubility parameter. *Pet. Sci. Technol.* **2017**, *35*, 1445–1450. [CrossRef]
11. Adeyemi, A.F.; Adebiyi, F.M.; Koya, O.A. Evaluation of physico-chemical properties of re-refined lubricating oils obtained from fabricated packed bed reactor. *Pet. Sci. Technol.* **2017**, *35*, 1712–1723. [CrossRef]
12. Ahmad, I.; Khan, R.; Ishaq, M.; Khan, H.; Ismail, M.; Gul, K.; Ahmad, W. Production of Lighter Fuels from Spent Lubricating Oil via Pyrolysis over Barium-Substituted Spinel Ferrite. *Energy Fuel* **2016**, *30*, 4781–4789. [CrossRef]
13. Martín, M.I.; López, F.A.; Torralba, J.M. Production of sponge iron powder by reduction of rolling mill scale. *Ironmak. Steelmak.* **2013**, *39*, 155–162. [CrossRef]

14. Liu, B.; Zang, S.-G.; Tian, J.-J.; Pan, D.-A.; Zhu, H.-X. Strontium ferrite powders prepared from oily cold rolling mill sludge by solid-state reaction method. *Rare Met.* **2013**, *32*, 518–523. [CrossRef]
15. Park, J.-W.; Ahn, J.-C.; Song, H.; Park, K.; Shin, H.; Ahn, J.-S. Reduction characteristics of oily hot rolling mill sludge by direct reduced iron method. *Resour. Conserv. Recycl.* **2002**, *34*, 129–140. [CrossRef]
16. Hocking, M.B. Petroleum Refining. In *Handbook of Chemical Technology and Pollution Control*; Academic Press: Cambridge, MA, USA, 2005.
17. Pujadó, P. Petroleum Refining Processes. *J. Pet. Sci. Eng.* **2002**, *45*, 295–296. [CrossRef]
18. Heck, R.M.; Farrauto, R.J.; Gulati, S.T. *Catalytic Air Pollution Control: Commercial Technology*, 3rd ed.; John Wiley & Sons: Hoboken, NJ, USA, 2009.
19. Freund, H.J.; Meijer, G.; Scheffler, M.; Schlögl, R.; Wolf, M. CO oxidation as a prototypical reaction for heterogeneous processes. *Angew. Chem. Int. Ed.* **2011**, *50*, 10064–10094. [CrossRef]
20. Liu, B.; Wu, H.; Li, S.; Xu, M.; Cao, Y.; Li, Y. Solid-state construction of $CuO_x/Cu_{1.5}Mn_{1.5}O_4$ nanocomposite with abundant surface CuO_x species and oxygen vacancies to promote CO oxidation activity. *Int. J. Mol. Sci.* **2022**, *23*, 6856. [CrossRef]
21. Yang, L.; Shan, S.; Loukrakpam, R.; Petkov, V.; Ren, Y.; Wanjala, B.N.; Engelhard, M.H.; Luo, J.; Yin, J.; Chen, Y. Role of support–nanoalloy interactions in the atomic-scale structural and chemical ordering for tuning catalytic sites. *J. Am. Chem. Soc.* **2012**, *134*, 15048–15060. [CrossRef]
22. Liu, J.; Wang, L.; Okejiri, F.; Luo, J.; Zhao, J.; Zhang, P.; Liu, M.; Yang, S.; Zhang, Z.; Song, W. Deep understanding of strong metal interface confinement: A journey of Pd/FeO x catalysts. *ACS Catal.* **2020**, *10*, 8950–8959. [CrossRef]
23. Laguna, O.H.; Centeno, M.A.; Boutonnet, M.; Odriozola, J.A. Fe-doped ceria solids synthesized by the microemulsion method for CO oxidation reactions. *Appl. Catal. B* **2011**, *106*, 621–629. [CrossRef]
24. Biabani-Ravandi, A.; Rezaei, M. Low temperature CO oxidation over Fe-Co mixed oxide nanocatalysts. *Chem. Eng. J.* **2012**, *184*, 141–146. [CrossRef]
25. Papaioannou, E.I.; Neagu, D.; Ramli, W.K.W.; Irvine, J.T.S.; Metcalfe, I.S. Sulfur-Tolerant, Exsolved Fe-Ni Alloy Nanoparticles for CO Oxidation. *Top. Catal.* **2018**, *62*, 1149–1156. [CrossRef]
26. Waqas, M.; Mountapmbeme Kouotou, P.; El Kasmi, A.; Wang, Y.; Tian, Z.-Y. Role of copper grid mesh in the catalytic oxidation of CO over one-step synthesized Cu-Fe-Co ternary oxides thin film. *Chin. Chem. Lett.* **2020**, *31*, 1201–1206. [CrossRef]
27. Ramakrishna, C.; Krishna, R.; Gopi, T.; Swetha, G.; Saini, B.; Chandra Shekar, S.; Srivastava, A. Complete oxidation of 1,4-dioxane over zeolite-13X-supported Fe catalysts in the presence of air. *Chin. J. Catal.* **2016**, *37*, 240–249. [CrossRef]
28. Li, X.; Wang, J.; Guo, Y.; Zhu, T.; Xu, W. Adsorption and desorption characteristics of hydrophobic hierarchical zeolites for the removal of volatile organic compounds. *Chem. Eng. J.* **2021**, *411*, 128558. [CrossRef]
29. Imran, M.; Affandi, A.M.; Alam, M.M.; Khan, A.; Khan, A.I. Advanced biomedical applications of iron oxide nanostructures based ferrofluids. *Nanotechnology* **2021**, *32*, 422001. [CrossRef]
30. Imran, M.; Abutaleb, M.; Ashraf Ali, M.; Ahamad, T.; Rahman Ansari, A.; Shariq, M.; Lolla, D.; Khan, A. UV light enabled photocatalytic activity of α-Fe_2O_3 nanoparticles synthesized via phase transformation. *Mater. Lett.* **2020**, *258*, 126748. [CrossRef]
31. Giecko, G.; Borowiecki, T.; Gac, W.; Kruk, J. Fe_2O_3/Al_2O_3 catalysts for the N_2O decomposition in the nitric acid industry. *Catal. Today* **2008**, *137*, 403–409. [CrossRef]
32. Sierra-Pereira, C.A.; Urquieta-González, E.A. Reduction of NO with CO on CuO or Fe_2O_3 catalysts supported on TiO_2 in the presence of O_2, SO_2 and water steam. *Fuel* **2014**, *118*, 137–147. [CrossRef]
33. Basińska, A.; Jóźwiak, W.K.; Góralski, J.; Domka, F. The behaviour of Ru/Fe_2O_3 catalysts and Fe_2O_3 supports in the TPR and TPO conditions. *Appl. Catal. A* **2000**, *190*, 107–115. [CrossRef]
34. Tanaka, S.-I.; Yuzaki, K.; Ito, S.-I.; Kameoka, S.; Kunimori, K. Mechanism of O_2 Desorption during N_2O Decomposition on an Oxidized Rh/USY Catalyst. *J. Catal.* **2001**, *200*, 203–208. [CrossRef]
35. Li, G.; Li, N.; Sun, Y.; Qu, Y.; Jiang, Z.; Zhao, Z.; Zhang, Z.; Cheng, J.; Hao, Z. Efficient defect engineering in Co-Mn binary oxides for low-temperature propane oxidation. *Appl. Catal. B* **2021**, *282*, 119512. [CrossRef]
36. Cai, Z.; Liu, Q.; Li, H.; Wang, J.; Tai, G.; Wang, F.; Han, J.; Zhu, Y.; Wu, G. Waste-to-resource strategy to fabricate functionalized MOFs composite material based on durian shell biomass carbon fiber and Fe_3O_4 for highly efficient and recyclable dye adsorption. *Int. J. Mol. Sci.* **2022**, *23*, 5900. [CrossRef]
37. Kaur Ubhi, M.; Kaur, M.; Singh, D.; Javed, M.; Oliveira, A.C.; Kumar Garg, V.; Sharma, V.K. Hierarchical nanoflowers of $MgFe_2O_4$, bentonite and B-,P- Co-doped graphene oxide as adsorbent and photocatalyst: Optimization of parameters by Box-Behnken methodology. *Int. J. Mol. Sci.* **2022**, *23*, 9678. [CrossRef]
38. Shekar, S.G.C.; Alkanad, K.; Hezam, A.; Alsalme, A.; Al-Zaqri, N.; Lokanath, N.K. Enhanced photo-Fenton activity over a sunlight-driven ignition synthesized α-Fe_2O_3-Fe_3O_4/CeO_2 heterojunction catalyst enriched with oxygen vacancies. *J. Mol. Liq.* **2021**, *335*, 116186. [CrossRef]
39. Tepluchin, M.; Casapu, M.; Boubnov, A.; Lichtenberg, H.; Wang, D.; Kureti, S.; Grunwaldt, J.-D. Fe and Mn-Based Catalysts Supported on γ-Al_2O_3 for CO Oxidation under O_2-Rich Conditions. *ChemCatChem* **2014**, *6*, 1763–1773. [CrossRef]
40. Schoch, R.; Bauer, M. Pollution Control Meets Sustainability: Structure-Activity Studies on New Iron Oxide-Based CO Oxidation Catalysts. *ChemSusChem* **2016**, *9*, 1996–2004. [CrossRef]
41. Liu, X.; Liu, J.; Chang, Z.; Sun, X.; Li, Y. Crystal plane effect of Fe_2O_3 with various morphologies on CO catalytic oxidation. *Catal. Commun.* **2011**, *12*, 530–534. [CrossRef]

42. Gao, Q.-X.; Wang, X.-F.; Di, J.-L.; Wu, X.-C.; Tao, Y.-R. Enhanced catalytic activity of α-Fe$_2$O$_3$ nanorods enclosed with {110} and {001} planes for methane combustion and CO oxidation. *Catal. Sci. Technol.* **2011**, *1*, 574–577. [CrossRef]
43. Tang, X.; Wang, J.; Li, J.; Zhang, X.; La, P.; Jiang, X.; Liu, B. In-situ growth of large-area monolithic Fe$_2$O$_3$/TiO$_2$ catalysts on flexible Ti mesh for CO oxidation. *J. Mater. Sci. Technol.* **2021**, *69*, 119–128. [CrossRef]
44. Schlicher, S.; Prinz, N.; Bürger, J.; Omlor, A.; Singer, C.; Zobel, M.; Schoch, R.; Lindner, J.K.N.; Schünemann, V.; Kureti, S.; et al. Quality or Quantity? How Structural Parameters Affect Catalytic Activity of Iron Oxides for CO Oxidation. *Catalysts* **2022**, *12*, 675. [CrossRef]
45. Li, P.; Miser, D.E.; Rabiei, S.; Yadav, R.T.; Hajaligol, M.R. The removal of carbon monoxide by iron oxide nanoparticles. *Appl. Catal. B* **2003**, *43*, 151–162. [CrossRef]
46. Carriazo, J.G.; Centeno, M.A.; Odriozola, J.A.; Moreno, S.; Molina, R. Effect of Fe and Ce on Al-pillared bentonite and their performance in catalytic oxidation reactions. *Appl. Catal. A* **2007**, *317*, 120–128. [CrossRef]

International Journal of
Molecular Sciences

Review

Research Progress on Graphitic Carbon Nitride/Metal Oxide Composites: Synthesis and Photocatalytic Applications

Hao Lin [†], Yao Xiao [†], Aixia Geng, Huiting Bi, Xiao Xu, Xuelian Xu and Junjiang Zhu *

Hubei Key Laboratory of Biomass Fibers and Eco-Dyeing & Finishing, College of Chemistry and Chemical Engineering, Wuhan Textile University, Wuhan 430200, China
* Correspondence: jjzhu@wtu.edu.cn; Tel.: +86-27-59367434
† These authors contributed equally to this work.

Abstract: Although graphitic carbon nitride (g-C_3N_4) has been reported for several decades, it is still an active material at the present time owing to its amazing properties exhibited in many applications, including photocatalysis. With the rapid development of characterization techniques, in-depth exploration has been conducted to reveal and utilize the natural properties of g-C_3N_4 through modifications. Among these, the assembly of g-C_3N_4 with metal oxides is an effective strategy which can not only improve electron–hole separation efficiency by forming a polymer–inorganic heterojunction, but also compensate for the redox capabilities of g-C_3N_4 owing to the varied oxidation states of metal ions, enhancing its photocatalytic performance. Herein, we summarized the research progress on the synthesis of g-C_3N_4 and its coupling with single- or multiple-metal oxides, and its photocatalytic applications in energy production and environmental protection, including the splitting of water to hydrogen, the reduction of CO_2 to valuable fuels, the degradation of organic pollutants and the disinfection of bacteria. At the end, challenges and prospects in the synthesis and photocatalytic application of g-C_3N_4-based composites are proposed and an outlook is given.

Keywords: graphitic carbon nitride; metal oxides; heterojunctions; synthesis; photocatalytic applications

1. Introduction

With the development of economies and the growth of populations, pressures on energy demand and environmental pollution continue to increase all over the world [1–4]. Fossil fuels, which currently account for a large amount of the world's energy, are increasingly consumed, resulting in negative impacts on the environment through the release of CO_2, which is a serious greenhouse gas. Solar-energy-based photocatalysis is a promising technology to solve energy and environment problems, and has received extensive attention recently [5–7]. The synthesis of efficient photocatalysts is a key factor in applying photocatalytic technology to solve energy and environmental issues, such as water splitting to produce H_2 and O_2 [8–12], tail gas treatment (NO, CO_2, etc.) [13,14], pollutant degradation [15–20], etc.

In the photocatalytic process, the electrons of photocatalysts are activated by absorbing photon energy [21]. Once the electrons have received enough energy, they will be excited to the valence band (VB), leaving holes at the conduction band (CB). The photogenerated electron–hole pairs (e^-/h^+) will then activate the reactants and promote the proceeding of a reaction [22,23]. In 1972, Fujishima and Honda reported the use of TiO_2 electrodes for photocatalytic water splitting under ultraviolet light, which can be regarded as the milestone of photocatalytic technology [24]. In 1979, Inoue reported the reduction of CO_2 into organic compounds in aqueous solution using TiO_2, ZnO, GaP and CdS semiconductors [25]. Since then, the development of efficient semiconductors for photocatalysis has become a hotspot. Traditional photocatalysts mainly contained inorganic compounds, including metal oxides [26,27], sulfides [28], nitrides [29] and their composites [30], etc.

The direct use of such materials was often restricted by their large band gap, which leads to low utilization efficiency for solar energy. Recently, graphite carbon nitride (g-C_3N_4) semiconductors have come into people's horizons and have become a research focus in the field of photocatalysis, owing to their abundance, simple synthesis, high visible-light utilization efficiency and excellent physicochemical stability.

The application of g-C_3N_4 to photocatalysis was reported in 2009 by Wang et al. [31]. This material received widespread attention in photocatalysis thereafter owing to its polymeric properties and good visible-light response [32,33]. The challenge of applying g-C_3N_4 to photocatalysis mainly lies in its small specific surface area, narrow light response range and high e^-/h^+ recombination rate. To this end, many strategies have been proposed in the literature, such as adjustment of the microstructure [34–36], the doping of heteroatoms [37–39], the coupling of semiconductors [11,30,31,40–42], etc. Among these, coupling with other semiconductors is an attractive strategy, which can not only compensate for the shortcomings of g-C_3N_4 with their own properties, but also produce synergistic effects by forming heterojunctions. Both metal-free polymeric materials [43–45] and metal-containing inorganic materials, such as CdS [46], Fe_2O_3 [47] Fe_3O_4 [48], ZnO [49], TiO_2 [50], Bi_2WO_6 [51] and $Ce_2(WO_4)_3$ [52], can couple with g-C_3N_4 and form heterojunctions. In particular, the coupling of materials with special properties can give the composites interesting advantages. For example, the coupling of magnetic materials, e.g., g-C_3N_4/Fe_3O_4 [53] and g-C_3N_4/$CoFe_2O_4$ [54], can facilitate the recycling of photocatalysts (as they can be simply separated by a magnet), in addition to the improving photocatalytic performance.

Among the coupling materials, metal oxides came into the eyes of researchers early, because of their low-cost, abundance and easy synthesis. Many works on the coupling of g-C_3N_4 and metal oxides have been reported and great achievements have been made. The secular growth of related publications commendably reveals the flourishment of g-C_3N_4/metal oxide heterojunction materials in photocatalytic applications (Figure 1). To date, numerous breakthroughs and advances have been made in the photocatalysis system based on g-C_3N_4-based heterojunction materials, but a comprehensive summary still needs to be further subdivided, especially regarding the g-C_3N_4/metal oxide composite system. In this context, it is of great significance to summarize the recent advances in the synthesis and photocatalytic application of g-C_3N_4/metal oxide composites to alleviate environmental pollution and energy shortage. In detail, this work reviewed the recent progress on (1) the synthesis of g-C_3N_4 and its coupling with single or double metal oxides; (2) the photocatalytic applications of the composites in energy production and environmental protection; and (3) the challenges and prospects of g-C_3N_4-based heterojunction materials in photocatalytic applications. This review enables a wide range of researchers to understand these important areas and prospects, and the challenges and potential of g-C_3N_4/metal oxide composites.

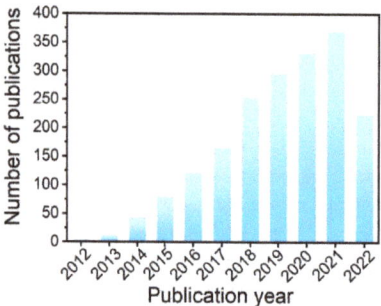

Figure 1. The number of publications on carbon nitride/metal oxide complexes for photocatalytic applications published in the last 10 years. Literature searched in Web of Science with the keywords: "carbon nitride" AND "metal oxide complex" AND "photocatalysis".

2. Synthesis of g-C$_3$N$_4$ and Metal Oxides/g-C$_3$N$_4$ Composites

2.1. Synthesis of g-C$_3$N$_4$

With the in-depth study of g-C$_3$N$_4$ year-by-year, various modification strategies have been proposed and applied to improve the catalytic properties of g-C$_3$N$_4$ materials, including plasma sputtering deposition [55], solvothermal synthesis [56], chemical vapor deposition [57], thermal condensation [58], etc. The thermal condensation method receives special attention owning to its convenience, low-cost and time-savings. Nitrogen-rich materials, such as cyanamide [59], dicyandiamide [60], melamine [61], thiourea [62], urea [63], ammonium thiocyanate [64] and their mixtures [65], are generally used as the precursors to g-C$_3$N$_4$. However, this method often results in materials with low surface area and structural defects, which hinder the exposure of active sites on the surface [66] and act as the recombination centers of photogenerated electron–hole pairs, thereby reducing the photocatalytic performance. To solve these problems, it is suggested that the band gap structure of g-C$_3$N$_4$ be optimized to improve the separation efficiency of photogenerated e^-/h^+ pairs, and to adjust the microstructure to increase its specific surface area.

In this section, we mainly focus on the influence of precursors and preparation conditions on the properties of g-C$_3$N$_4$. In the case of precursors, cyanamide is first used to synthesize g-C$_3$N$_4$. In 2005, Antonietti et al. [67] prepared g-C$_3$N$_4$ via the thermal polymerization of cyanamide. In the process, cyanamide is first self-condensed to dicyandiamide at 150 °C, which then transforms to melamine, melem and, finally, g-C$_3$N$_4$ at 240 °C, 390 °C and 520 °C, respectively, accompanying the release of NH$_3$. However, the high price, high toxicity and special transportation limit its wide use. Therefore, intermediate products with low cost, low toxicity and chemical stability, e.g., dicyandiamide and melamine, are generally used instead of cyanamide.

Ge et al. [68] used melamine as precursor to producing g-C$_3$N$_4$ at a temperature of 500~600 °C. They found that samples prepared at 520 °C showed the best performance for the photodegradation of phenol. This indicates that the properties of g-C$_3$N$_4$ depend intimately on the synthesis temperature, which promotes the modification of samples in, for example, the degree of crystallinity. Additionally, they also found that an increase in temperature can introduce nanostructures to the material, due to the exfoliation caused by the high temperature. This provides a way to control the structure and surface area of g-C$_3$N$_4$ with secondary thermal treatment, as is widely reported in the literature [69,70].

In addition to cyanamide and its derivates, other nitrogen-containing organics can also be used as precursors to g-C$_3$N$_4$. For example, Schaber et al. found that the thermal decomposition of urea in an open reaction vessel can yield g-C$_3$N$_4$, through the transformation of biuret, cyanuric acid, ammelide, ammeline and melamine intermediates [71]. Later, Liu et al. used urea as precursor to producing g-C$_3$N$_4$ without adding auxiliary agents, finding that the obtained material can show excellent activity for the photocatalytic degradation of methylene blue (MB) [72].

Zhang et al. investigated the reaction mechanism of transforming urea to produce a g-C$_3$N$_4$ network at high temperature, and found that the oxygen-containing groups of urea promote the condensation process [73]. In addition to urea, thiourea is also employed to fabricate g-C$_3$N$_4$, and it was found that the sulfur existing in thiourea changes the traditional monomer condensation pathway and plays a crucial role in optimizing the structure. In particular, no signal of sulfur was detected in the X-ray Photoelectron Spectroscopy (XPS) spectrum (Figure 2), which suggests that the sulfur acts as a medium rather than a component of the final material. In addition to the types of precursor, the polymerization temperature also affects the formation process of g-C$_3$N$_4$. The same authors found that the condensation of thiourea to g-C$_3$N$_4$ is insufficient at 450 °C, but could be completed at 500 °C. When the temperature continues increasing to 550 °C and 600 °C, the structure is optimized. However, the g-C$_3$N$_4$ starts to thermally decompose once the temperature is raised to 650 °C. One of the advantages of using thiourea as precursor is that it can induce the formation of nanostructured g-C$_3$N$_4$, as the oxygen in the structure gradually escapes at high temperatures. This further results in the exposure of surface sites and the localization of light-induced electrons in the conjugated systems, thereby improving the photocatalytic performance [74].

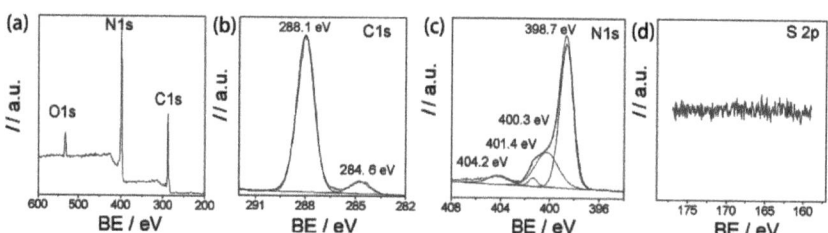

Figure 2. (a) XPS survey spectrum and the corresponding high-resolution spectra of (b) C 1s, (c) N 1s and (d) S 2p obtained from the CN-T$_{500}$ sample. Used with permission from [73]. Copyright 2012 Royal Society of Chemistry.

As mentioned above, urea serving as precursor can accelerate the production of large amounts of gases at high temperatures owing to the presence of oxygen in the structure, thereby improving the surface area of the product. For this reason, urea is often used as porogen in the preparation of g-C$_3$N$_4$ to increase the porosity, as well as the nitrogen content [75]. However, the excess addition of urea would produce a large number of fragments, which tend to agglomerate during the reaction, reducing the surface free energy and decreasing the photocatalytic activity. Therefore, the amount of urea added during synthesis is of great importance and worth being optimized.

For the improvement of surface area, Wu et al. [76] reported that the addition of NH$_4$Cl additives during the synthesis procedure is also highly efficient, as they can be decomposed into HCl and NH$_3$ gases during the heat-treatment process, promoting the delamination and depolymerization of g-C$_3$N$_4$, and thus, improving the surface area. Moreover, the presence of NH$_4$Cl can lower the temperature of g-C$_3$N$_4$ formation to 400 °C and introduce numerous surface amino groups, which are beneficial to, for example, the photocatalytic H$_2$ evolution reaction, with a reaction rate twice that of bulk g-C$_3$N$_4$. Similar cooperative effects are also observed for other multi-component systems, e.g., urea-mixed imidazole [77], or melamine and urea mixed with thiourea [78].

Pretreatment of the precursor is also an effective way to improve the surface area of g-C$_3$N$_4$. Sun et al. [79] prepared protonated g-C$_3$N$_4$ using HCl-treated melamine as a precursor and compared the effects of treatment time on the properties of the material. They found that the reaction of melamine with HCl changes the crystal structure and vibration bands of g-C$_3$N$_4$. Compared with g-C$_3$N$_4$ originating from untreated melamine, the material obtained from HCl-treated melamine exhibits smaller grain size and a bigger surface area. Powder X-ray diffraction (XRD) patterns shows that pretreatment with acid changes the structure of melamine, and shortens the formation process to 1 h (Figure 3a). The structure of the samples after treatment is similar, except for a slight shift in peak position due to the formation of nanosheets in the samples, which facilitates the strengthening of stacking between layers and reduction in the spacing distance [80]. Indeed, scanning electron microscope (SEM) images show that g-C$_3$N$_4$ obtained from the untreated melamine exhibited a particle size of 7.5 μm, which is larger than that obtained from HCl-treated melamine (Figure 3b). This verifies that acid-treated melamine prevents the thermal condensation of melamine into large-sized g-C$_3$N$_4$, by releasing HCl and NH$_3$ gases in the heating process. Similar results are also reported for nitric acid-treated melamine [81] and sulfuric acid-treated melamine [82]. These results suggest that acid treatment of the precursor is beneficial to improve the surface area of g-C$_3$N$_4$, by generating cracks during the heating process. Moreover, the samples obtained from the acid-treated precursor possess the advantages of rich surface defects, excellent electron–hole separation efficiency and strong light absorption ability.

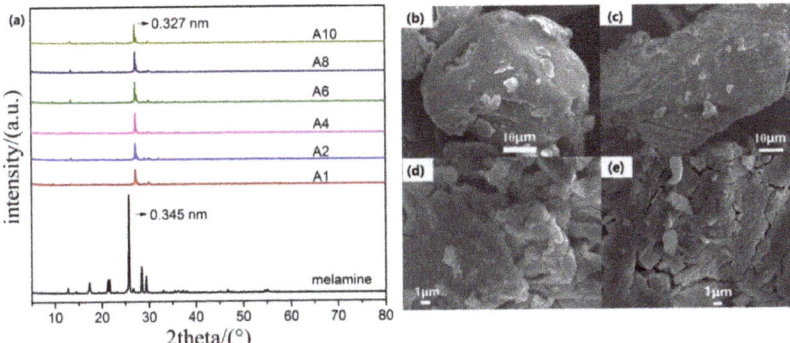

Figure 3. (**a**) XRD patterns of melamine and acid-treated melamine samples; SEM images of g-C_3N_4 prepared with (**b**) melamine and acid-treated melamine for (**c**) 2 h, (**d**) 6 h and (**e**) 8 h. Used with permission from [79]. Copyright 2019 IOP Publishing.

In addition to the acid pretreatment, the hydrothermal treatment of dicyandiamide also yields g-C_3N_4 with a high surface area and various surface morphologies, e.g., flower-like [83–85], hollow spheres [86–88], needle-like and rod-like [89–91], depending on the solvents. Such materials exhibit more attracting properties than the bulk one, for example: (1) the lamellar and porous structure is conducive to gas permeation; (2) the large surface area facilitates the reactant's adsorption; (3) the special morphology provides the benefit of widening the visible-light response range and improving the light absorption ability.

As well as the precursor and temperature, the reaction atmosphere is also crucial in affecting the properties of g-C_3N_4, through generating carbon and nitrogen vacancies, for example. Wang et al. [66] fabricated nanorod g-C_3N_4/metal oxide composites by heating a copper–melamine supramolecular framework, [Cu(μ-OAc)(μ-OCH$_3$)(MA)](Cu-MA1), under an argon atmosphere, which shows 94% Rhodamine B (RhB) conversion within 20 min under visible-light irradiation. Niu et al. [92] reported the generation of nitrogen vacancies by heating g-C_3N_4 in a hydrogen environment and foresaw the importance of self-modification and vacancies to completely modify the electronic structure of the layered g-C_3N_4 structure. Liang et al. [93] prepared porous g-C_3N_4 with abundant carbon vacancies by heating bulk g-C_3N_4 in a NH_3 atmosphere. The obtained material showed a surface area of 196 m^2/g, and exposed additional active edges, which significantly accelerated the transfer of photoinduced electron–hole pairs through a cross-plane diffusion pathway. The in-plane pores and wrinkled structures of g-C_3N_4 greatly enhance mass transfer and promote the dynamics of photoactivity. Xu et al. [94] reported that g-C_3N_4 prepared by pyrolyzing 3-amino-1,2,4-triazole in a CO_2 atmosphere shows excellent activity for hydrogen production, which was 2.4 and 1.7 times higher than that prepared in air and N_2 atmospheres, respectively. This could be because treatment in a CO_2 atmosphere causes a reduction in nitrogen vacancies (V_n) and the formation of NH_x groups on the surface of g-C_3N_4, generating hydrogen bond interactions between the layers, which facilitate the transfer of electrons from the heptazine ring to the g-C_3N_4 layer. Transient photocurrent response measurement confirms that the g-C_3N_4 prepared in a CO_2 atmosphere produces the largest current density under visible-light driving (Figure 4a), which implies improvement in the separation efficiency of electron–hole pairs. Additionally, electrochemical impedance spectroscopy (EIS) shows that this sample has a smaller EIS arc radius than the others (Figure 4b) which confirms again that the g-C_3N_4 prepared in a CO_2 atmosphere has lower charge transfer resistance and higher charge transfer efficiency.

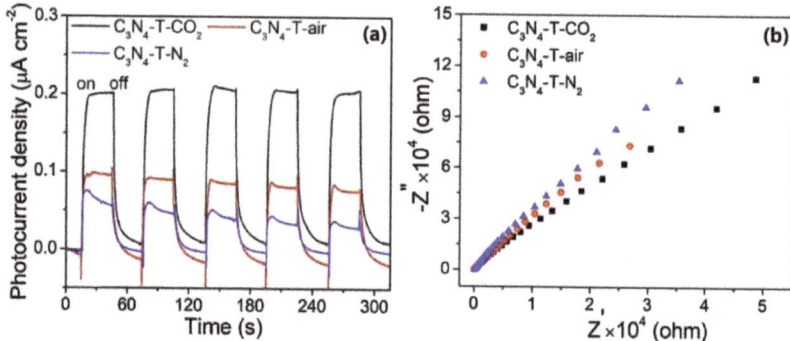

Figure 4. (a) Photocurrent responses and (b) EIS Nyquist plots of C_3N_4-T-Y under light irradiation. Used with permission from [94]. Copyright 2019 Elsevier.

Table 1 summarizes the band gap and surface area of g-C_3N_4 prepared under different reaction conditions, which shows that the selection of precursors and the proper control of reaction conditions are effective strategies to optimize the electronic structure and surface area of g-C_3N_4.

Table 1. Surface area and band gap of g-C_3N_4 synthesized under different preparation conditions.

Precursor	Reaction Conditions	Band Gap [eV]	Surface Area [m²/g]	Ref.
Cyanamide	550 °C, 4 h, N_2	2.62	10	[95]
Dicyandiamide	550 °C, 3 h, air	2.64	40.5	[38]
Melamine	550 °C, 3 h, air	2.66	28.2	[38]
Urea	550 °C, 3 h, air	2.72	67.1	[38]
Urea	550 °C, 2 h, air	2.76	58	[39]
Thiourea	550 °C, 2 h, air	2.58	18	[39]
3-amino-1, 2, 4-triazole	550 °C, 4 h, CO_2	2.05	7.2	[56]
Ammonium thiocyanate	550 °C, 2 h, NH_3	2.87	46	[55]
Guanidine hydrochlorides	550 °C, 3 h, air	2.70	16.08	[58]
Guanidine thiocyanate	550 °C, 2 h, N_2	2.74	8	[96]
Urea Melamine	520 °C, 4 h, air	2.47	39.06	[97]
Imidazole-mixed urea	550 °C, 4 h, air	2.26	105.28	[77]
Sulfur-mixed melamine	650 °C, 2 h, N_2	2.65	26	[98]
Melamine–cyanuric acid	550 °C, 10 h, air	2.72	142.8	[99]
H_2SO_4-treated melamine	600 °C, 4 h, Ar	2.69	15.6	[82]
HCl-treated melamine	550 °C, 2 h, air	2.66	24.7	[79]
HNO_3-treated melamine	550 °C, 2 h, air	2.65	59.3	[81]

2.2. Synthesis of Single-Metal Oxide/g-C_3N_4 Heterojunctions

It is known that a single semiconductor often encounters problems such as low quantum yields, a narrow light absorption spectrum and low e^-/h^+ separation efficiency in photocatalysis, due to the contradiction between light absorption capability and e^-/h^+ recombination rate. Thus, modifications such as heteroatom doping, morphological control and semiconductor combination are often adopted in order to fully exhibit the photocatalytic properties of semiconductors. Metal oxides are one class of semiconductor and their inorganic characteristics can largely compensate for the shortcomings of polymeric g-C_3N_4. For example, the good redox ability of metal oxides can compensate for that of g-C_3N_4 when conducting redox reactions. Therefore, it would be of great interest to combine metal oxides with g-C_3N_4 to form inorganic–polymeric heterojunctions, which produce synergistic effects not only in the band gaps, but also in redox and other properties.

The construction of heterojunctions requires unequal band levels between the semiconductors to create interface band arrangement, which can result in a built-in electric field to drive the opposite migration of photogenerated electrons and holes, improving e^-/h^+ separation efficiency. Type II and Z-scheme are two typical heterojunctions and their formation mechanisms are shown in Figure 5. The former requires two coupled semiconductors with an interleaved band structure. Thereby, the photo-generated electrons transfer from semiconductor 1 to semiconductor 2, and the holes transfer in the opposite direction (Figure 5a). The electrons accumulated on semiconductor 2 are used for a reduction reaction and the holes accumulated on semiconductor 1 are used for an oxidation reaction. This process can separate photo-generated electrons and holes in space, but sacrifices the redox capacity of the materials. Hence, both the oxidation potential and the reduction potential of the heterojunction are reduced compared to those of the semiconductor alone.

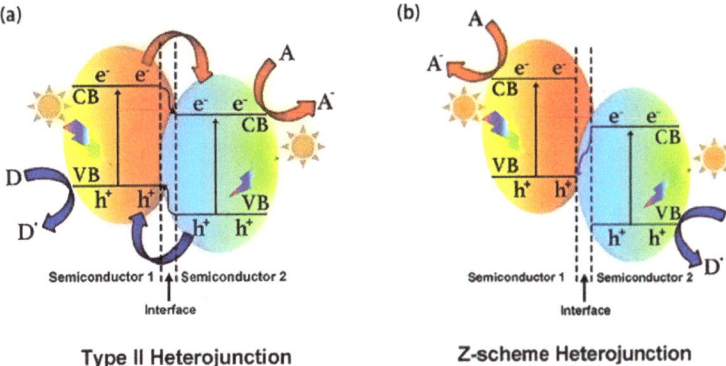

Figure 5. Band structure of (**a**) type II and (**b**) Z−scheme heterojunctions. "A" and "D" represent electron acceptor and electron donor, respectively. Used with permission from [100]. Copyright 2016 American Chemical Society.

A Z−scheme heterojunction requires the components to have staggered energy-band configurations, with the electron transfer in a zigzag mode. Typically, the photogenerated electrons at the CB of semiconductor 2 transfer and combine with the holes at the VB of semiconductor 1. The retained electrons at the CB of semiconductor 1 and holes at the VB of semiconductor 2 participate in the reduction and oxidation reactions, respectively (Figure 5b). This charge transfer mode enables the system to have not only improved charge-separation efficiency, but also stronger redox capability compared to that of the sole semiconductor.

Among the metal oxides, TiO_2 is well known for its first application to photocatalysis. Generally, it has three polymorphs in nature, including anatase, rutile and brookite [101]. Rutile TiO_2, with a band gap of 3.0 eV, has the most stable and compact structure, while anatase TiO_2, with a band gap of 3.2 eV, is better facilitates photocatalysis owing to its good e^-/h^+ separation efficiency and high adsorption capacity [102]. In the fabrication of TiO_2 and $g\text{-}C_3N_4$ heterojunctions, Fang et al. [103] prepared an anatase/rutile $TiO_2/g\text{-}C_3N_4$ (A/R/CN) multi-heterostructure using a facile thermoset hybrid method, finding that the combination of two type II heterostructures (i.e., A/R and R/CN) greatly improved the separation and transfer efficiency of e^-/h^+. As a result, the heterostructures showed activity that was eight and four times higher for the photocatalytic hydrolysis of hydrogen than $g\text{-}C_3N_4$ and TiO_2 (P25) alone, respectively. Similarly, other TiO_2 polymorphs, e.g., brookite TiO_2, can combine with $g\text{-}C_3N_4$ to prepare heterojunctions with improved photocatalytic activity [104]. Zhu et al. prepared $g\text{-}C_3N_4/TiO_2$ hybrids via a ball-milling method, finding that the composites possess a wider light-absorption range and higher photocatalytic activity than the respective component, with the activity for MB degradation being 3.0 and 1.3 times higher than that of $g\text{-}C_3N_4$ and TiO_2, respectively [105].

In addition to TiO_2, the combination of $g\text{-}C_3N_4$ with other metal oxides is also widely reported. Liu et al. reported that the coupling of $g\text{-}C_3N_4$ with ZnO prolongs the lifetime and separation efficiency of photogenerated e^-/h^+, therefore improving its photocatalytic activity for phenol degradation [106]. Moreover, the introduction of a silicate group to the ZnO/g-C_3N_4 composites further improves the lifetime and separation efficiency of e^-/h^+ pairs, and thereby, the photocatalytic activity. This indicates that the built-in silicate group in the composites acts as a bridge to link ZnO and g-C_3N_4, promoting the transfer and separation efficiency of e^-/h^+ pairs. Consequently, the electrons and holes have a longer lifetime to interact with the reactants and contribute to the reaction.

Guo et al. [107] coupled oxygen-deficient molybdenum oxide (MoO_3) nanoplates with g-C_3N_4 nanoplates using a one-step hydrothermal method, and found that MoO_3 particles grew well on the surface of g-C_3N_4 (Figure 6). MoO_3 is a chemically inert semiconductor with a large work function, which is suitable to couple with g-C_3N_4 and form a Z-scheme heterojunction. Moreover, the oxygen vacancies facilitate the promotion of plasmon resonance and expand the range of spectral absorption, and their concentrations can be adjusted via annealing in air. Combined with surface plasmon resonance and the synergistic effects of Z-scheme heterojunctions, it is expected that the composites will exhibit efficient performance for photocatalytic reactions, e.g., the H_2 evolution reaction.

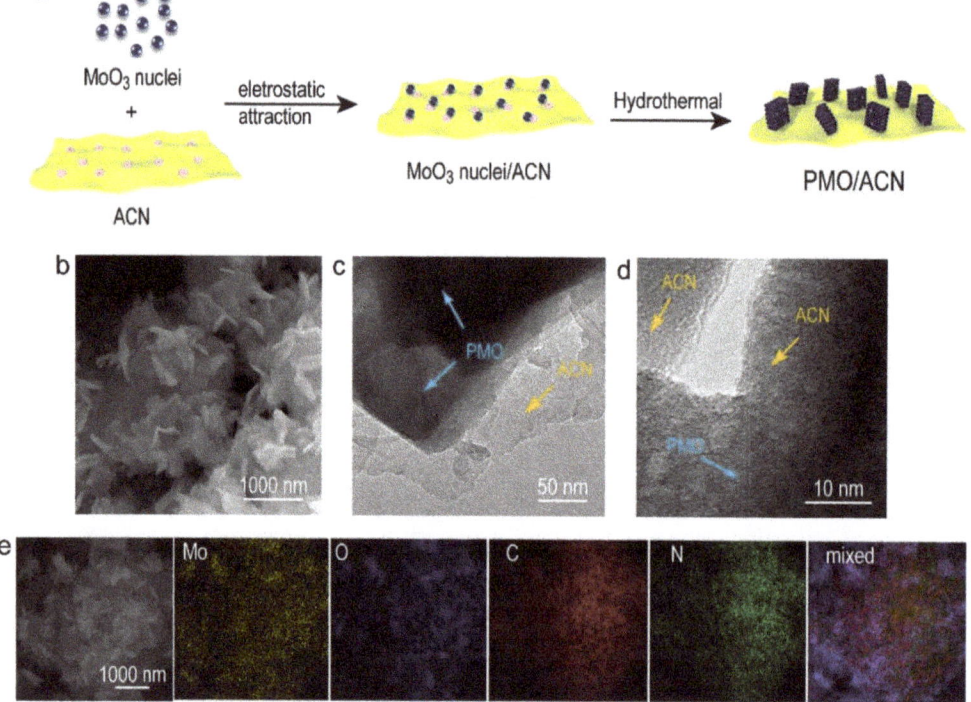

Figure 6. Synthesis of the PMO/ACN nanohybrid. (**a**) Schematic illustration for the synthesis of PMO/ACN, (**b**) SEM, (**c**) TEM, (**d**) HRTEM images of the PMO/ACN sample and (**e**) SEM elemental mapping for the PMO/ACN nanohybrid. Used with permission from [107]. Copyright 2020 Elsevier.

The morphology, structure and contacting patterns are also crucial factors affecting the electron transfer and photocatalytic activity of g-C_3N_4. Using seed-induced solvent heat treatment, 0D nanoparticles, 1D nanowires, 2D nanosheets and 3D mesoporous crystals can be loaded on the surface of g-C_3N_4. For example, 3D/2D MnO_2/g-C_3N_4 nanocomposites can be prepared via a calcination process using MnO_2 polyhedron and 2D g-C_3N_4 nanosheets as precursors [108], as shown in Figure 7. The 3D polyhedral morphology and multi-phase polycrystalline structure of MnO_2 are beneficial as they strengthen the interaction between MnO$_2$ and g-C$_3$N$_4$, owing to the presence of low-valence Mn species, graphitic N species and oxygen vacancy. Like the lamellar structure, materials with other structures can increase the interfacial area and surface area of the resulting composites. Liu et al. [109] reported that core-shell CeO@g-C$_3$N$_4$ exhibits high efficiency for the photocatalytic degradation of doxycycline, owing to the high surface area (82.37 m^2/g) and low e$^-$/h$^+$ recombination rate. They also found that shuttle-like CeO$_2$/g-C$_3$N$_4$ is efficient for the degradation of norfloxacin under visible light using persulfate as an oxidant, during which the norfloxacin is degraded into small molecules via gradual shedding of the functional groups.

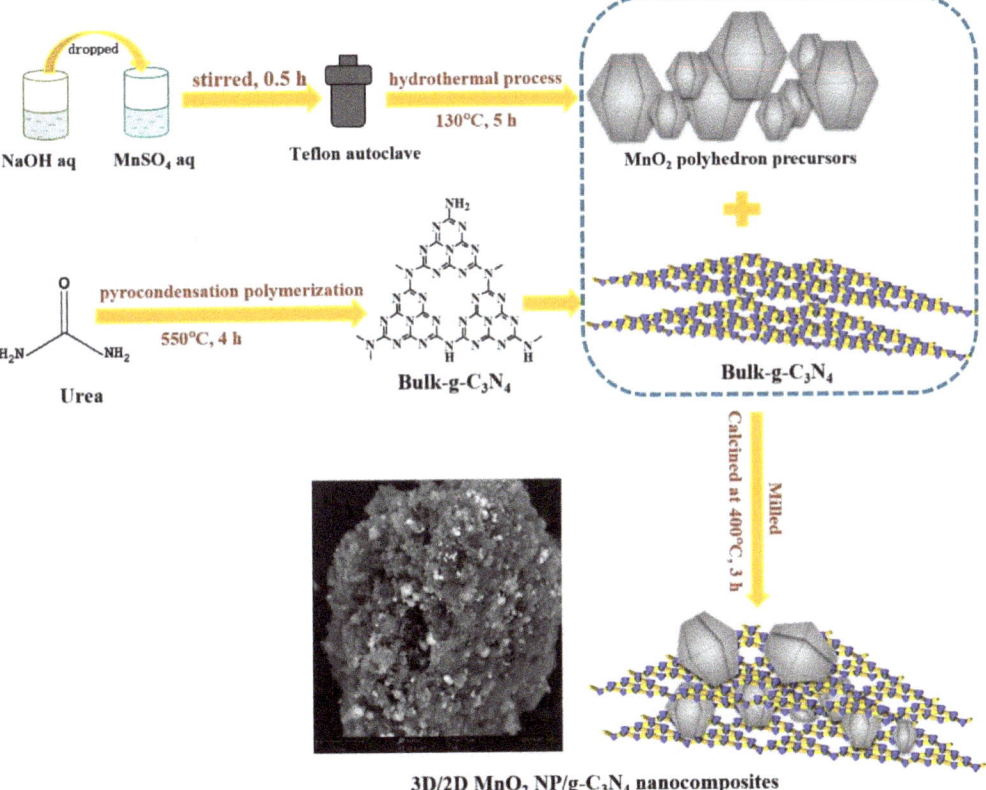

Figure 7. Synthesis procedures of 3D/2D MnO$_2$ NP/g-C$_3$N$_4$ nanocomposites. Used with permission from [108]. Copyright 2019 American Chemical Society.

The unique properties of metal oxides also give the composites special functions. Ye et al. [110] loaded magnetic Fe$_2$O$_3$ on g-C$_3$N$_4$ to introduce magnetization to the sample, which makes it easy to separate from the reaction liquid, and hence, reduces the cost of the recycling process. Mou et al. [111] used amorphous ZrO$_2$ as a cocatalyst of g-C$_3$N$_4$ for ammonia synthesis to improve its activity. The introduction of ZrO$_2$ not only restrains the hydrogen generation rate, but also improves the electron transfer rate and the e$^-$/h$^+$ separation efficiency. These results demonstrate that the combination with metal oxide is efficient in improving the photocatalytic performance of g-C$_3$N$_4$. Tables 2 and 3 summarize the recent advances in metal oxide/g-C$_3$N$_4$ heterojunctions in photocatalytic applications.

Table 2. Summary of recent advances in photocatalytic degradation using metal oxide/g-C_3N_4 composites.

Sample	Model Reaction	Reaction Activity (mol/g/min)	Refs.
CaO/g-C_3N_4	Degradation of MB	2.6×10^{-5}	[112]
SrO_2/g-C_3N_4	Degradation of RhB	4.5×10^{-7}	[113]
MnO_2/g-C_3N_4	Degradation of MO	9.4×10^{-7}	[108]
ZnO/g-C_3N_4	Degradation of MO	3.3×10^{-7}	[114]
MoO_3/g-C_3N_4	Degradation of MB	9.7×10^{-7}	[115]
AgO/g-C_3N_4	Degradation of RhB	5.2×10^{-6}	[116]
CdO/g-C_3N_4	Degradation of RhB	1.1×10^{-7}	[117]
In_2O_3/g-C_3N_4	Degradation of RhB	5.2×10^{-10}	[118]
SnO_2/g-C_3N_4	Degradation of MO	7.9×10^{-8}	[119]
TiO_2/g-C_3N_4	Degradation of RhB	9.3×10^{-9}	[120]
Bi_2O_3/g-C_3N_4	Degradation of Amido black 10B dye	3.3×10^{-7}	[121]
Nb_2O_5/g-C_3N_4	Degradation of tetracycline	5.3×10^{-7}	[122]
CeO_2/g-C_3N_4	Degradation of Norfloxacin	4.6×10^{-7}	[123]
ZnO/$NiFe_2O_4$/g-C_3N_4	Degradation of LVX	1.7×10^{-7}	[124]
TiO_2/ZnO/g-C_3N_4	Degradation of MB	4.9×10^{-7}	[125]
Fe_3O_4/BiOBr/g-C_3N_4	Degradation of TC	8.8×10^{-7}	[126]
ZnO/CuO/g-C_3N_4	Degradation of MB	6.1×10^{-6}	[127]
WO_3/Fe_3O_4/g-C_3N_4	Degradation of diazinon	6.5×10^{-7}	[128]
$Ni_3(VO_4)_2$/$ZnCr_2O_4$/g-C_3N_4	Degradation of p-CP	4.8×10^{-4}	[129]

MO: methyl orange; MB: methylene blue; RhB: rhodamine B; LVX: levofloxacin; TC: tetracycline; p-CP: p-chlorophenol.

Table 3. Summary of recent advances in photocatalytic hydrogen evolution using metal oxide/g-C_3N_4 composites.

Sample	Reaction Activity (mol/g/min)	Refs.
Al_2O_3/g-C_3N_4	Al_2O_3/g-C_3N_4: 8.7×10^{-6} g-C_3N_4: 3.5×10^{-6}	[130]
CoO/g-C_3N_4	CoO/g-C_3N_4: 8.4×10^{-7} CoO: 4.9×10^{-8} g-C_3N_4: 8.3×10^{-8}	[131]
NiO/g-C_3N_4	NiO/g-C_3N_4: 2.5×10^{-7} g-C_3N_4: 2.7×10^{-9}	[132]
Cu_2O/g-C_3N_4	Cu_2O/g-C_3N_4: 4.0×10^{-6} g-C_3N_4: 2.4×10^{-6}	[133]
MgO/g-C_3N_4	MgO/g-C_3N_4: 5.0×10^{-6} g-C_3N_4: 9.7×10^{-7}	[134]
FEO_X/G-C_3N_4	FEO_x/g-C_3N_4: 1.8×10^{-5} g-C_3N_4: 4.3×10^{-6}	[110]

2.3. Multiple-Metal Oxide/g-C_3N_4 Heterojunctions

The achievements in combining single-metal oxide with g-C_3N_4 have stimulated researchers to use multiple-metal oxides to upgrade the materials. The construction of multi-component composites can induce multi-step charge transfer and charge separation, and hence, better photocatalytic performance could be expected when compared to single ones. However, more attention should be paid to the matching of the energy-band potential of each component, so that the photo-generated electrons can transfer at the phase interface, reaching the goal of constructing heterojunctions.

Bajiri et al. constructed ternary and double Z-scheme CuO/ZnO/g-C$_3$N$_4$ heterojunctions using a solvothermal method [135], which consisted of g-C$_3$N$_4$ flakes decorated with small nanoparticles (<5 nm) (Figure 8). It is interesting to find that the gases released from the solute combustion process build up a porous structure in the material, similar to the function of porogens. The porous and sheet-like structure increases the capability of the material to absorb reactants on the surface, thereby improving the photodegradation efficiency. Indeed, the material exhibits activity of 98% (45 min) and 91% (6 h) for the degradation of MB and ammonia nitrogen, respectively, under visible-light irradiation.

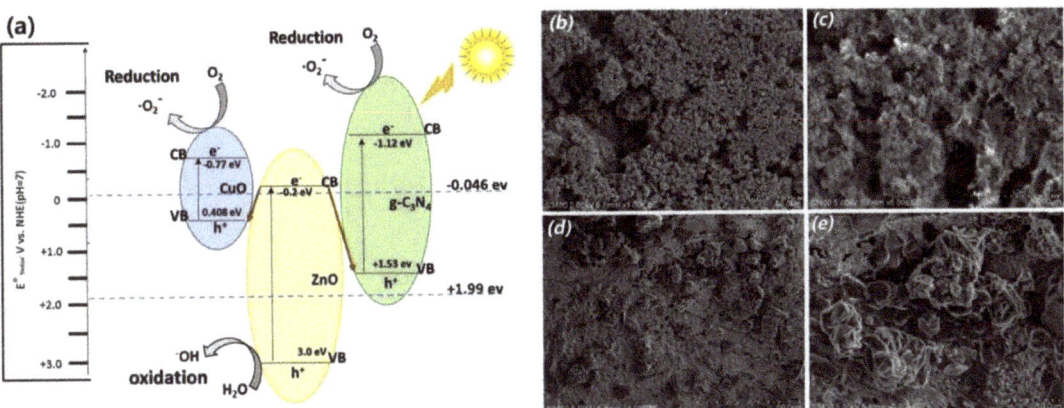

Figure 8. (**a**) Schematic diagram of the charge migration pathway in CuO/ZnO/g-C$_3$N$_4$; SEM images of (**b**,**c**) CuO/ZnO and (**d**,**e**) CuO/ZnO/g-C$_3$N$_4$ with different magnifications. Used with permission from [135]. Copyright 2019 Elsevier.

Jiang et al. [136] found that in addition to acting as photocatalyst, g-C$_3$N$_4$ can be an intermediate for charge transfer, by constructing a WO$_3$/g-C$_3$N$_4$/Bi$_2$O$_3$ (WCB) catalyst. Compared to the single or binary materials, ternary WCB exhibits moderate surface area and the highest photocatalytic activity. This indicates that the high surface area facilitated the reaction but was not the key factor determining the reaction. Optical characterizations from the UV-vis and PL spectra showed that the light absorption edge is red-shifted and the e$^-$/h$^+$ recombination rate is inhibited for WCB, when compared to the single or binary counterparts, due to the interactions between WO$_3$, g-C$_3$N$_4$ and Bi$_2$O$_3$ (Figure 9a). Consequently, the WCB exhibits enhanced optical properties and improved photocatalytic activity for tetracycline (TC) degradation under visible-light irradiation, with TC conversion of 80.2% at 60 min, which is much higher than that of g-C$_3$N$_4$ (22.1%), WO$_3$ (7.17%) and Bi$_2$O$_3$ (28.6%), and the binary CW (g-C$_3$N$_4$/WO$_3$), CB (g-C$_3$N$_4$/Bi$_2$O$_3$) and WB (WO$_3$/Bi$_2$O$_3$) (Figure 9b–f).

Yuan et al. [137] constructed ternary g-C$_3$N$_4$/CeO$_2$/ZnO composites with multiple heterogeneous interfaces. Binary g-C$_3$N$_4$/CeO$_2$ nanosheets were first prepared via pyrolysis and exfoliation. Thereafter, spherical ZnO nanoparticles were anchored on the g-C$_3$N$_4$/CeO$_2$ surface to form a ternary heterojunction structure. Because of the formation of the type II staggered belt arrangement between the components, the g-C$_3$N$_4$/CeO$_2$/ZnO shows efficient three-level transfer of electrons and holes, resulting in the effective separation of photo-excited carriers, as shown in Figure 10.

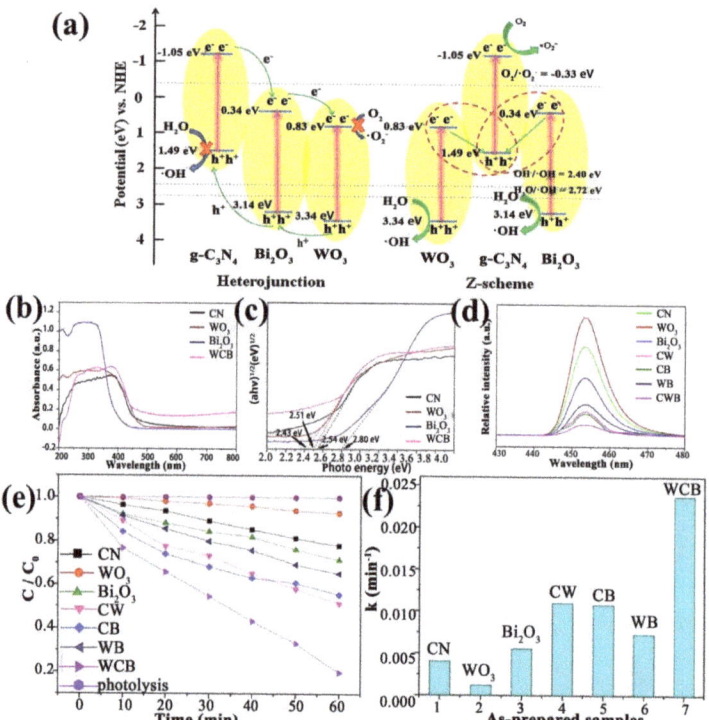

Figure 9. (**a**) Schematic diagram of charge separation in WO$_3$/g-C$_3$N$_4$/Bi$_2$O$_3$, (**b**) UV-vis spectra, (**c**) band gap, (**d**) photoluminescence spectra, (**e**) photocatalytic activities for TC degradation under visible-light and (**f**) apparent rate constants for TC degradation obtained from the various samples. Used with permission from [136]. Copyright 2018 Elsevier.

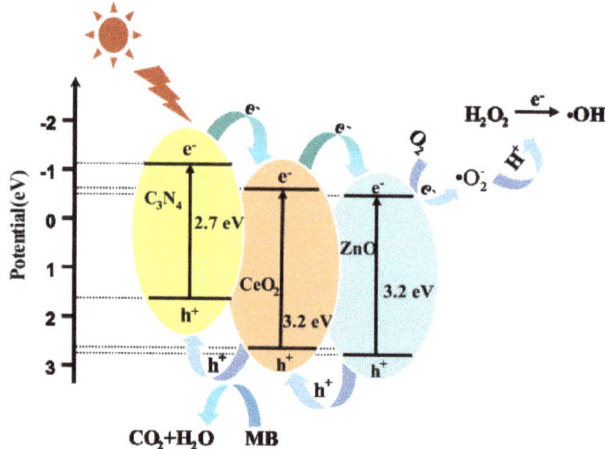

Figure 10. Schematic diagram of the photoexcited e$^-$/h$^+$ separation process in the g-C$_3$N$_4$/CeO$_2$/ZnO composite under visible-light irradiation. Used with permission from [137]. Copyright 2017 Elsevier.

Morphology control is also effective in improving the photocatalytic performance of materials, by enhancing the interactions and enlarging the contact areas of the heterogeneous interfaces, which are beneficial to electron transfer and separation (from the holes). For example, Jiang et al. [138] fabricated g-C_3N_4, TiO_2 and ZnO nanoflakes, and then, assembled them to form g-C_3N_4/TiO_2/ZnO Z-scheme heterojunctions. A high-resolution TEM image shows that the TiO_2 (101) plane and ZnO (002) plane are stacked on the g-C_3N_4 surface to form heterojunctions (Figure 11a–f). The similar morphologies of g-C_3N_4/TiO_2/ZnO and g-C_3N_4 indicate that TiO_2 and ZnO are uniformly dispersed on the g-C_3N_4 surface. The 2D/2D nanosheet/nanosheet structure not only increases the surface area of the material (g-C_3N_4: 8.18 m^2/g, g-C_3N_4/TiO_2/ZnO: 27.21 m^2/g), but also improves the e^-/h^+ separation efficiency by facilitating electron transfer through the abundant interfaces (Figure 11g–i).

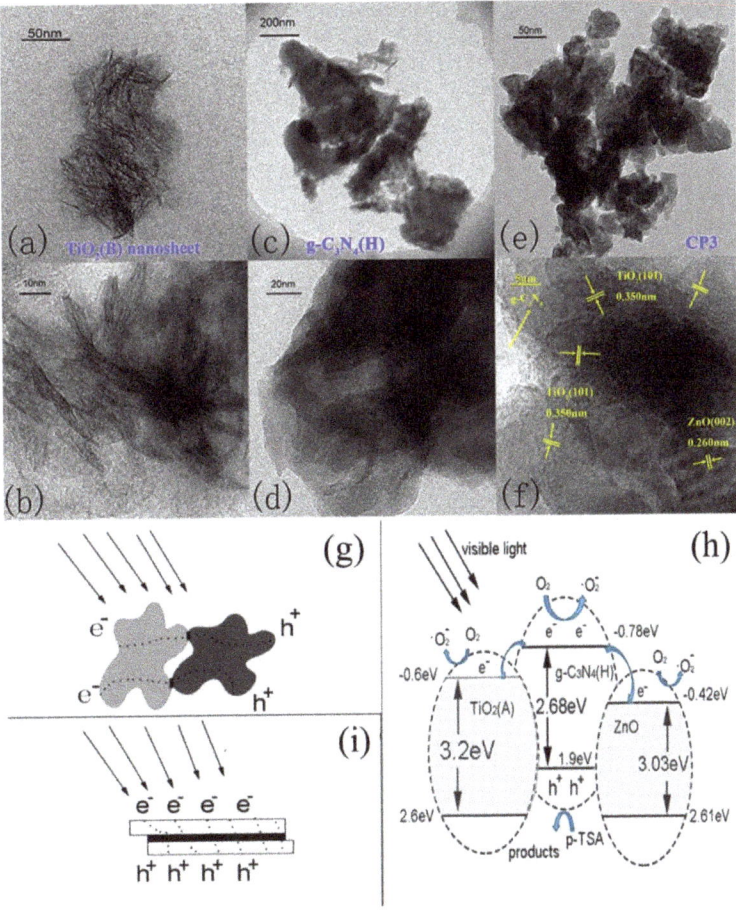

Figure 11. TEM images of (**a**,**b**) TiO_2 nanosheets, (**c**,**d**) g-C_3N_4 nanosheets and (**e**,**f**) g-C_3N_4/TiO_2/ZnO nanocomposites; (**g**,**i**) comparison of the electron transfer routes between the granule/granule and the nanosheet/nanosheet composites with different heterojunction areas; (**h**) the mechanism of facet-coupled ternary nanocomposites for p-TSA degradation under visible light. Used with permission from [138]. Copyright 2017 Elsevier.

The formation mechanism of the composites has been explored to reveal how the heterogeneous interfaces affect the electron transfer process. Liu et al. [139] proposed a lattice-matching assumption of amorphous materials in the structural hybridization process and clarified a coordination effect in the unoccupied d orbitals of N atoms of g-C_3N_4. Because of the different crystal structures and lattice parameters of metal oxides (e.g., ZnO) and g-C_3N_4, lattice matching between them is difficult. Amorphous materials (e.g., Al_2O_3) have disordered atomic distribution and unfixed lattice parameters; hence, they can easily accept the charge of g-C_3N_4. Therefore, amorphous Al_2O_3 can be an intermediary to improving electron transfer efficiency between g-C_3N_4 and ZnO [140]. As shown in Figure 12a, the lattice fringes of ZnO and Al_2O_3 are entangled with that of g-C_3N_4, which proves that the two components are in close contact. The tight contact interface provides a step to transfer the induced carrier (Figure 12b). XPS spectra show that the binding energies of Al atoms in g-C_3N_4/Al_2O_3 and g-C_3N_4/Al_2O_3/ZnO shifted to a higher position compared to that of the original Al_2O_3 (Figure 12c). This indicates that a chemical force between Al and g-C_3N_4 is formed, due to the coordination of the unoccupied 3p or 3d orbital of Al ions with the lone electron pair of N atoms of g-C_3N_4, as verified by the shift in the binding energy of N 1s (Figure 12d–f). These results provide ideas to correct the lattice mismatch between g-C_3N_4 and metal oxides, and promote the application of amorphous materials to fabricate heterojunctions.

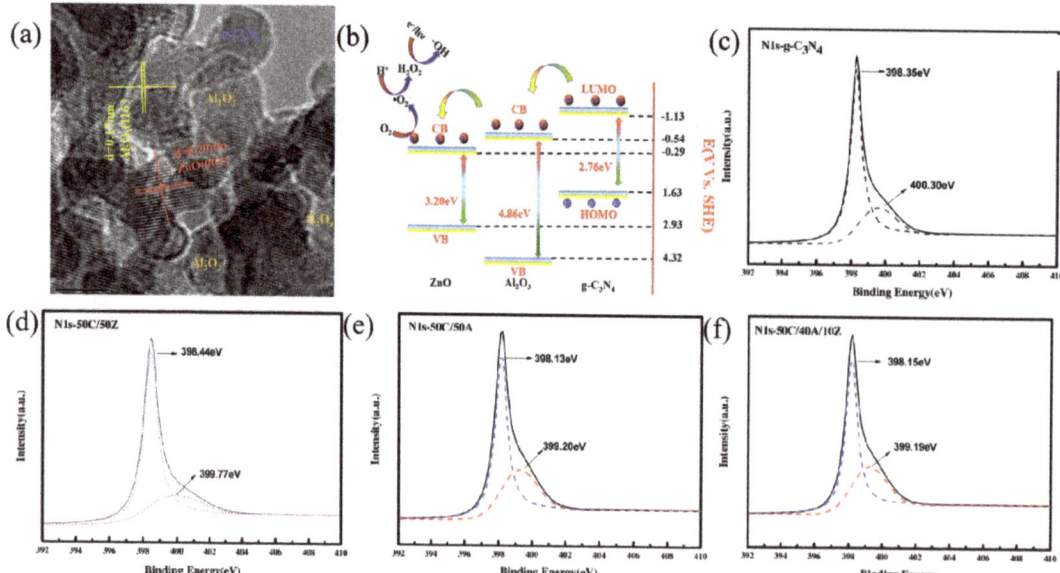

Figure 12. (a) HRTEM image of ternary g-C_3N_4/Al_2O_3/ZnO heterojunctions. (b) Cascade of electron transfer in ternary g-C_3N_4/Al_2O_3/ZnO heterojunctions under visible-light irradiation; high–resolution XPS spectra of (c) Al2p. (d–f) N1s of g-C_3N_4, 50C/50Z, 50C/50A and 50C/40A/10Z. Used with permission from [139]. Copyright 2017 Elsevier.

Fe_3O_4 is an attractive material in the synthesis of multi-component heterojunctions, owing to its good photocatalytic and especially magnetic properties, which promote not only the reaction activity but also the separation efficiency of catalysts from liquid solutions. Adil Raza et al. [141] prepared a Fe_3O_4/TiO_2/g-C_3N_4 composite using a hydrothermal method, finding that anatase TiO_2 and magnetic Fe_3O_4 can enter the g-C_3N_4 frame if treated at 200 °C. The composites show efficient activity for RhB and MO degradation under visible-light irradiation, with degradation conversions of 96.4% and 90%, respectively, which are 3.73 and 2.74 times higher than that of g-C_3N_4. Amir Mirzaei [142] prepared petal-like

Fe$_3$O$_4$-ZnO@g-C$_3$N$_4$ composites using an in situ growth method, finding that the hydrolysis of urea (precursor) produces stable and continuous OH$^-$ ions, which can react with zinc ions and control the growth of nuclei. The coating of g-C$_3$N$_4$ corrodes the surface of Fe$_3$O$_4$-ZnO and creates pores in the structure, benefiting electron transfer, while the presence of Fe$_3$O$_4$ not only reduces the e$^-$/h$^+$ recombination rate by accepting useless electrons, but also improves the separation efficiency of catalysts from solution (via a magnet) owing to its magnetic properties (Figure 13a). Moreover, the composite is stable and no leaching of Zn^{2+} and Fe^{2+} ions is observed in the reaction of photocatalytic SMX degradation (Figure 13b,c). This suggests that g-C$_3$N$_4$ acts not only as a semiconductor contributing to the photocatalytic reaction, but also a protective layer against photo-corrosion of the Fe-ZnO surface. Similar phenomena are observed for other composites, e.g., Ag$_2$O/g-C$_3$N$_4$/Fe$_3$O$_4$ [53], ZnO/Fe$_3$O$_4$/g-C$_3$N$_4$ [143] and α-Fe$_2$O$_3$/g-C$_3$N$_4$/ZnO [144].

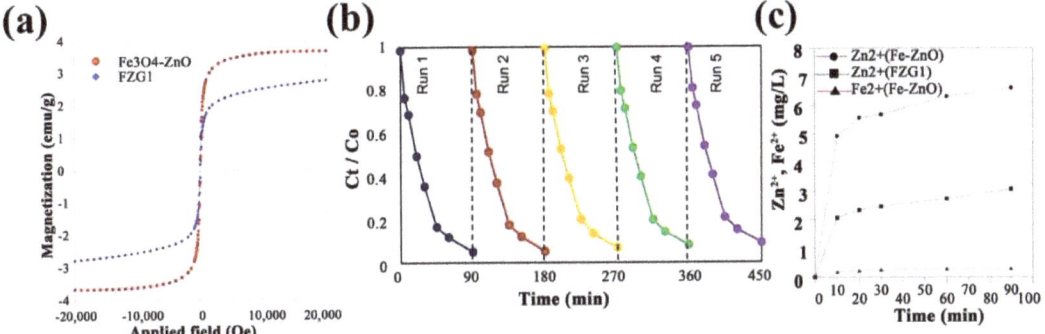

Figure 13. (a) Magnetization curves of Fe$_3$O$_4$-ZnO and FZG1; (b) release of zinc and iron ions into the solution as a function of time for Fe-ZnO and FZG1; (c) recyclability of FZG1 for photocatalytic degradation of SMX. Used with permission from [142]. Copyright 2018 Elsevier.

In addition to simple metal oxide, compound oxides are also interesting materials in catalysis, and they usually exhibit different electronic and chemical properties relative to their parent materials. Many compound metal oxides have been used to couple with g-C$_3$N$_4$ and form heterojunctions. Among them, perovskite oxides with an ABO$_3$ structure attract much attention owing to their unique physical and chemical properties, such as variable ion valences, controllable oxygen vacancies, adjustable redox properties and the ability to accommodate foreign ions [145–147]. The typical perovskite oxide, CaTiO$_3$ (CT), has a band gap of ~3.5 eV, which means that its photocatalytic activity is limited to ultraviolet excitation. However, when it is coupled with narrow-band-gap semiconductors such as g-C$_3$N$_4$ to form binary heterojunctions, the large band gap of CT can efficiently enhance the photocatalytic activity of g-C$_3$N$_4$ under visible light by promoting the charge-separation efficiency. Kumar et al. [148] found that the combination of 2D CT nanosheets with g-C$_3$N$_4$ flakes, to form 2D/2D composite nanoflake (CT/CN), greatly increased the BET surface area to 50.7 m^2/g, which is larger than that of CT nanosheets (29.3 m^2/g) and g-C$_3$N$_4$ flakes (41.0 m^2/g). Hence, more active sites can be exposed on the surface, shortening the bulk diffusion length and reducing the e$^-$/h$^+$ recombination rate. Ye et al. [149] reported that the fabrication of 1D CoTiO$_3$ rod–2D g-C$_3$N$_4$ flake Z-scheme heterojunctions (CT-U) improves not only e$^-$/h$^+$ separation efficiency, but also redox ability. SEM images show that the CoTiO$_3$ rods are fully wrapped with g-C$_3$N$_4$, forming heterogeneous interfaces that are beneficial to electron transfer (Figure 14a–d). As a result, CT-U exhibited a reaction rate of 858 μmol/h/g for the hydrogen evolution reaction, which is about two times higher than that obtained from g-C$_3$N$_4$.

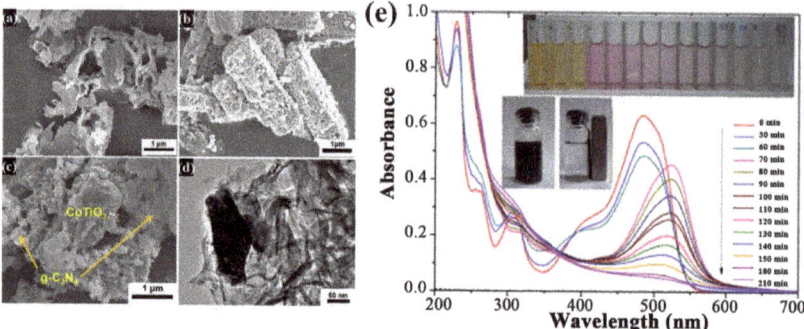

Figure 14. SEM images of (**a**) g-C$_3$N$_4$, (**b**) CoTiO$_3$, (**c**) CT-U and (**d**) 0.15% CT-U. Used with permission from [149]. Copyright 2016 American Chemical Society; (**e**) the photodegradation of Orange II by CuFe$_2$O$_4$@C$_3$N$_4$/H$_2$O$_2$/Vis system (inset: the solution before and after magnetic separation using an external magnet). Used with permission from [150]. Copyright 2015 Elsevier.

The effects of surface morphology on photocatalytic activity are documented by Zhang et al. [151], who used KNbO$_3$ as a model catalyst and found that its efficiency for the photocatalytic conversion of methanol to hydrogen depends intimately on the morphology, with an order of cubic > orthogonal > tetragonal. On this basis, a cubic KNbO$_3$/g-C$_3$N$_4$ composite was synthesized and it exhibited excellent activity for photocatalytic hydrogen production, owing to the close contact between KNbO$_3$ cubes and g-C$_3$N$_4$ nanosheets, which forms active heterojunction interfaces and effectively inhibits the e$^-$/h$^+$ recombination rate in the system [152]. With the same principle, many ABO$_3$/g-C$_3$N$_4$ composites are prepared and reported in the literature, such as LaFeO$_3$/g-C$_3$N$_4$ [153], g-C$_3$N$_4$/SrTiO$_3$ [154] and LaMnO$_3$/g-C$_3$N$_4$ [155].

As well as perovskite oxides, spinel oxides with an AB$_2$O$_4$ structure are also promising materials in catalysis [156,157]. Compared to ABO$_3$ perovskites, AB$_2$O$_4$ spinels have a narrower band gap and stronger responses to visible light. Moreover, the AB$_2$O$_4$ spinels can accommodate transitional metals at both the A- and B-sites; thus, the metals at both the A- and B-sites can contribute to the reactions. For example, Chang et al. [158] reported that Z-scheme NiCo$_2$O$_4$/g-C$_3$N$_4$ heterojunctions exhibit not only a larger surface area (141.7 m^2/g) than g-C$_3$N$_4$ (89.2 m^2/g) and NiCo$_2$O$_4$ (98.8 m^2/g), but also higher photoactivity for water splitting than Co$_3$O$_4$/g-C$_3$N$_4$ and NiO/g-C$_3$N$_4$, owing to their abundant active sites and good photoelectric properties.

Some spinel oxides (e.g., CuFe$_2$O$_4$) also have magnetic properties, exhibiting the advantages of easy separation. In the research of Yao et al. [150], they constructed a type II CuFe$_2$O$_4$@g-C$_3$N$_4$ heterojunction, in which the CuFe$_2$O$_4$ and g-C$_3$N$_4$ are intertwined to form a three-dimensional hybrid structure that is beneficial to electron transfer. As a result, the material shows improved e$^-$/h$^+$ separation efficiency and photocatalytic activity compared to the respective g-C$_3$N$_4$, CuFe$_2$O$_4$ and g-C$_3$N$_4$/CuFe$_2$O$_4$ mixtures. Moreover, the material exhibits good easy-to-separate magnetism owing to its magnetic properties (Figure 14e), and thus, can be well recycled in the reaction.

3. Applications of g-C$_3$N$_4$-Based Photocatalysts

Its promising optical and physicochemical properties enable g-C$_3$N$_4$, utilizing sunlight, to solve the problems of environmental pollution and energy crises, while avoiding secondary pollution. In the following, we briefly introduce the application of g-C$_3$N$_4$-based materials in photocatalysis [159–162], including water splitting to generate H$_2$ and O$_2$, the degradation of pollutants, CO$_2$ reduction and bacterial disinfection.

3.1. Photocatalytic Water Splitting for H_2

Because of the decreasing storage of fossil fuels and their negative impacts on the environment (releasing CO_2 for example), the use of green and renewable hydrogen fuels attracts much attention from scientists. The photocatalytic splitting of water is an ideal way to generate hydrogen and has become a hot topic in recent years. Figure 15 presents a simplified diagram of splitting water into hydrogen and oxygen over g-C_3N_4 under light irradiation. First, g-C_3N_4 is excited by photons to generate electrons, which then jump to the CB, leaving holes at the VB. The photogenerated e^- and h^+ flow to the surface of g-C_3N_4, reducing and oxidizing the adsorbed water to hydrogen and oxygen, respectively. However, the generated e^-/h^+ will rapidly recombine each other due to the Coulombic attraction, losing activity. The improvement in the separation efficiency of the photogenerated e^-/h^+ pairs, thus, is a challenging topic in the field of g-C_3N_4 photocatalysis.

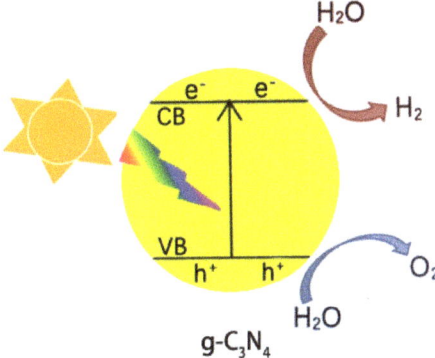

Figure 15. Scheme of photocatalytic water splitting into H_2 and O_2 over g-C_3N_4 under light irradiation.

To achieve this, the coupling of g-C_3N_4 with metal oxide is a solution, which can separate e^-/h^+ pairs in space by forming an opposite flow of e^- and h^+ (for type II heterojunctions), or by inducing the recombination of unused e^- and h^+ (for Z-Scheme heterojunctions), as reported in the literature [163,164]. Shi et al. [165] reported the in situ synthesis of MoO_3/g-C_3N_4, via co-pyrolysis of MoS_2 and melamine, for photocatalytic water splitting to hydrogen, finding that the activity of g-C_3N_4 was significantly enhanced with the increase in MoO_3 content. It is possible that the use of layered MoS_2 as a precursor not only improves the dispersion of MoO_3 on g-C_3N_4, but also enhances the interactions between them. Li et al. [166] synthesized $W_{18}O_{49}$/g-C_3N_4 composites by roasting a g-C_3N_4-impregnated ammonium tungstate solution. The loading of $W_{18}O_{49}$ greatly improves the surface area (by about five times) and exhibits excellent activity for a photocatalytic hydrogen evolution reaction, with a reaction rate of 912.3 $\mu mol \cdot g^{-1} \cdot h^{-1}$, which is 9.7 times higher than that of g-C_3N_4.

The coupling of g-C_3N_4 with two metal oxides could be more interesting when compared to that with single-metal oxide, as multiple heterojunctions can be established, exhibiting rich optical properties, and hence, better photocatalytic activities. This is observed in many studies [167–169]. For example, Wang et al. [170] found that Fe_2O_3@MnO_2 core-shell g-C_3N_4 ternary composites can form double heterojunctions, which provide abundant channels for electrons transfer, exhibit enhanced optical properties and allow the two half-reactions (the production of hydrogen and oxygen) to occur on the opposite surfaces of the semiconductor (Figure 16a–c); this results in improved activity for both hydrogen and oxygen production, with an optimal reaction rate of 124 $\mu mol \cdot h^{-1}$ and 60 $\mu mol \cdot h^{-1}$, respectively (Figure 16d).

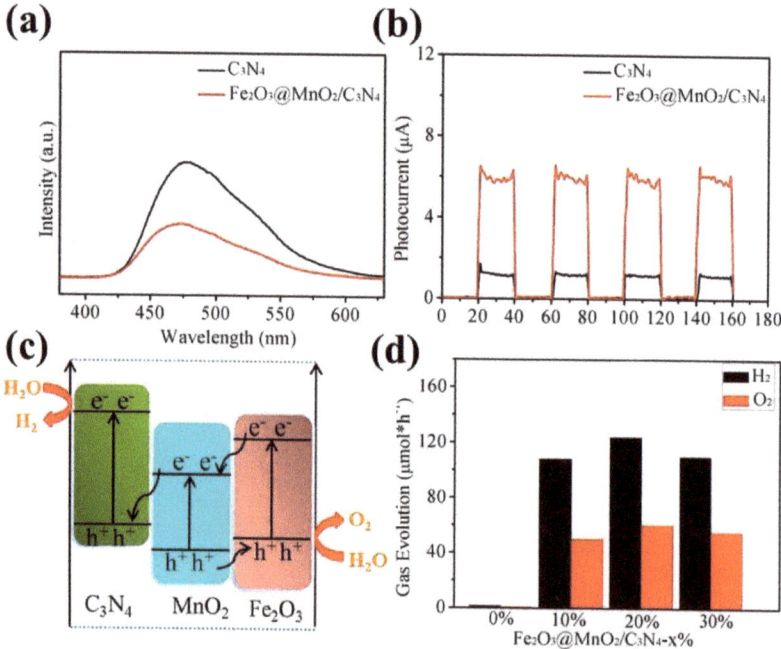

Figure 16. (**a**) PL spectra and (**b**) photocurrent response of C_3N_4 and $Fe_2O_3@MnO_2/C_3N_4$ samples; (**c**) schematic diagram of electron transfer; and (**d**) activity for splitting of water into H_2 and O_2 for the $Fe_2O_3@MnO_2/C_3N_4$ photocatalyst. Used with permission from [170]. Copyright 2020 Elsevier.

3.2. Photocatalytic Reduction of CO_2 to Renewable Hydrocarbon Fuels

With increasing global warming, it is critical to find effective ways to deal with greenhouse gases. Carbon dioxide (CO_2) is not only a typical greenhouse gas but also a valuable C1 resource. Hence, utilizing solar energy to reduce CO_2 into higher-value chemicals shows great advantages in solving the problems of both global warming and energy crises. In the past few years, g-C_3N_4 has been employed as a photocatalyst for CO_2 reduction owing to its high CB potential, which can activate CO_2 by donating electrons to the unoccupied orbits of CO_2. The photocatalytic CO_2 reduction involves a proton-assisted multi-electron process, as shown in Equations (1)–(5) below [171]. From the viewpoint of thermodynamics, CO_2 is gradually reduced to HCOOH, CO, HCHO, CH_3OH and CH_4 by receiving multiple (2, 2, 4, 6 and 8) electrons and protons, accompanying the increase in reduction potential. This means that the photocatalyst used to reduce CO_2 should have strong redox capability in order to supply sufficient driving force for the reaction.

$$CO_2 + 2H^+ + 2e^- \rightarrow HCOOH \qquad E^0_{redox} = -0.61 \text{V (vs. NHE at pH 7)} \tag{1}$$

$$CO_2 + 2H^+ + 2e^- \rightarrow CO + H_2O \qquad E^0_{redox} = -0.53 \text{V (vs. NHE at pH 7)} \tag{2}$$

$$CO_2 + 4H^+ + 4e^- \rightarrow HCHO + H_2O \qquad E^0_{redox} = -0.48 \text{V (vs. NHE at pH 7)} \tag{3}$$

$$CO_2 + 6H^+ + 6e^- \rightarrow CH_3OH + H_2O \qquad E^0_{redox} = -0.38 \text{V (vs. NHE at pH 7)} \tag{4}$$

$$CO_2 + 8H^+ + 8e^- \rightarrow CH_4 + 2H_2O \qquad E^0_{redox} = -0.24 \text{V (vs. NHE at pH 7)} \tag{5}$$

ZnO can absorb CO_2 and has a CB potential (E_{CB}) of −0.44 eV, which is more negative than the reduction potential of CO_2. Therefore, the combination of ZnO and g-C_3N_4 would benefit the CO_2 reduction reaction. Indeed, it is found that although the deposition of ZnO has negligible effects on the light absorption capacity and surface area of g-C_3N_4, the ZnO/g-C_3N_4 composite shows better photocatalytic activity for CO_2 reduction than individual ZnO and g-C_3N_4, due to the formation of heterojunctions that facilitate the separation of e^-/h^+ pairs [172]. The CO_2 conversion rate obtained from ZnO/g-C_3N_4 reaches 45.6 μmol/g/h, which is 4.9 times and 6.4 times higher than that obtained from g-C_3N_4 and P25, respectively. Additionally, based on the fact that the zeta potential of ZnO is positive and that of g-C_3N_4 is negative, Nie et al. [173] constructed a ZnO/g-C_3N_4 composite using an electrostatic self-assembly method, as shown in Figure 17a,b. The combination of them induces synergistic effects that are conducive to photocatalytic reactions, in which the ZnO microsphere prevents falling g-C_3N_4 nano flakes from gathering, and the g-C_3N_4 improves light utilization efficiency through the multi-scattering effect (Figure 17c).

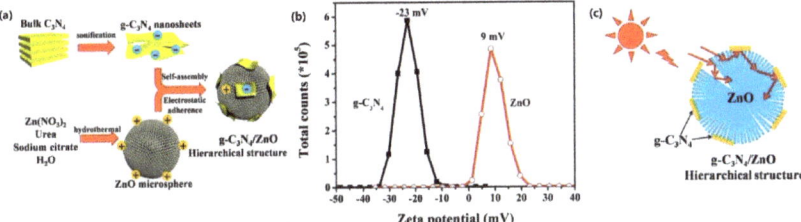

Figure 17. (**a**) Schematic diagram of synthesizing g-C_3N_4/ZnO microspheres; (**b**) zeta potential of ZnO and g-C_3N_4 (pH = 7); (**c**) illustration of enhanced reflections within the g-C_3N_4/ZnO photocatalyst. Used with permission from [173]. Copyright 2018 Elsevier.

In addition to ZnO, many other metal oxides can couple with g-C_3N_4 and contribute to the CO_2 reduction reaction. For example, Bhosale et al. [174] employed a wet chemical method to couple $FeWO_4$ with g-C_3N_4, forming a Z-scheme g-C_3N_4/$FeWO_4$ photocatalyst; it showed good activity for the reduction of CO_2 to CO without any medium, with a CO production rate of 6 μmol/g/h, which is 6 and 15 times higher than that of individual g-C_3N_4 and $FeWO_4$.

3.3. Photocatalytic Degradation of Pollutants

With the rapid development of the economy, various toxic pollutants emitted from industrial plants have been discharged to the environment and have seriously destroyed the ecological system. The removal of pollutants and the remediation of the environment have thus become essential topics and have attracted broad attention in recent years. Photocatalysis is a prospective technology for pollutant removal, and is able to mineralize organic pollutants into CO_2 and H_2O by producing oxidizing intermediates (such as •O_2^-, •OH and h^+). Depending on the properties of the pollutants, three reaction types can be classified: (1) the removal of organic pollutants in aqueous solution, such as dye [166,175] and antibiotic degradation [176]; (2) the removal of heavy-metal cations in aqueous solution, such as the reduction of chromium (VI) [177]; and (3) the removal of organic or inorganic pollutants in gas phase, such as the degradation of ortho-dichlorobenzene [178], acetaldehyde [179] and nitric oxide [180].

The Fenton advanced oxidation process (with an Fe^{2+} and H_2O_2 system) is a traditional technology used to treat industrial wastewater, but it is limited to a narrow pH range (<3) and causes secondary pollution due to the production of iron sludge. For this reason, it is proposed that a photocatalyst should be used instead of Fe^{2+}, to activate H_2O_2 into •OH radicals under light irradiation conditions, which can be achieved in a wide pH range without producing secondary pollutants. Hence, it is a green route to removing organic pollutants in aqueous solution and has good prospects for industrial use.

In this respect, Xu et al. [181] recently reported that the LFO@CN photocatalyst is highly efficient for the oxidative degradation of RhB with H_2O_2 under visible-light irradiation, with 98% conversion obtained within 25 min, and the material can be recycled for four cycles with no appreciable deactivation. Moreover, when applying a ternary $LaFe_{0.5}Co_{0.5}O_3/Ag/g\text{-}C_3N_4$ heterojunction that consists of a redox part $LaFe_{0.5}Co_{0.5}O_3$ (LFCO), photo part $g\text{-}C_3N_4$ and plasmonic part (Ag), for the degradation of tetracycline hydrochloride (TC), in the presence of H_2O_2 and light irradiation, the system exhibits good activity due to a photo-Fenton effect induced in the reaction, as shown in Figure 18 [182]. In this system, H_2O_2 is first activated into •OH radicals and OH^- anions over the LFCO, and the OH^- anions subsequently react with holes (h^+) produced at the VB band of LFCO to form more •OH radicals. Hence, H_2O_2 can be fully utilized to oxidize TC in the reaction. Meanwhile, the O_2 dissolved in the solution can react with the electrons (e^-) generated at the CB band of $g\text{-}C_3N_4$ and form •O_2^-, which is also a strong oxidant that is able to oxidize TC into CO_2 and H_2O. These results support that $g\text{-}C_3N_4$-based catalysts have good chemical stability and can be an effective substitute for Fenton catalysts in environmental purification.

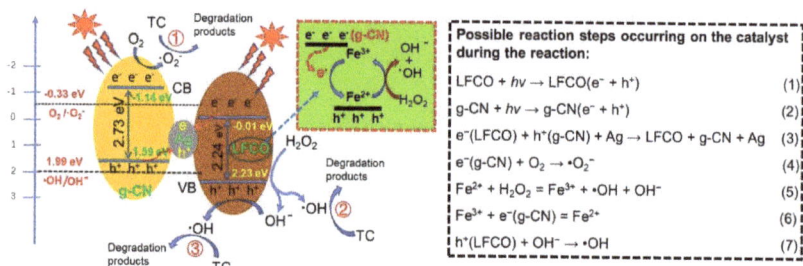

Figure 18. Mechanism of photo—Fenton degradation of tetracycline hydrochloride over the ternary LFCO/Ag/g-CN heterojunctions under visible-light irradiation. Used with permission from [182]. Copyright 2022 John Wiley and Sons.

In addition to the direct addition of H_2O_2, the photocatalytic in situ generation of H_2O_2 in the reaction for pollutant oxidation, which is a more promising way but a more challenging topic, is also possible. For example, Xu et al. reported that ternary $g\text{-}C_3N_4/Co_3O_4/Ag_2O$ heterojunctions can accelerate the mineralization of RhB due to the presence of H_2O_2 in situ, produced from O_2 reduction [183]. Through studying the catalytic behavior of the composites in the electrochemical oxygen reduction reaction (ORR), they found that the average number of electrons transferred in the reaction is 2.07, which indicates that the two-electron O_2 reduction process is the dominant step in the reaction.

The morphology of metal oxide, the interface interaction between metal oxide and $g\text{-}C_3N_4$ and the method of coupling metal oxide with $g\text{-}C_3N_4$ are also crucial factors affecting the photocatalytic performance of $g\text{-}C_3N_4$ for pollutant removal. For instance, the coupling of cubic CeO_2 (3~10 nm) with $g\text{-}C_3N_4$ using a hydrothermal method can greatly improve the activity of $g\text{-}C_3N_4$ for methyl orange degradation, with the reaction rate reaching 1.27 min^{-1}, which is 7.8 times higher than that of $g\text{-}C_3N_4$ alone (0.16 min^{-1}) [184]. The hybridization of NiO with $g\text{-}C_3N_4$ causes a red shift in the UV absorption edge and boosts the ability of light response; hence, it exhibits improved activity for methylene blue degradation, which is about 2.3 times higher than that of $g\text{-}C_3N_4$ [185]. Similar phenomena are also observed for other materials, e.g., $TiO_2\text{-}In_2O_3@g\text{-}C_3N_4$ [186].

The heavy-metal ions produced in electroplating, metallurgy, printing and dyeing, medicine and other industries cause serious damage to the ecological environment. Cr(VI) is a typical heavy metal in wastewater and its removal receives wide attention. The photocatalytic reduction of Cr(VI) to Cr(III) is an efficient way to treat Cr(VI)-containing wastewater, due to its simple process, energy savings, high efficiency and lower levels of secondary pollution [187]. It has been reported that the in situ self-assembly of g-C$_3$N$_4$/WO$_3$ in different organic acid media can lead to various surface morphologies and catalytic activities for Cr(VI) removal, as the number of carboxyl groups in organic acid greatly affects the shape and performance of g-C$_3$N$_4$/WO$_3$. Its synthesis in ethanedioic acid medium, which contains two carboxyl groups, yields a disc shape and has the best activity for nitroaromatic reduction (Figure 19a,b). Furthermore, the material has good stability for the reaction, with no appreciable activity loss within four cycles, as shown in Figure 19c [188].

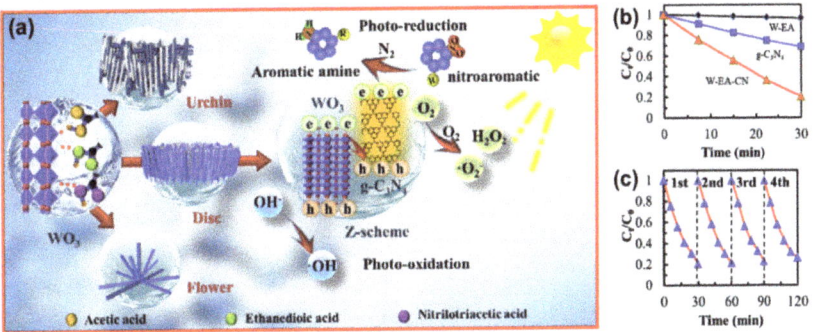

Figure 19. (a) The mechanism of organic acid inducing the growth of WO$_3$ with different shapes and the photocatalytic process occurring over g-C$_3$N$_4$/WO$_3$; (b) photocatalytic activity of different samples; and (c) recycle stability for the reduction of Cr(VI) and (m) cyclic experiments of W-EA-CN for photoreduction of Cr(VI). Used with permission from [188]. Copyright 2012 Royal Society of Chemistry.

Bi$_2$WO$_6$ is a promising semiconductor that can couple with g-C$_3$N$_4$ and form a heterojunction for the photocatalytic treatment of Cr(VI)-containing wastewater. Song et al. [189] found that a C$_3$N$_4$/Bi$_2$WO$_6$ composite prepared using a hydrothermal method exhibits a surface area up to 46.3 m^2/g and shows a rate constant of 0.0414 min^{-1} for the photocatalytic reduction of Cr(VI), as the high surface area of the catalyst facilitates not only the reactant's adsorption, but also the visible-light absorption.

Photocatalysis is also effective for removing gas-phase pollutants and receives great interest from scientists. It is known that air pollution is a big problem for the environment, and causes serious harm to the human body and ecological systems by forming acid rain, chemical smog, particulate matter, etc. Hence, seeking an effective and feasible technology for its removal is a challenging topic. Photocatalysis provides a way to remove air pollutants (e.g., NOx) by installing catalysts either inside the exhaust pipe or on the road surface [1]. As a typical photocatalyst, g-C$_3$N$_4$-based materials are also widely investigated in this aspect. Zhu et al. reported that g-C$_3$N$_4$ is active in NO removal via thermal catalysis, and proposed that the N atoms of g-C$_3$N$_4$, with a lone electron pair, serve as the active site of NO by donating electrons to weaken the N-O bond order [190]. This lays the foundation or using photocatalysis for NO removal, as electrons can be effectively excited from g-C$_3$N$_4$ under light irradiation.

However, it is known that the surface area of g-C$_3$N$_4$ prepared using the thermal condensation method is small, which grfieatly limits the light absorption capacity, the e$^-$/h$^+$ separation efficiency and other physicochemical properties; thus, many strategies have been adopted to overcome this problem. For example, Sano et al. [191] reported that pretreating melamine with NaOH solution before the condensation process favors the hydrolysis of unstable domains and the generation of mesopores in the structure of g-C$_3$N$_4$, leading to an increase in surface area from 7.7 m^2/g to 65 m^2/g, and the NO oxidation activity is accordingly increased 8.6 times. Duan et al. [180] found that flower-like g-C_3N_4 prepared using the self-assembly method can notably improve photocatalytic activity for NO oxidation compared to bulk g-C_3N_4, owing to the enlargement of the BET surface area, the formation of nitrogen vacancies, the condensation of π–π layer stacking, and the improvement in e^-/h^+ separation efficiency. The alternation of the precursor, e.g., urea [192] and guanidine hydrochloride [193] is also efficient in preparing g-C_3N_4 with a large surface area and improving photocatalytic performance.

3.4. Sterilization and Disinfection

In addition to the above applications, photocatalysis is also widely applied to inactivate pathogens in surface water owing to its broad compatibility, long durability, anti-drug resistance and thorough sterilization [194]. Bacteria, such as salmonella, staphylococcus aureus and bacillus anthracis, are commonly used as model pathogens to evaluate photocatalytic disinfection efficiency. Since the first work of Matsunaga et al. [195] on photochemical sterilization in 1985, this technique has rapidly developed and receives great interest from scientists. The principle of photocatalytic sterilization is to excite and separate the e^-/h^+ pairs via illumination; the photoinduced electrons and/or holes then inactivate the bacteria by directly or indirectly inflicting oxidative damage on their organs (through the formation of •O_2^-, •OH, etc.). Hence, the disinfection efficiency of materials closely depends on the properties that influence the generation and separation of e^-/h^+ pairs, e.g., the surface area, the band gap and the surface morphology, as reported for other photocatalytic processes.

In the case of g-C_3N_4, Huang et al. [196] found that mesoporous g-C_3N_4 synthesized using the hard template method can inactivate most of the bacteria (e.g., *E. coli* K-12) within 4 h, owing to its large surface area, which allows more active sites exposed on the surface to produce h^+ for bacterial disinfection. To support that the inactivation of bacteria is caused by photocatalysis, Xu et al. [197] conducted a dark contrasting experiment using a porous g-C_3N_4 nanosheet (PCNS) as the photocatalyst and *E. coli* as the model bacteria; they found that the adsorption of *E. coli* on PCNS reaches equilibrium within 1 h and about 85.5% of *E. coli* survive after 4 h, while nearly 100% of *E. coli* are killed by PCNS within 4 h under visible-light irradiation (Figure 20a). This demonstrates that the PCNS has little toxic effect on *E. coli* and the disinfection is mainly caused by the electrons or holes induced from PCNS under light irradiation. Figure 20b–g display the morphology of *E. coli* before and after photocatalytic disinfection, observed from TEM images, showing that the bacterial cells are tightly bound to PCNS and the outer membrane is partially damaged after 4 h of irradiation.

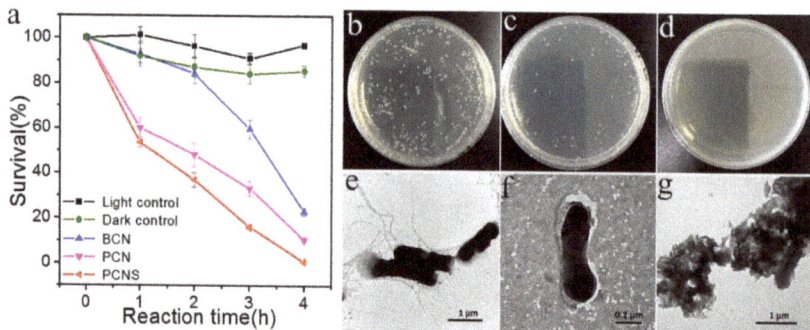

Figure 20. (**a**) Visible-light-driven photocatalytic disinfection performance against *E. coli* over BCN (bulk g-C_3N_4), PCN (porous g-C_3N_4) and PCNS (porous g-C_3N_4 nanosheets). Images of *E. coli* on solid culture medium before (**b**) and after ((**c**) 2 h; (**d**) 4 h) light irradiation on PCNS. TEM images of *E. coli* cells (**e**) before irradiation and (**f,g**) after disinfection for 4 h on PCNS. Used with permission from [197]. Copyright 2017 American Chemical Society.

In addition to bacterial infection, viral outbreaks, including SARS, bird flu, Ebola and the recent COVID-19, are also important events related to human health, and they are generally more resistant than bacteria to conventional disinfection due to their small size. Thus, the inactivation of viruses normally requires strong oxidative agents. $g\text{-}C_3N_4$-based materials have good photocatalytic reactivity to produce strong oxidative agents, e.g., $\bullet O_2^-$ and $\bullet OH$; hence, they are potential photocatalysts for virus inactivation. It has been reported that phage MS_2 can be completely inactivated by $g\text{-}C_3N_4$ under visible-light irradiation within 360 min [198], and the main active species for the reaction are $\bullet O_2^-$ and $\bullet OH$. Figure 21 shows that the phage MS2 in contact with $g\text{-}C_3N_4$ maintains integrity before irradiation, and its structure is severely damaged after 6 h of visible-light irradiation. The loss of protein triggers the leakage and rapid destruction of internal components, and ultimately leads to the death of the virus without regrowth.

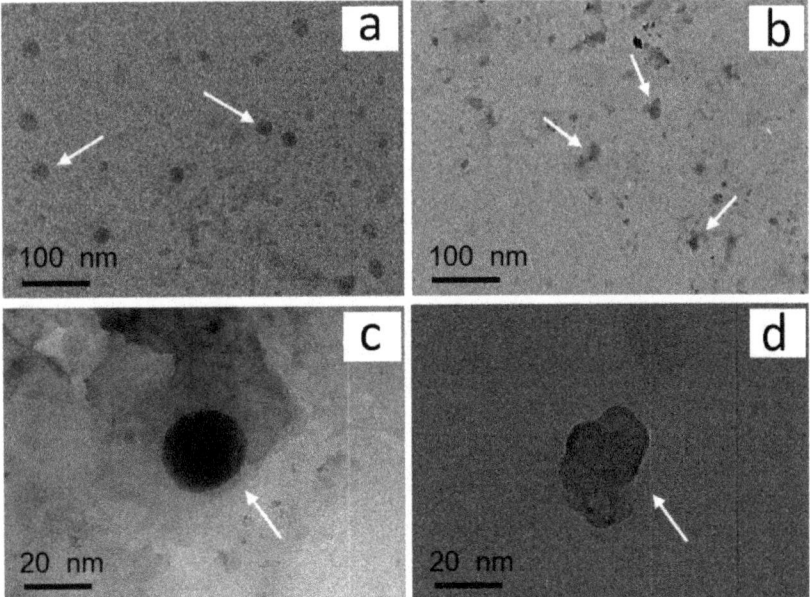

Figure 21. TEM images of phage MS2 before (**a**,**b**) and after (**c**,**d**) treatment with $g\text{-}C_3N_4$ for 6 h under visible-light irradiation. Used with permission from [198]. Copyright 2016 Elsevier.

4. Summary and Outlook

We provide an overview of the synthesis and photocatalytic applications of $g\text{-}C_3N_4$ and its coupling with single- or multi-metal oxides. Currently, the improvement in the photocatalytic performance of $g\text{-}C_3N_4$ mainly focuses on three aspects: (1) enhancing the adsorption capacity for target reactants, (2) broadening the absorption range to visible light, and (3) improving the e^-/h^+ pair separation efficiency. Generally, coupling with metal oxide can almost accomplish these three aspects, by increasing the surface area, narrowing the band gap and forming heterojunctions, for example.

Coupling metal oxide semiconductors with suitable energy levels is a promising strategy to improve the activity of $g\text{-}C_3N_4$ for photocatalytic reactions, by forming type II or Z-scheme heterojunctions, which facilitate the separation, and hence, the utilization of e^-/h^+ pairs. Moreover, the alterable valence of transition metals enables the composites to exhibit redox properties that benefit the proceeding of reactions undergoing electron transfer steps. Hence, the coupling of metal oxide semiconductors can widen the applications of $g\text{-}C_3N_4$ and may result in synergistic effects (e.g., photo-redox) that facilitate the proceeding of complex reactions.

However, because metal oxides are easy to sinter and reduce with g-C_3N_4 (which can be regarded as a type of reducing agent) at high temperature, developing a suitable method to obtain desirable effects on the composites is of great importance. This can be related to (1) the surface morphology, such as surface area and pore size, which affects, for example, the ability to absorb light, the spatial separation efficiency of e^-/h^+ pairs and the capability to adsorb the reactant; (2) the interface interaction, which influences the mobility of electrons and/or holes between g-C_3N_4 and the metal oxide, either for the formation of the newly balanced band gap (for type II heterojunctions) or for the recombination of unused electrons and holes (for Z-Scheme heterojunctions); or (3) the type of heterojunction (type II or Z-Scheme) formed, which depends on the properties of metal oxide, such as the Fermi level, the band level or the band gap, and the semiconductor type (*p*- or n-type). Hence, it is essential to consider the preparation method, the properties of the metal oxide and the integrating degree between g-C_3N_4 and the metal oxide, to obtain the best synergistic effect and fully exhibit the photocatalytic performance of the g-C_3N_4/metal oxide composites.

Photocatalytic applications of g-C_3N_4 and g-C_3N_4/metal oxide composites for energy synthesis and environmental protection have been widely reported for gas-, liquid- and gas–liquid-phase reactions. Because of the redox capability of metal oxides, redox and photocatalytic processes can simultaneously occur in the reaction, and a synergistic effect may be induced between them, for instance, the photo-Fenton reaction process. This improves the reaction rate while making the reaction process complex. Hence, the reaction process deserves further exploration and investigation in order to reveal and understand the photocatalytic mechanism, such as the manners of charge transfer, the internal force-field adjustment, the electron interaction between g-C_3N_4 and the metal oxide, etc. The collaboration of advanced characterizations and theoretic simulation calculations would be useful in this respect and could be a development tendency in photocatalysis in future.

Funding: Financial support provided by the National Natural Science Foundation of China (21976141, 52002292, 22102123, 42277485), the Department of Science and Technology of Hubei Province (2021CFA034), the Department of Education of Hubei Province (T2020011, Q20211712) and the Opening Project of the Hubei Key Laboratory of Biomass Fibers and Eco-Dyeing and Finishing (STRZ202101) is gratefully acknowledged.

Conflicts of Interest: The authors declare no conflict of interest.

References

1. Ali, T.; Muhammad, N.; Qian, Y.; Liu, S.; Wang, S.; Wang, M.; Qian, T.; Yan, C. Recent advances in material design and reactor engineering for electrocatalytic ambient nitrogen fixation. *Mater. Chem. Front.* **2022**, *6*, 843–879. [CrossRef]
2. Ali, T.; Qiao, W.; Zhang, D.; Liu, W.; Sajjad, S.; Yan, C.; Su, R. Surface Sulfur Vacancy Engineering of Metal Sulfides Promoted Desorption of Hydrogen Atoms for Enhanced Electrocatalytic Hydrogen Evolution. *J. Phys. Chem. C* **2021**, *125*, 12707–12712. [CrossRef]
3. Ali, T.; Wang, X.; Tang, K.; Li, Q.; Sajjad, S.; Khan, S.; Farooqi, S.A.; Yan, C. SnS_2 quantum dots growth on MoS_2: Atomic-level heterostructure for electrocatalytic hydrogen evolution. *Electrochim. Acta* **2019**, *300*, 45–52. [CrossRef]
4. Khan, S.; Ali, T.; Wang, X.; Iqbal, W.; Bashir, T.; Chao, W.; Sun, H.; Lu, H.; Yan, C.; Muhammad Irfan, R. $Ni_3S_2@Ni_5P_4$ nanosheets as highly productive catalyst for electrocatalytic oxygen evolution. *Chem. Eng. Sci.* **2022**, *247*, 117020. [CrossRef]
5. Xu, X.; Liu, H.; Wang, J.; Chen, T.; Ding, X.; Chen, H. Insight into surface hydroxyl groups for environmental purification: Characterizations, applications and advances. *Surf. Interfaces* **2021**, *25*, 101272. [CrossRef]
6. Yeganeh, M.; Sobhi, H.R.; Esrafili, A. Efficient photocatalytic degradation of metronidazole from aqueous solutions using Co/g-C_3N_4/Fe_3O_4 nanocomposite under visible light irradiation. *Environ. Sci. Pollut. Res.* **2022**, *29*, 25486–25495. [CrossRef] [PubMed]
7. Palanivel, B.; Jayaraman, V.; Ayyappan, C.; Alagiri, M. Magnetic binary metal oxide intercalated g-C3N4: Energy band tuned p-n heterojunction towards Z-scheme photo-Fenton phenol reduction and mixed dye degradation. *J. Water Process. Eng.* **2019**, *32*, 100968. [CrossRef]
8. Hsieh, M.C.; Wu, G.C.; Liu, W.G.; Goddard, W.A.; Yang, C.M. Nanocomposites of Tantalum-Based Pyrochlore and Indium Hydroxide Showing High and Stable Photocatalytic Activities for Overall Water Splitting and Carbon Dioxide Reduction. *Angew. Chem. Int. Edit.* **2014**, *53*, 14216–14220. [CrossRef]

9. Yang, X.; Xu, X.; Wang, J.; Chen, T.; Wang, S.; Ding, X.; Chen, H. Insights into the Surface/Interface Modifications of Bi_2MoO_6: Feasible Strategies and Photocatalytic Applications. *Sol. RRL* **2021**, *5*, 2000442. [CrossRef]
10. Oshima, T.; Nishioka, S.; Kikuchi, Y.; Hirai, S.; Yanagisawa, K.I.; Eguchi, M.; Miseki, Y.; Yokoi, T.; Yui, T.; Kimoto, K.; et al. An Artificial Z-Scheme Constructed from Dye-Sensitized Metal Oxide Nanosheets for Visible Light-Driven Overall Water Splitting. *J. Am. Chem. Soc.* **2020**, *142*, 8412–8420. [CrossRef] [PubMed]
11. Yan, C.Z.; Xue, X.L.; Zhang, W.J.; Li, X.J.; Liu, J.; Yang, S.Y.; Hu, Y.; Chen, R.P.; Yan, Y.P.; Zhu, G.Y.; et al. Well-designed $Te/SnS_2/Ag$ artificial nanoleaves for enabling and enhancing visible-light driven overall splitting of pure water. *Nano Energy* **2017**, *39*, 539–545. [CrossRef]
12. Sajjad, S.; Wang, C.; Wang, X.; Ali, T.; Qian, T.; Yan, C. In situ evolved $NiMo/NiMoO_4$ nanorods as a bifunctional catalyst for overall water splitting. *Nanotechnology* **2020**, *31*, 495404. [CrossRef]
13. Li, Y.H.; Gu, M.L.; Shi, T.; Cui, W.; Zhang, X.M.; Dong, F.; Cheng, J.S.; Fan, J.J.; Lv, K.L. Carbon vacancy in C_3N_4 nanotube: Electronic structure, photocatalysis mechanism and highly enhanced activity. *Appl. Catal. B Environ.* **2020**, *262*, 118281. [CrossRef]
14. Ismael, M. A review on graphitic carbon nitride ($g-C_3N_4$) based nanocomposites: Synthesis, categories, and their application in photocatalysis. *J. Alloys. Compd.* **2020**, *846*, 156446. [CrossRef]
15. Sohrabnezhad, S.; Pourahmad, A.; Radaee, E. Photocatalytic degradation of basic blue 9 by CoS nanoparticles supported on AlMCM-41 material as a catalyst. *J. Hazard. Mater.* **2009**, *170*, 184–190. [CrossRef] [PubMed]
16. Bhunia, S.K.; Jana, N.R. Reduced Graphene Oxide-Silver Nanoparticle Composite as Visible Light Photocatalyst for Degradation of Colorless Endocrine Disruptors. *ACS Appl. Mater. Interfaces* **2014**, *6*, 20085–20092. [CrossRef] [PubMed]
17. Li, C.M.; Yu, S.Y.; Zhang, X.X.; Wang, Y.; Liu, C.B.; Chen, G.; Dong, H.J. Insight into photocatalytic activity, universality and mechanism of copper/chlorine surface dual-doped graphitic carbon nitride for degrading various organic pollutants in water. *J. Colloid Interf. Sci.* **2019**, *538*, 462–473. [CrossRef]
18. Vadaei, S.; Faghihian, H. Enhanced visible light photodegradation of pharmaceutical pollutant, warfarin by nano-sized SnTe, effect of supporting, catalyst dose, and scavengers. *Environ. Toxicol. Phar.* **2018**, *58*, 45–53. [CrossRef] [PubMed]
19. Danish, M.; Saud Athar, M.; Ahmad, I.; Warshagha, M.Z.A.; Rasool, Z.; Muneer, M. Highly efficient and stable $Fe_2O_3/g-C_3N_4/GO$ nanocomposite with Z-scheme electron transfer pathway: Role of photocatalytic activity and adsorption isotherm of organic pollutants in wastewater. *Appl. Surf. Sci.* **2022**, *604*, 154604. [CrossRef]
20. Palanivel, B.; Lallimathi, M.; Arjunkumar, B.; Shkir, M.; Alshahrani, T.; Al-Namshah, K.S.; Hamdy, M.S.; Shanavas, S.; Venkatachalam, M.; Ramalingam, G. rGO supported $g-C_3N_4/CoFe_2O_4$ heterojunction: Visible-light-active photocatalyst for effective utilization of H_2O_2 to organic pollutant degradation and OH radicals production. *J. Environ. Chem. Eng.* **2021**, *9*, 104698. [CrossRef]
21. Xu, X.; Ding, X.; Yang, X.; Wang, P.; Li, S.; Lu, Z.; Chen, H. Oxygen vacancy boosted photocatalytic decomposition of ciprofloxacin over Bi_2MoO_6: Oxygen vacancy engineering, biotoxicity evaluation and mechanism study. *J. Hazard. Mater.* **2018**, *364*, 691–699. [CrossRef] [PubMed]
22. Xu, X.; Yang, N.; Wang, P.; Wang, S.; Xiang, Y.; Zhang, X.; Ding, X.; Chen, H. Highly Intensified Molecular Oxygen Activation on $Bi@Bi_2MoO_6$ via a Metallic Bi-Coordinated Facet-Dependent Effect. *ACS Appl. Mater. Interfaces* **2020**, *12*, 1867–1876. [CrossRef] [PubMed]
23. Xu, X.; Wang, J.; Chen, T.; Yang, N.; Wang, S.; Ding, X.; Chen, H. Deep insight into ROS mediated direct and hydroxylated dichlorination process for efficient photocatalytic sodium pentachlorophenate mineralization. *Appl. Catal. B Environ.* **2021**, *296*, 120352. [CrossRef]
24. Fujishima, A.; Honda, K. Electrochemical Photolysis of Water at a Semiconductor Electrode. *Nature* **1972**, *238*, 37–38. [CrossRef] [PubMed]
25. Akira, F.; Tooru, I.; Tadashi, W.; Kenichi, H. Stabilization of photoanodes in electrochemical photocells for solar energy conversion. *Chem. Lett.* **1978**, *7*, 357–360.
26. Gao, M.; Zhu, L.; Ong, W.L.; Wang, J.; Ho, G. Structural Design of TiO_2-based Photocatalyst for H_2 Production and Degradation Applications. *Catal. Sci. Technol.* **2015**, *5*, 4703–4726. [CrossRef]
27. Zhou, X.; Xu, Q.; Lei, W.; Zhang, T.; Qi, X.; Liu, G.; Deng, K.; Yu, J. Origin of Tunable Photocatalytic Selectivity of Well-Defined $\alpha-Fe_2O_3$ Nanocrystals. *Small* **2014**, *10*, 674–679. [CrossRef]
28. Chen, J.; Wu, X.-J.; Yin, L.; Li, B.; Hong, X.; Fan, Z.; Chen, B.; Xue, C.; Zhang, H. One-pot Synthesis of CdS Nanocrystals Hybridized with Single-Layer Transition-Metal Dichalcogenide Nanosheets for Efficient Photocatalytic Hydrogen Evolution. *Angew. Chem. Int. Edit.* **2015**, *54*, 1210–1214. [CrossRef]
29. Chen, S.; Shen, S.; Liu, G.; Qi, Y.; Zhang, F.; Li, C. Interface Engineering of a CoO_x/Ta_3N_5 Photocatalyst for Unprecedented Water Oxidation Performance under Visible-Light-Irradiation. *Angew. Chem. Int. Edit.* **2015**, *54*, 3047–3051. [CrossRef]
30. Inoue, T.; Fujishima, A.; Konishi, S.; Honda, K. Photoelectrocatalytic reduction of carbon dioxide in aqueous suspensions of semiconductor powders. *Nature* **1979**, *277*, 637–638. [CrossRef]
31. Wang, X.; Maeda, K.; Thomas, A.; Takanabe, K.; Xin, G.; Carlsson, J.M.; Domen, K.; Antonietti, M. A metal-free polymeric photocatalyst for hydrogen production from water under visible light. *Nat. Mater.* **2009**, *8*, 76–80. [CrossRef] [PubMed]
32. Lallimathi, M.; Kalisamy, P.; Suryamathi, M.; Alshahrani, T.; Shkir, M.; Venkatachalam, M.; Palanivel, B. Carbon Dot Loaded Integrative $CoFe_2O_4/g-C_3N_4$ P-N Heterojunction: Direct Solar Light-Driven Photocatalytic H_2 Evolution and Organic Pollutant Degradation. *ChemistrySelect* **2020**, *5*, 10607–10617. [CrossRef]

33. Palanivel, B.; Hossain, M.S.; Macadangdang, R.R.; Ayappan, C.; Krishnan, V.; Marnadu, R.; Kalaivani, T.; Alharthi, F.A.; Sreedevi, G. Activation of Persulfate for Improved Naproxen Degradation Using FeCo$_2$O$_4$@g-C$_3$N$_4$ Heterojunction Photocatalysts. *ACS Omega* **2021**, *6*, 34563–34571. [CrossRef] [PubMed]
34. Cui, L.F.; Song, J.L.; McGuire, A.F.; Kang, S.F.; Fang, X.Y.; Wang, J.J.; Yin, C.C.; Li, X.; Wang, Y.G.; Cui, B.X. Constructing Highly Uniform Onion-Ring-like Graphitic Carbon Nitride for Efficient Visible-Light-Driven Photocatalytic Hydrogen Evolution. *ACS Nano* **2018**, *12*, 5551–5558. [CrossRef] [PubMed]
35. Zhang, Y.Z.; Huang, Z.X.; Shi, J.W.; Guan, X.J.; Cheng, C.; Zong, S.C.; Huangfu, Y.L.; Ma, L.J.; Guo, L.J. Maleic hydrazide-based molecule doping in three-dimensional lettuce-like graphite carbon nitride towards highly efficient photocatalytic hydrogen evolution. *Appl. Catal. B Environ.* **2020**, *272*, 119009. [CrossRef]
36. Zhang, M.L.; Yang, Y.; An, X.Q.; Zhao, J.J.; Bao, Y.P.; Hou, L.A. Exfoliation method matters: The microstructure-dependent photoactivity of g-C$_3$N$_4$ nanosheets for water purification. *J. Hazard. Mater.* **2022**, *424*, 127424. [CrossRef]
37. Zhang, Y.Z.; Shi, J.W.; Huang, Z.X.; Guan, X.J.; Zong, S.C.; Cheng, C.; Zheng, B.T.; Guo, L.J. Synchronous construction of CoS$_2$ in-situ loading and S doping for g-C$_3$N$_4$: Enhanced photocatalytic H$_2$-evolution activity and mechanism insight. *Chem. Eng. J.* **2020**, *401*, 126135. [CrossRef]
38. Xiong, T.; Cen, W.L.; Zhang, Y.X.; Dong, F. Bridging the g-C$_3$N$_4$ Interlayers for Enhanced Photocatalysis. *ACS Catal.* **2016**, *6*, 2462–2472. [CrossRef]
39. Luo, L.; Gong, Z.; Ma, J.; Wang, K.; Zhu, H.; Li, K.; Xiong, L.; Guo, X.; Tang, J. Ultrathin sulfur-doped holey carbon nitride nanosheets with superior photocatalytic hydrogen production from water. *Appl. Catal. B Environ.* **2021**, *284*, 119742. [CrossRef]
40. Guo, H.T.; Huang, H.; Li, Y.; Lu, S.K.; Xue, M.H.; Weng, W.; Zheng, T. Stepwise preparation of Ti-doped functionalized carbon nitride nanoparticles and hybrid TiO$_2$/graphitic-C$_3$N$_4$ for detection of free residual chlorine and visible-light photocatalysis. *Chem. Commun.* **2019**, *55*, 13848–13851. [CrossRef]
41. Liu, J.; Han, D.D.; Chen, P.J.; Zhai, L.P.; Wang, Y.J.; Chen, W.H.; Mi, L.W.; Yang, L.P. Positive roles of Br in g-C$_3$N$_4$/PTCDI-Br heterojunction for photocatalytic degrading chlorophenols. *Chem. Eng. J.* **2021**, *418*, 129492. [CrossRef]
42. Palanivel, B.; Mani, A. Conversion of a Type-II to a Z-Scheme Heterojunction by Intercalation of a 0D Electron Mediator between the Integrative NiFe$_2$O$_4$/g-C$_3$N$_4$ Composite Nanoparticles: Boosting the Radical Production for Photo-Fenton Degradation. *ACS Omega* **2020**, *5*, 19747–19759. [CrossRef] [PubMed]
43. Sun, L.; Du, T.; Hu, C.; Chen, J.; Lu, J.; Lu, Z.; Han, H. Antibacterial Activity of Graphene Oxide/g-C$_3$N$_4$ Composite through Photocatalytic Disinfection under Visible Light. *ACS Sustain. Chem. Eng.* **2017**, *5*, 8693–8701. [CrossRef]
44. Christoforidis, K.C.; Syrgiannis, Z.; La Parola, V.; Montini, T.; Petit, C.; Stathatos, E.; Godin, R.; Durrant, J.R.; Prato, M.; Fornasiero, P. Metal-free dual-phase full organic carbon nanotubes/g-C$_3$N$_4$ heteroarchitectures for photocatalytic hydrogen production. *Nano Energy* **2018**, *50*, 468–478. [CrossRef]
45. Yu, F.T.; Wang, Z.Q.; Zhang, S.C.; Ye, H.N.; Kong, K.Y.; Gong, X.Q.; Hua, J.L.; Tian, H. Molecular Engineering of Donor-Acceptor Conjugated Polymer/g-C$_3$N$_4$ Heterostructures for Significantly Enhanced Hydrogen Evolution Under Visible-Light Irradiation. *Adv. Funct. Mater.* **2018**, *28*, 1804512. [CrossRef]
46. Jo, W.-K.; Selvam, N.C.S. Z-scheme CdS/g-C$_3$N$_4$ composites with RGO as an electron mediator for efficient photocatalytic H$_2$ production and pollutant degradation. *Chem. Eng. J.* **2017**, *317*, 913–924. [CrossRef]
47. Mullakkattuthodi, S.; Haridas, V.; Sugunan, S.; Narayanan, B.N. Z-scheme mechanism for methylene blue degradation over Fe$_2$O$_3$/g-C$_3$N$_4$ nanocomposite prepared via one-pot exfoliation and magnetization of g-C$_3$N$_4$. *Front. Mater. Sci.* **2022**, *16*, 220612. [CrossRef]
48. Zhu, D.; Liu, S.; Chen, M.; Zhang, J.; Wang, X. Flower-like-flake Fe$_3$O$_4$/g-C$_3$N$_4$ nanocomposite: Facile synthesis, characterization, and enhanced photocatalytic performance. *Colloid. Surf. A* **2018**, *537*, 372–382. [CrossRef]
49. Hu, S.; Ouyang, W.; Guo, L.; Lin, Z.; Jiang, X.; Qiu, B.; Chen, G. Facile synthesis of Fe$_3$O$_4$/g-C$_3$N$_4$/HKUST-1 composites as a novel biosensor platform for ochratoxin A. *Bios. Bioelectron.* **2017**, *92*, 718–723. [CrossRef]
50. Li, K.; Gao, S.; Wang, Q.; Xu, H.; Wang, Z.; Huang, B.; Dai, Y.; Lu, J. In-Situ-Reduced Synthesis of Ti^{3+} Self-Doped TiO$_2$/g-C$_3$N$_4$ Heterojunctions with High Photocatalytic Performance under LED Light Irradiation. *ACS Appl. Mater. Interfaces* **2015**, *7*, 9023–9030. [CrossRef]
51. Tang, L.; Liu, Y.; Wang, J.; Zeng, G.; Deng, Y.; Dong, H.; Feng, H.; Wang, J.; Peng, B. Enhanced activation process of persulfate by mesoporous carbon for degradation of aqueous organic pollutants: Electron transfer mechanism. *Appl. Catal. B Environ.* **2018**, *231*, 1–10. [CrossRef]
52. Ahilandeswari, G.; Arivuoli, D. Investigation of Ce$_2$(WO$_4$)$_3$/g-C$_3$N$_4$ nanocomposite for degradation of industrial pollutants through sunlight-driven photocatalysis. *Appl. Phys. A* **2022**, *128*, 705. [CrossRef]
53. Zhang, D.; Cui, S.; Yang, J. Preparation of Ag$_2$O/g-C$_3$N$_4$/Fe$_3$O$_4$ composites and the application in the photocatalytic degradation of Rhodamine B under visible light. *J. Alloy. Compd.* **2017**, *708*, 1141–1149. [CrossRef]
54. Huang, S.; Xu, Y.; Xie, M.; Xu, H.; He, M.; Xia, J.; Huang, L.; Li, H. Synthesis of magnetic CoFe$_2$O$_4$/g-C$_3$N$_4$ composite and its enhancement of photocatalytic ability under visible-light. *Colloid. Surf. A* **2015**, *478*, 71–80. [CrossRef]
55. Xu, Z.; Guan, L.; Li, H.; Sun, J.; Ying, Z.; Wu, J.; Xu, N. Structure Transition Mechanism of Single-Crystalline Silicon, g-C$_3$N$_4$, and Diamond Nanocone Arrays Synthesized by Plasma Sputtering Reaction Deposition. *J. Phys. Chem. C* **2015**, *119*, 29062–29070. [CrossRef]

56. Cui, Y.; Tang, Y.; Wang, X. Template-free synthesis of graphitic carbon nitride hollow spheres for photocatalytic degradation of organic pollutants. *Mater. Lett.* **2015**, *161*, 197–200. [CrossRef]
57. Fang, W.; Xing, M.; Zhang, J. Modifications on reduced titanium dioxide photocatalysts: A review. *J. Photoch. Photobio. C* **2017**, *32*, 21–39. [CrossRef]
58. Wang, Y.; Wang, F.; Zuo, Y.; Zhang, X.; Cui, L.F. Simple synthesis of ordered cubic mesoporous graphitic carbon nitride by chemical vapor deposition method using melamine. *Mater. Lett.* **2014**, *136*, 271–273. [CrossRef]
59. Shiraishi, Y.; Kofuji, Y.; Sakamoto, H.; Tanaka, S.; Ichikawa, S.; Hirai, T. Effects of Surface Defects on Photocatalytic H_2O_2 Production by Mesoporous Graphitic Carbon Nitride under Visible Light Irradiation. *ACS Catal.* **2015**, *5*, 3058–3066. [CrossRef]
60. Yang, L.; Liu, X.; Liu, Z.; Wang, C.; Liu, G.; Li, Q.; Feng, X. Enhanced photocatalytic activity of $g-C_3N_4$ 2D nanosheets through thermal exfoliation using dicyandiamide as precursor. *Ceram. Int.* **2018**, *44*, 20613–20619. [CrossRef]
61. Papailias, I.; Giannakopoulou, T.; Todorova, N.; Demotikali, D.; Vaimakis, T.; Trapalis, C. Effect of processing temperature on structure and photocatalytic properties of $g-C_3N_4$. *Appl. Surf. Sci.* **2015**, *358*, 278–286. [CrossRef]
62. Hong, Y.; Liu, E.; Shi, J.; Lin, X.; Sheng, L.; Zhang, M.; Wang, L.; Chen, J. A direct one-step synthesis of ultrathin $g-C_3N_4$ nanosheets from thiourea for boosting solar photocatalytic H_2 evolution. *Int. J. Hydrogen Energy* **2019**, *44*, 7194–7204. [CrossRef]
63. Paul, D.R.; Sharma, R.; Nehra, S.P.; Sharma, A. Effect of calcination temperature, pH and catalyst loading on photodegradation efficiency of urea derived graphitic carbon nitride towards methylene blue dye solution. *RSC Adv.* **2019**, *9*, 15381–15391. [CrossRef]
64. Cui, Y.; Zhang, G.; Lin, Z.; Wang, X. Condensed and low-defected graphitic carbon nitride with enhanced photocatalytic hydrogen evolution under visible light irradiation. *Appl. Catal. B: Environ.* **2016**, *181*, 413–419. [CrossRef]
65. Kumru, B.; Antonietti, M. Colloidal properties of the metal-free semiconductor graphitic carbon nitride. *Adv. Colloid Interface Sci.* **2020**, *283*, 102229. [CrossRef]
66. Hasija, V.; Raizada, P.; Sudhaik, A.; Sharma, K.; Kumar, A.; Singh, P.; Jonnalagadda, S.B.; Thakur, V.K. Recent advances in noble metal free doped graphitic carbon nitride based nanohybrids for photocatalysis of organic contaminants in water: A review. *Appl. Mater. Today* **2019**, *15*, 494–524. [CrossRef]
67. Groenewolt, M.; Antonietti, M. Synthesis of $g-C_3N_4$ Nanoparticles in Mesoporous Silica Host Matrices. *Adv. Mater.* **2005**, *17*, 1789–1792. [CrossRef]
68. Ge, L. Synthesis and photocatalytic performance of novel metal-free $g-C_3N_4$ photocatalysts. *Mater. Lett.* **2011**, *65*, 2652–2654. [CrossRef]
69. Samanta, S.; Yadav, R.; Kumar, A.; Kumar Sinha, A.; Srivastava, R. Surface modified C, O co-doped polymeric $g-C_3N_4$ as an efficient photocatalyst for visible light assisted CO_2 reduction and H_2O_2 production. *Appl. Catal. B Environ.* **2019**, *259*, 118054. [CrossRef]
70. Wang, T.; Nie, C.; Ao, Z.; Wang, S.; An, T. Recent progress in $g-C_3N_4$ quantum dots: Synthesis, properties and applications in photocatalytic degradation of organic pollutants. *J. Mater. Chem. A* **2020**, *8*, 485–502. [CrossRef]
71. Schaber, P.M.; Colson, J.; Higgins, S.; Thielen, D.; Anspach, B.; Brauer, J. Thermal decomposition (pyrolysis) of urea in an open reaction vessel. *Thermochim. Acta* **2004**, *424*, 131–142. [CrossRef]
72. Liu, J.; Zhang, T.; Wang, Z.; Dawson, G.; Chen, W. Simple pyrolysis of urea into graphitic carbon nitride with recyclable adsorption and photocatalytic activity. *J. Mater. Chem.* **2011**, *21*, 14398–14401. [CrossRef]
73. Zhang, G.; Zhang, J.; Zhang, M.; Wang, X. Polycondensation of thiourea into carbon nitride semiconductors as visible light photocatalysts. *J. Mater. Chem.* **2012**, *22*, 8083–8091. [CrossRef]
74. Chen, X.; Jun, Y.S.; Takanabe, K.; Maeda, K.; Domen, K.; Fu, X.; Antonietti, M.; Wang, X. Ordered Mesoporous SBA-15 Type Graphitic Carbon Nitride: A Semiconductor Host Structure for Photocatalytic Hydrogen Evolution with Visible Light. *Chem. Mater.* **2009**, *21*, 4093–4095. [CrossRef]
75. Tian, N.; Xiao, K.; Zhang, Y.; Lu, X.; Ye, L.; Gao, P.; Ma, T.; Huang, H. Reactive sites rich porous tubular yolk-shell $g-C_3N_4$ via precursor recrystallization mediated microstructure engineering for photoreduction. *Appl. Catal. B Environ.* **2019**, *253*, 196–205. [CrossRef]
76. Wu, X.; Gao, D.; Wang, P.; Yu, H.; Yu, J. NH_4Cl-induced low-temperature formation of nitrogen-rich $g-C_3N_4$ nanosheets with improved photocatalytic hydrogen evolution. *Carbon* **2019**, *153*, 757–766. [CrossRef]
77. Qi, H.; Liu, Y.; Li, C.; Zou, X.; Huang, Y.; Wang, Y. Precursor-reforming protocol to synthesis of porous N-doped $g-C_3N_4$ for highly improved photocatalytic water treatments. *Mater. Lett.* **2020**, *264*, 127329. [CrossRef]
78. Zhang, D.; Tan, G.; Wang, M.; Li, B.; Dang, M.; Ren, H.; Xia, A. The modulation of $g-C_3N_4$ energy band structure by excitons capture and dissociation. *Mater. Res. Bull.* **2020**, *122*, 110685. [CrossRef]
79. Sun, S.; Fan, E.; Xu, H.; Cao, W.; Shao, G.; Fan, B.; Wang, H.; Zhang, R. Enhancement of photocatalytic activity of $g-C_3N_4$ by hydrochloric acid treatment of melamine. *Nanotechnology* **2019**, *30*, 315601. [CrossRef]
80. Yin, J.-T.; Li, Z.; Cai, Y.; Zhang, Q.-F.; Chen, W. Ultrathin graphitic carbon nitride nanosheets with remarkable photocatalytic hydrogen production under visible LED irradiation. *Chem. Commun.* **2017**, *53*, 9430–9433. [CrossRef]
81. Zhong, Y.; Wang, Z.; Feng, J.; Yan, S.; Zhang, H.; Li, Z.; Zou, Z. Improvement in photocatalytic H_2 evolution over $g-C_3N_4$ prepared from protonated melamine. *Appl. Surf. Sci.* **2014**, *295*, 253–259. [CrossRef]
82. Yan, H.; Chen, Y.; Xu, S. Synthesis of graphitic carbon nitride by directly heating sulfuric acid treated melamine for enhanced photocatalytic H_2 production from water under visible light. *Int. J. Hydrogen Energy* **2012**, *37*, 125–133. [CrossRef]

83. Zheng, Y.J.; Cao, L.Y.; Xing, G.X.; Bai, Z.Q.; Huang, J.F.; Zhang, Z.P. Microscale flower-like magnesium oxide for highly efficient photocatalytic degradation of organic dyes in aqueous solution. *RSC Adv.* **2019**, *9*, 7338–7348. [CrossRef] [PubMed]
84. Xu, J.; Huang, Z.F.; Ji, H.; Tang, H.; Tang, G.G.; Jiang, H.B. g-C_3N_4 anchored with MoS_2 ultrathin nanosheets as high performance anode material for supercapacitor. *Mater. Lett.* **2019**, *241*, 35–38. [CrossRef]
85. Wu, J.; Xie, Y.; Ling, Y.; Dong, Y.Y.; Li, J.; Li, S.Q.; Zhao, J.S. Synthesis of Flower-Like g-C_3N_4/BiOBr and Enhancement of the Activity for the Degradation of Bisphenol A Under Visible Light Irradiation. *Front. Chem.* **2019**, *7*, 649. [CrossRef]
86. Bai, X.J.; Li, J.; Cao, C.B.; Hussain, S. Solvothermal synthesis of the special shape (deformable) hollow g-C_3N_4 nanospheres. *Mater. Lett.* **2011**, *65*, 1101–1104. [CrossRef]
87. Lv, S.; Ng, Y.H.; Zhu, R.; Li, S.; Wu, C.; Liu, Y.; Zhang, Y.; Jing, L.; Deng, J.; Dai, H. Phosphorus vapor assisted preparation of P-doped ultrathin hollow g-C3N4 sphere for efficient solar-to-hydrogen conversion. *Appl. Catal. B Environ.* **2021**, *297*, 120438. [CrossRef]
88. Huang, Z.W.; Zhang, Y.W.; Dai, H.Y.; Wang, Y.Y.; Qin, C.C.; Chen, W.X.; Zhou, Y.M.; Yuan, S.H. Highly dispersed Pd nanoparticles hybridizing with 3D hollow-sphere g-C_3N_4 to construct 0D/3D composites for efficient photocatalytic hydrogen evolution. *J. Catal.* **2019**, *378*, 331–340. [CrossRef]
89. Li, Y.N.; Chen, Z.Y.; Wang, M.Q.; Zhang, L.Z.; Bao, S.J. Interface engineered construction of porous g-C_3N_4/TiO_2 heterostructure for enhanced photocatalysis of organic pollutants. *Appl. Surf. Sci.* **2018**, *440*, 229–236. [CrossRef]
90. Wang, Q.Z.; Shi, Y.B.; Du, Z.Y.; He, J.J.; Zhong, J.B.; Zhao, L.C.; She, H.D.; Liu, G.; Su, B.T. Synthesis of Rod-Like g-C_3N_4/ZnS Composites with Superior Photocatalytic Activity for the Degradation of Methyl Orange. *Eur. J. Inorg. Chem.* **2015**, *2015*, 4108–4115. [CrossRef]
91. Wu, M.; Li, Q.; Chen, C.; Su, G.; Song, M.; Sun, B.; Meng, J.; Shi, B. Constructed palladium-anchored hollow-rod-like graphitic carbon nitride created rapid visible-light-driven debromination of hexabromocyclododecane. *Appl. Catal. B Environ.* **2021**, *297*, 120409. [CrossRef]
92. Niu, P.; Liu, G.; Cheng, H.M. Nitrogen Vacancy-Promoted Photocatalytic Activity of Graphitic Carbon Nitride. *J. Phys. Chem. C* **2012**, *116*, 11013–11018. [CrossRef]
93. Liang, Q.; Li, Z.; Huang, Z.-H.; Kang, F.; Yang, Q.-H. Holey Graphitic Carbon Nitride Nanosheets with Carbon Vacancies for Highly Improved Photocatalytic Hydrogen Production. *Adv. Funct. Mater.* **2015**, *25*, 6885–6892. [CrossRef]
94. Xu, J.; Fujitsuka, M.; Kim, S.; Wang, Z.; Majima, T. Unprecedented effect of CO_2 calcination atmosphere on photocatalytic H_2 production activity from water using g-C_3N_4 synthesized from triazole polymerization. *Appl. Catal. B Environ.* **2019**, *241*, 141–148. [CrossRef]
95. Maeda, K.; Wang, X.; Nishihara, Y.; Lu, D.; Antonietti, M.; Domen, K. Photocatalytic Activities of Graphitic Carbon Nitride Powder for Water Reduction and Oxidation under Visible Light. *J. Phys. Chem. C* **2009**, *113*, 4940–4947. [CrossRef]
96. Long, B.; Lin, J.; Wang, X. Thermally-induced desulfurization and conversion of guanidine thiocyanate into graphitic carbon nitride catalysts for hydrogen photosynthesis. *J. Mater. Chem. A* **2014**, *2*, 2942–2951. [CrossRef]
97. Tian, N.; Zhang, Y.; Li, X.; Xiao, K.; Huang, H. Precursor-reforming protocol to 3D mesoporous g-C_3N_4 established by ultrathin self-doped nanosheets for superior hydrogen evolution. *Nano Energy* **2017**, *38*, 72–81. [CrossRef]
98. Zhang, J.; Zhang, M.; Zhang, G.; Wang, X. Synthesis of Carbon Nitride Semiconductors in Sulfur Flux for Water Photoredox Catalysis. *ACS Catal.* **2012**, *2*, 940–948. [CrossRef]
99. Duan, Y.; Li, X.; Lv, K.; Zhao, L.; Liu, Y. Flower-like g-C_3N_4 assembly from holy nanosheets with nitrogen vacancies for efficient NO abatement. *Appl. Surf. Sci.* **2019**, *492*, 166–176. [CrossRef]
100. Ong, W.-J.; Tan, L.-L.; Ng, Y.H.; Yong, S.-T.; Chai, S.-P. Graphitic Carbon Nitride (g-C_3N_4)-Based Photocatalysts for Artificial Photosynthesis and Environmental Remediation: Are We a Step Closer To Achieving Sustainability? *Chem. Rev.* **2016**, *116*, 7159–7329. [CrossRef]
101. Wu, L.; Fu, C.; Huang, W. Surface chemistry of TiO_2 connecting thermal catalysis and photocatalysis. *Phys. Chem. Chem. Phys.* **2020**, *22*, 9875–9909. [CrossRef] [PubMed]
102. Ko, K.C.; Bromley, S.T.; Lee, J.Y.; Illas, F. Size-Dependent Level Alignment between Rutile and Anatase TiO_2 Nanoparticles: Implications for Photocatalysis. *J. Phys. Chem. Lett.* **2017**, *8*, 5593–5598. [CrossRef] [PubMed]
103. Fang, Y.; Huang, W.; Yang, S.; Zhou, X.; Ge, C.; Gao, Q.; Fang, Y.; Zhang, S. Facile synthesis of anatase/rutile TiO_2/g-C_3N_4 multi-heterostructure for efficient photocatalytic overall water splitting. *Int. J. Hydrogen Energy* **2020**, *45*, 17378–17387. [CrossRef]
104. Chen, H.; Xie, Y.; Sun, X.; Lv, M.; Xu, X.X. Efficient charge separation based on type-II g-C_3N_4/TiO_2-B nanowire/tube heterostructure photocatalysts. *Dalton Trans.* **2015**, *44*, 13030–13039. [CrossRef] [PubMed]
105. Zhou, J.; Zhang, M.; Zhu, Y. Photocatalytic enhancement of hybrid C_3N_4/TiO_2 prepared via ball milling method. *Phys. Chem.* **2015**, *17*, 3647–3652. [CrossRef] [PubMed]
106. Liu, C.; Li, C.; Fu, X.; Raziq, F.; Qu, Y.; Jing, L. Synthesis of silicate-bridged ZnO/g-C_3N_4 nanocomposites as efficient photocatalysts and its mechanism. *RSC Adv.* **2015**, *5*, 37275–37280. [CrossRef]
107. Guo, Y.; Chang, B.; Wen, T.; Zhang, S.; Yang, B. A Z-scheme photocatalyst for enhanced photocatalytic H_2 evolution, constructed by growth of 2D plasmonic MoO_{3-x} nanoplates onto 2D g-C_3N_4 nanosheets. *J. Colloid Interf. Sci.* **2020**, *567*, 213–223. [CrossRef]
108. Zhang, Y.; Li, H.; Zhang, L.; Gao, R.; Dai, W.-L. Construction of Highly Efficient 3D/2D MnO_2/g-C_3N_4 Nanocomposite in the Epoxidation of Styrene with TBHP. *ACS Sustain. Chem. Eng.* **2019**, *7*, 17008–17019. [CrossRef]

109. Liu, W.; Zhou, J.; Hu, Z. Nano-sized g-C_3N_4 thin layer @ CeO_2 sphere core-shell photocatalyst combined with H_2O_2 to degrade doxycycline in water under visible light irradiation. *Sep. Purif. Technol.* **2019**, *227*, 115665. [CrossRef]
110. Cheng, R.; Zhang, L.; Fan, X.; Wang, M.; Shi, J. One-step construction of FeO_x modified g-C_3N_4 for largely enhanced visible-light photocatalytic hydrogen evolution. *Carbon* **2016**, *101*, 62–70. [CrossRef]
111. Mou, H.; Wang, J.; Yu, D.; Zhang, D.; Chen, W.; Wang, Y.; Wang, D.; Mu, T. Fabricating Amorphous g-C_3N_4/ZrO_2 Photocatalysts by One-Step Pyrolysis for Solar-Driven Ambient Ammonia Synthesis. *ACS Appl. Mater. Interfaces* **2019**, *11*, 44360–44365. [CrossRef] [PubMed]
112. Ramacharyulu, P.V.R.K.; Abbas, S.J.; Ke, S.-C. Enhanced charge separation and photoactivity in heterostructured g-C_3N_4: A synergistic interaction in environmental friendly CaO/g-C_3N_4. *Catal. Sci. Technol.* **2017**, *7*, 4940–4943. [CrossRef]
113. Prakash, K.; Senthil, K.P.; Latha, P.; Shanmugam, R.; Karuthapandian, S. Dry synthesis of water lily flower like SrO_2/g-C_3N_4 nanohybrids for the visible light induced superior photocatalytic activity. *Mater. Res. Bull.* **2017**, *93*, 112–122. [CrossRef]
114. Guan, R.; Li, J.; Zhang, J.; Zhao, Z.; Wang, D.; Zhai, H.; Sun, D. Photocatalytic Performance and Mechanistic Research of ZnO/g-C_3N_4 on Degradation of Methyl Orange. *ACS Omega* **2019**, *4*, 20742–20747. [CrossRef]
115. Huang, L.; Xu, H.; Zhang, R.; Cheng, X.; Xia, J.; Xu, Y.; Li, H. Synthesis and characterization of g-C_3N_4/MoO_3 photocatalyst with improved visible-light photoactivity. *Appl. Surf. Sci.* **2013**, *283*, 25–32. [CrossRef]
116. Shen, W.; Wang, X.; Ge, Y.; Feng, H.; Wang, L. Synthesis and characterization of AgO/g-C_3N_4 hybrids with enhanced visible-light photocatalytic activity for Rhodamine B degradation and bactericidal inactivation. *Colloid. Surf. A* **2019**, *575*, 102–110. [CrossRef]
117. Munusamy, T.D.; Yee, C.S.; Khan, M.M.R. Construction of hybrid g-C_3N_4/CdO nanocomposite with improved photodegradation activity of RhB dye under visible light irradiation. *Adv. Powder Technol.* **2020**, *31*, 2921–2931. [CrossRef]
118. Chen, L.-Y.; Zhang, W.-D. In_2O_3/g-C_3N_4 composite photocatalysts with enhanced visible light driven activity. *Appl Surf Sci* **2014**, *301*, 428–435. [CrossRef]
119. Zhang, T.; Chang, F.; Qi, Y.; Zhang, X.; Yang, J.; Liu, X.; Li, S. A facile one-pot and alkali-free synthetic procedure for binary SnO_2/g-C_3N_4 composites with enhanced photocatalytic behavior. *Mat. Sci. Semicon. Proc.* **2020**, *115*, 105112. [CrossRef]
120. Zhou, D.; Chen, Z.; Yang, Q.; Dong, X.; Zhang, J.; Qin, L. In-situ construction of all-solid-state Z-scheme g-C_3N_4/TiO_2 nanotube arrays photocatalyst with enhanced visible-light-induced properties. *Sol. Energ. Mat. Sol. C.* **2016**, *157*, 399–405. [CrossRef]
121. Li, B.; Nengzi, L.-C.; Guo, R.; Cui, Y.; Zhang, Y.; Cheng, X. Novel synthesis of Z-scheme α-Bi_2O_3/g-C_3N_4 composite photocatalyst and its enhanced visible light photocatalytic performance: Influence of calcination temperature. *Chin. Chem. Lett.* **2020**, *31*, 2705–2711. [CrossRef]
122. Idrees, F.; Dillert, R.; Bahnemann, D.; Butt, F.K.; Tahir, M. In-Situ Synthesis of Nb_2O_5/g-C_3N_4 Heterostructures as Highly Efficient Photocatalysts for Molecular H_2 Evolution under Solar Illumination. *Catalysts* **2019**, *9*, 169. [CrossRef]
123. Liu, W.; Zhou, J.; Yao, J. Shuttle-like CeO_2/g-C_3N_4 composite combined with persulfate for the enhanced photocatalytic degradation of norfloxacin under visible light. *Ecotox. Environ. Safe.* **2020**, *190*, 110062. [CrossRef] [PubMed]
124. Garg, T.; Renu; Kaur, J.; Kaur, P.; Nitansh; Kumar, V.; Tikoo, K.; Kaushik, A.; Singhal, S. An innovative Z-scheme g-C_3N_4/ZnO/$NiFe_2O_4$ heterostructure for the concomitant photocatalytic removal and real-time monitoring of noxious fluoroquinolones. *Chem. Eng. J.* **2022**, *443*, 136441. [CrossRef]
125. Khoshnevisan, B.; Boroumand, Z. Synthesis and characterization of a g-C_3N_4/TiO_2-ZnO nanostructure for Photocatalytic degradation of methylene blue. *Nano Futures* **2022**, *6*, 035001.
126. Preeyanghaa, M.; Dhileepan, M.D.; Madhavan, J.; Neppolian, B. Revealing the charge transfer mechanism in magnetically recyclable ternary g-C_3N_4/BiOBr/Fe_3O_4 nanocomposite for efficient photocatalytic degradation of tetracycline antibiotics. *Chemosphere* **2022**, *303*, 135070. [CrossRef]
127. Hanif, M.A.; Akter, J.; Kim, Y.S.; Kim, H.G.; Hahn, J.R.; Kwac, L.K. Highly efficient and sustainable ZnO/CuO/g-C_3N_4 photocatalyst for wastewater treatment under visible light through heterojunction development. *Catalysts* **2022**, *12*, 151. [CrossRef]
128. Pirsaheb, M.; Hossaini, H.; Asadi, A.; Jafari, Z. Enhanced degradation of diazinon with WO_3-Fe_3O_4/g-C_3N_4-persulfate system under visible light: Pathway, intermediates toxicity and mechanism. *Process Saf. Environ. Prot.* **2022**, *162*, 1107–1123. [CrossRef]
129. Swedha, M.; Alatar, A.A.; Okla, M.K.; Alaraidh, I.A.; Mohebaldin, A.; Aufy, M.; Raju, L.L.; Thomas, A.M.; Abdel-Maksoud, M.A.; Sudheer Khan, S. Graphitic carbon nitride embedded $Ni_3(VO_4)_2$/$ZnCr_2O_4$ Z-scheme photocatalyst for efficient degradation of p-chlorophenol and 5-fluorouracil, and genotoxic evaluation in Allium cepa. *J. Ind. Eng. Chem.* **2022**, *112*, 244–257. [CrossRef]
130. Li, F.-T.; Liu, S.-J.; Xue, Y.-B.; Wang, X.-J.; Hao, Y.-J.; Zhao, J.; Liu, R.-H.; Zhao, D. Structure Modification Function of g-C_3N_4 for Al_2O_3 in the In Situ Hydrothermal Process for Enhanced Photocatalytic Activity. *Chem. Eur. J.* **2015**, *21*, 10149–10159. [CrossRef]
131. Guo, F.; Shi, W.; Zhu, C.; Li, H.; Kang, Z. CoO and g-C_3N_4 complement each other for highly efficient overall water splitting under visible light. *Appl. Catal. B Environ.* **2018**, *226*, 412–420. [CrossRef]
132. Liu, J.; Jia, Q.; Long, J.; Wang, X.; Gao, Z.; Gu, Q. Amorphous NiO as co-catalyst for enhanced visible-light-driven hydrogen generation over g-C_3N_4 photocatalyst. *Appl. Catal. B Environ.* **2018**, *222*, 35–43. [CrossRef]
133. Chen, J.; Shen, S.; Guo, P.; Wang, M.; Wu, P.; Wang, X.; Guo, L. In-situ reduction synthesis of nano-sized Cu_2O particles modifying g-C_3N_4 for enhanced photocatalytic hydrogen production. *Appl. Catal. B: Environ.* **2014**, *152–153*, 335–341. [CrossRef]
134. Mao, N.; Jiang, J.X. MgO/g-C_3N_4 Nanocomposites as Efficient Water Splitting Photocatalysts under Visible Light Irradiation. *Appl. Surf. Sci.* **2019**, *476*, 144–150. [CrossRef]
135. Bajiri, M.A.; Hezam, A.; Namratha, K.; Viswanath, R.; Drmosh, Q.A.; Bhojya Naik, H.S.; Byrappa, K. CuO/ZnO/g-C_3N_4 heterostructures as efficient visible light-driven photocatalysts. *J. Environ. Chem. Eng.* **2019**, *7*, 103412. [CrossRef]

136. Jiang, L.; Yuan, X.; Zeng, G.; Liang, J.; Chen, X.; Yu, H.; Wang, H.; Wu, Z.; Zhang, J.; Xiong, T. In-situ synthesis of direct solid-state dual Z-scheme $WO_3/g-C_3N_4/Bi_2O_3$ photocatalyst for the degradation of refractory pollutant. *Appl. Catal. B Environ.* **2018**, *227*, 376–385. [CrossRef]
137. Yuan, Y.; Huang, G.-F.; Hu, W.-Y.; Xiong, D.-N.; Zhou, B.-X.; Chang, S.; Huang, W.-Q. Construction of $g-C_3N_4/CeO_2/ZnO$ ternary photocatalysts with enhanced photocatalytic performance. *J. Phys. Chem. Solids* **2017**, *106*, 1–9. [CrossRef]
138. Jiang, D.; Yu, H.; Yu, H. Modified $g-C_3N_4/TiO_2$ nanosheets/ZnO ternary facet coupled heterojunction for photocatalytic degradation of p-toluenesulfonic acid (p-TSA) under visible light. *Physica E* **2017**, *85*, 1–6. [CrossRef]
139. Liu, S.J.; Li, F.T.; Li, Y.L.; Hao, Y.J.; Wang, X.J.; Li, B.; Liu, R.H. Fabrication of ternary $g-C_3N_4/Al_2O_3/ZnO$ heterojunctions based on cascade electron transfer toward molecular oxygen activation. *Appl. Catal. B Environ.* **2017**, *212*, 115–128. [CrossRef]
140. Ren, Y.; Zhan, W.; Tang, L.; Zheng, H.; Liu, H.; Tang, K. Constructing a ternary $H_2SrTa_2O_7/g-C_3N_4/Ag_3PO_4$ heterojunction based on cascade electron transfer with enhanced visible light photocatalytic activity. *CrystEngComm* **2020**, *22*, 6485–6494. [CrossRef]
141. Raza, A.; Shen, H.; Haidry, A.A.; Cui, S. Hydrothermal synthesis of $Fe_3O_4/TiO_2/g-C_3N_4$: Advanced photocatalytic application. *Appl. Surf. Sci.* **2019**, *488*, 887–895. [CrossRef]
142. Mirzaei, A.; Chen, Z.; Haghighat, F.; Yerushalmi, L. Hierarchical magnetic petal-like Fe_3O_4-ZnO@$g-C_3N_4$ for removal of sulfamethoxazole, suppression of photocorrosion, by-products identification and toxicity assessment. *Chemosphere* **2018**, *205*, 463. [CrossRef] [PubMed]
143. Wu, Z.; Chen, X.; Liu, X.; Yang, X.; Yang, Y. A Ternary Magnetic Recyclable $ZnO/Fe_3O_4/g-C_3N_4$ Composite Photocatalyst for Efficient Photodegradation of Monoazo Dye. *Nanoscale Res. Lett.* **2019**, *14*, 147. [CrossRef] [PubMed]
144. Balu, S.; Velmurugan, S.; Palanisamy, S.; Chen, S.-W.; Velusamy, V.; Yang, T.C.K.; El-Shafey, E.-S.I. Synthesis of α-Fe_2O_3 decorated $g-C_3N_4/ZnO$ ternary Z-scheme photocatalyst for degradation of tartrazine dye in aqueous media. *J. Taiwan Inst. Chem. E.* **2019**, *99*, 258–267. [CrossRef]
145. Kumar, A.; Kumar, A.; Krishnan, V. Perovskite Oxide Based Materials for Energy and Environment-Oriented Photocatalysis. *ACS Catal.* **2020**, *10*, 10253–10315. [CrossRef]
146. Grabowska, E. Selected perovskite oxides: Characterization, preparation and photocatalytic properties-A review. *Appl. Catal. B Environ.* **2016**, *186*, 97–126. [CrossRef]
147. Shi, R.; Waterhouse, G.I.N.; Zhang, T.R. Recent Progress in Photocatalytic CO_2 Reduction Over Perovskite Oxides. *Sol. RRL* **2017**, *1*, 1700126. [CrossRef]
148. Kumar, A.; Schuerings, C.; Kumar, S.; Kumar, A.; Krishnan, V. Perovskite-structured $CaTiO_3$ coupled with $g-C_3N_4$ as a heterojunction photocatalyst for organic pollutant degradation. *Beilstin. J. Nanotechnol.* **2018**, *9*, 671–685. [CrossRef]
149. Ye, R.; Fang, H.; Zheng, Y.-Z.; Li, N.; Wang, Y.; Tao, X. Fabrication of $CoTiO_3/g-C_3N_4$ Hybrid Photocatalysts with Enhanced H_2 Evolution: Z-Scheme Photocatalytic Mechanism Insight. *ACS Appl. Mater. Interfaces* **2016**, *8*, 13879–13889. [CrossRef]
150. Yao, Y.; Lu, F.; Zhu, Y.; Wei, F.; Liu, X.; Lian, C.; Wang, S. Magnetic core–shell $CuFe_2O_4$@C_3N_4 hybrids for visible light photocatalysis of Orange II. *J. Hazard. Mater.* **2015**, *297*, 224–233. [CrossRef]
151. Zhang, T.; Zhao, K.; Yu, J.; Jin, J.; Qi, Y.; Li, H.; Hou, X.; Liu, G. Photocatalytic water splitting for hydrogen generation on cubic, orthorhombic, and tetragonal $KNbO_3$ microcubes. *Nanoscale* **2013**, *5*, 8375–8383. [CrossRef] [PubMed]
152. Yu, J.; Chen, Z.; Wang, Y.; Ma, Y.; Feng, Z.; Lin, H.; Wu, Y.; Zhao, L.; He, Y. Synthesis of $KNbO_3/g-C_3N_4$ composite and its new application in photocatalytic H2 generation under visible light irradiation. *J. Mater. Sci.* **2018**, *53*, 7453–7465. [CrossRef]
153. Wu, Y.; Wang, H.; Tu, W.; Liu, Y.; Tan, Y.Z.; Yuan, X.; Chew, J.W. Quasi-polymeric construction of stable perovskite-type $LaFeO_3$/$g-C_3N_4$ heterostructured photocatalyst for improved Z-scheme photocatalytic activity via solid p-n heterojunction interfacial effect. *J. Hazard. Mater.* **2018**, *347*, 412–422. [CrossRef] [PubMed]
154. Luo, J.; Deng, B.; Pu, Y.; Liu, A.; Wang, J.; Ma, K.; Gao, F.; Gao, B.; Zou, W.; Dong, L. Interfacial coupling effects in $g-C_3N_4/SrTiO_3$ nanocomposites with enhanced H_2 evolution under visible light irradiation. *Appl. Catal. B Environ.* **2019**, *247*, 1–9. [CrossRef]
155. Luo, J.; Chen, J.; Guo, R.; Qiu, Y.; Li, W.; Zhou, X.; Ning, X.; Zhan, L. Rational construction of direct Z-scheme $LaMnO_3/g-C_3N_4$ hybrid for improved visible-light photocatalytic tetracycline degradation. *Sep. Purif. Technol.* **2019**, *211*, 882–894. [CrossRef]
156. Wang, D.F.; Zou, Z.G.; Ye, J.H. A new spinel-type photocatalyst $BaCr_2O_4$ for H_2 evolution under UV and visible light irradiation. *Chem. Phys. Lett.* **2003**, *373*, 191–196. [CrossRef]
157. Jia, Y.F.; Ma, H.X.; Zhang, W.B.; Zhu, G.Q.; Yang, W.; Son, N.; Kang, M.; Liu, C.L. Z-scheme $SnFe_2O_4$-graphitic carbon nitride: Reusable, magnetic catalysts for enhanced photocatalytic CO_2 reduction. *Chem. Eng. J.* **2020**, *383*, 123172. [CrossRef]
158. Chang, W.; Xue, W.; Liu, E.; Fan, J.; Zhao, B. Highly efficient H_2 production over $NiCo_2O_4$ decorated $g-C_3N_4$ by photocatalytic water reduction. *Chem. Eng. J.* **2019**, *362*, 392–401. [CrossRef]
159. Fu, J.; Xu, Q.; Low, J.; Jiang, C.; Yu, J. Ultrathin 2D/2D $WO_3/g-C_3N_4$ step-scheme H_2-production photocatalyst. *Appl. Catal. B: Environ.* **2019**, *243*, 556–565. [CrossRef]
160. Sivasakthi, S.; Gurunathan, K. Graphitic carbon nitride bedecked with CuO/ZnO hetero-interface microflower towards high photocatalytic performance. *Renew. Energy* **2020**, *159*, 786–800. [CrossRef]
161. Jiang, Z.; Wan, W.; Li, H.; Yuan, S.; Zhao, H.; Wong, P.K. A Hierarchical Z-Scheme α-$Fe_2O_3/g-C_3N_4$ Hybrid for Enhanced Photocatalytic CO_2 Reduction. *Adv. Mater.* **2018**, *30*, 1706108. [CrossRef] [PubMed]
162. Zhang, C.; Li, Y.; Shuai, D.; Shen, Y.; Xiong, W.; Wang, L. Graphitic carbon nitride ($g-C_3N_4$)-based photocatalysts for water disinfection and microbial control: A review. *Chemosphere* **2019**, *214*, 462–479. [PubMed]
163. Xu, Q.; Zhang, L.; Cheng, B.; Fan, J.; Yu, J. S-Scheme Heterojunction Photocatalyst. *Chem* **2020**, *6*, 1543–1559. [CrossRef]

164. Low, J.; Yu, J.; Jaroniec, M.; Wageh, S.; Al-Ghamdi, A.A. Heterojunction Photocatalysts. *Adv. Mater.* **2017**, *29*, 1601694. [CrossRef] [PubMed]
165. Shi, J.; Zheng, B.; Mao, L.; Cheng, C.; Hu, Y.; Wang, H.; Li, G.; Jing, D.; Liang, X. $MoO_3/g-C_3N_4$ Z-scheme (S-scheme) system derived from MoS_2/melamine dual precursors for enhanced photocatalytic H_2 evolution driven by visible light. *Int. J. Hydrogen Energy* **2021**, *46*, 2927–2935. [CrossRef]
166. Li, W.; Da, P.; Zhang, Y.; Wang, Y.; Zheng, G. WO nanoflakes for enhanced photoelectrochemical conversion. *ACS Nano* **2014**, *8*, 11770–11777. [CrossRef]
167. Bai, Y.; Ye, L.; Wang, L.; Shi, X.; Wang, P.; Bai, W. A dual-cocatalyst-loaded Au/BiOI/MnOx system for enhanced photocatalytic greenhouse gas conversion into solar fuels. *Environ. Sci. Nano.* **2016**, *3*, 902–909.
168. Chen, X.; Zhu, K.; Wang, P.; Sun, G.; Yao, Y.; Luo, W.; Zou, Z. Reversible Charge Transfer and Adjustable Potential Window in Semiconductor/Faradaic Layer/Liquid Junctions. *iScience* **2020**, *23*, 100949. [CrossRef]
169. Jahurul Islam, M.; Amaranatha Reddy, D.; Han, N.S.; Choi, J.; Song, J.K.; Kim, T.K. An oxygen-vacancy rich 3D novel hierarchical MoS_2/BiOI/AgI ternary nanocomposite: Enhanced photocatalytic activity through photogenerated electron shuttling in a Z-scheme manner. *Phys. Chem. Chem. Phys.* **2016**, *18*, 24984–24993.
170. Wang, N.; Wu, L.; Li, J.; Mo, J.; Peng, Q.; Li, X. Construction of hierarchical $Fe_2O_3@MnO_2$ core/shell nanocube supported C_3N_4 for dual Z-scheme photocatalytic water splitting. *Sol. Energy Mater. Sol. Cells* **2020**, *215*, 110624.
171. Chang, X.; Wang, T.; Gong, J. CO_2 photo-reduction: Insights into CO_2 activation and reaction on surfaces of photocatalysts. *Energy Environ. Sci.* **2016**, *9*, 2177–2196. [CrossRef]
172. He, Y.; Wang, Y.; Zhang, L.; Teng, B.; Fan, M. High-efficiency conversion of CO_2 to fuel over $ZnO/g-C_3N_4$ photocatalyst. *Appl. Catal. B Environ.* **2015**, *168–169*, 1–8. [CrossRef]
173. Nie, N.; Zhang, L.; Fu, J.; Cheng, B.; Yu, J. Self-assembled hierarchical direct Z-scheme $g-C_3N_4$/ZnO microspheres with enhanced photocatalytic CO_2 reduction performance. *Appl. Surf. Sci.* **2018**, *441*, 12–22. [CrossRef]
174. Bhosale, R.; Jain, S.; Vinod, C.P.; Kumar, S.; Ogale, S. Direct Z-Scheme $g-C_3N_4$/$FeWO_4$ Nanocomposite for Enhanced and Selective Photocatalytic CO_2 Reduction under Visible Light. *ACS Appl. Mater. Interfaces* **2019**, *11*, 6174–6183. [CrossRef] [PubMed]
175. Ge, L.; Peng, Z.; Wang, W.; Tan, F.; Wang, X.; Su, B.; Qiao, X.; Wong, P.K. $g-C_3N_4$/MgO nanosheets: Light-independent, metal-poisoning-free catalysts for the activation of hydrogen peroxide to degrade organics. *J. Mater. Chem. A* **2018**, *6*, 16421–16429. [CrossRef]
176. Fan, J.; Qin, H.; Jiang, S. Mn-doped $g-C_3N_4$ composite to activate peroxymonosulfate for acetaminophen degradation: The role of superoxide anion and singlet oxygen. *Chem. Eng. J.* **2019**, *359*, 723–732. [CrossRef]
177. Ding, X.; Xiao, D.; Ji, L.; Jin, D.; Dai, K.; Yang, Z.; Wang, S.; Chen, H. Simple fabrication of $Fe_3O_4/C/g-C_3N_4$ two-dimensional composite by hydrothermal carbonization approach with enhanced photocatalytic performance under visible light. *Catal. Sci. Technol.* **2018**, *8*, 3484–3492. [CrossRef]
178. Zou, X.; Dong, Y.; Li, S.; Ke, J.; Cui, Y.; Ou, X. Fabrication of $V_2O_5/g-C_3N_4$ heterojunction composites and its enhanced visible light photocatalytic performance for degradation of gaseous ortho-dichlorobenzene. *J. Taiwan Inst. Chem. E.* **2018**, *93*, 158–165. [CrossRef]
179. Katsumata, K.-i.; Motoyoshi, R.; Matsushita, N.; Okada, K. Preparation of graphitic carbon nitride ($g-C_3N_4$)/WO_3 composites and enhanced visible-light-driven photodegradation of acetaldehyde gas. *J. Hazard. Mater.* **2013**, *260*, 475–482. [CrossRef]
180. Wang, S.; Ding, X.; Zhang, X.; Pang, H.; Hai, X.; Zhan, G.; Zhou, W.; Song, H.; Zhang, L.; Chen, H.; et al. In Situ Carbon Homogeneous Doping on Ultrathin Bismuth Molybdate: A Dual-Purpose Strategy for Efficient Molecular Oxygen Activation. *Adv. Funct. Mater.* **2017**, *27*, 1703923. [CrossRef]
181. Xu, X.; Geng, A.; Yang, C.; Carabineiro, S.A.C.; Lv, K.; Zhu, J.; Zhao, Z. One-pot synthesis of La–Fe–O@CN composites as photo-Fenton catalysts for highly efficient removal of organic dyes in wastewater. *Ceram. Int.* **2020**, *46*, 10740–10747. [CrossRef]
182. Xu, X.; Lin, H.; Xiao, P.; Zhu, J.; Bi, H.; Carabineiro, S.A.C. Construction of Ag-Bridged Z-Scheme $LaFe_{0.5}Co_{0.5}O_3/Ag_{10}$/Graphitic Carbon Nitride Heterojunctions for Photo-Fenton Degradation of Tetracycline Hydrochloride: Interfacial Electron Effect and Reaction Mechanism. *Adv. Mater. Interfaces* **2022**, *9*, 2101902. [CrossRef]
183. Xu, Q.; Zhao, P.; Shi, Y.-K.; Li, J.-S.; You, W.-S.; Zhang, L.-C.; Sang, X.-J. Preparation of a $g-C_3N_4/Co_3O_4/Ag_2O$ ternary heterojunction nanocomposite and its photocatalytic activity and mechanism. *New J. Chem.* **2020**, *44*, 6261–6268. [CrossRef]
184. She, X.; Xu, H.; Wang, H.; Xia, J.; Song, Y.; Yan, J.; Xu, Y.; Zhang, Q.; Du, D.; Li, H. Controllable synthesis of $CeO_2/g-C_3N_4$ composites and their applications in the environment. *Dalton T.* **2015**, *44*, 7021–7031. [CrossRef]
185. Chen, H.Y.; Qiu, L.G.; Xiao, J.D.; Ye, S.; Jiang, X.; Yuan, Y.P. Inorganic–organic hybrid $NiO–g-C_3N_4$ photocatalyst for efficient methylene blue degradation using visible light. *RSC Adv.* **2014**, *4*, 22491. [CrossRef]
186. Jiang, Z.; Jiang, D.; Yan, Z.; Liu, D.; Qian, K.; Xie, J. A new visible light active multifunctional ternary composite based on $TiO_2–In_2O_3$ nanocrystals heterojunction decorated porous graphitic carbon nitride for photocatalytic treatment of hazardous pollutant and H_2 evolution. *Appl. Catal. B Environ.* **2015**, *170–171*, 195–205. [CrossRef]
187. Zhang, Y.; Xu, M.; Li, H.; Ge, H.; Bian, Z. The enhanced photoreduction of Cr(VI) to Cr(III) using carbon dots coupled TiO_2 mesocrystals. *Appl. Catal. B Environ.* **2017**, *226*, 213–219. [CrossRef]
188. Liang, Z.-Y.; Wei, J.-X.; Wang, X.; Yu, Y.; Xiao, F.-X. Elegant Z-scheme-dictated $g-C_3N_4$ enwrapped WO_3 superstructures: A multifarious platform for versatile photoredox catalysis. *J. Mater. Chem. A* **2017**, *5*, 15601–15612. [CrossRef]

189. Song, X.-Y.; Chen, Q.-L. Facile preparation of g-C$_3$N$_4$/Bi$_2$WO$_6$ hybrid photocatalyst with enhanced visible light photoreduction of Cr(VI). *J. Nanopart. Res.* **2019**, *21*, 183. [CrossRef]
190. Zhu, J.; Wei, Y.; Chen, W.; Zhao, Z.; Thomas, A. Graphitic carbon nitride as a metal-free catalyst for NO decomposition. *Chem. Commun.* **2010**, *46*, 6965–6967. [CrossRef]
191. Sano, T.; Tsutsui, S.; Koike, K.; Hirakawa, T.; Teramoto, Y.; Negishi, N.; Takeuchi, K. Activation of graphitic carbon nitride (g-C$_3$N$_4$) by alkaline hydrothermal treatment for photocatalytic NO oxidation in gas phase. *J. Mater. Chem. A* **2013**, *1*, 6489–6496. [CrossRef]
192. Wang, Z.; Guan, W.; Sun, Y.; Dong, F.; Zhou, Y.; Ho, W.-K. Water-assisted production of honeycomb-like g-C$_3$N$_4$ with ultralong carrier lifetime and outstanding photocatalytic activity. *Nanoscale* **2015**, *7*, 2471–2479. [CrossRef] [PubMed]
193. Shi, L.; Liang, L.; Wang, F.; Ma, J.; Sun, J. Polycondensation of guanidine hydrochloride into a graphitic carbon nitride semiconductor with a large surface area as a visible light photocatalyst. *Catal. Sci. Technol.* **2014**, *4*, 3235–3243. [CrossRef]
194. Wang, W.; Huang, G.; Yu, J.C.; Wong, P.K. Advances in photocatalytic disinfection of bacteria: Development of photocatalysts and mechanisms. *J. Environ. Sci.* **2015**, *34*, 232–247. [CrossRef] [PubMed]
195. Tadashi, M.; Ryozo, T.; Toshiaki, N.; Hitoshi, W. Photoelectrochemical sterilization of microbial cells by semiconductor powders. *FEMS Microbiol. Lett.* **1985**, *29*, 211–214.
196. Huang, J.; Ho, W.; Wang, X. Metal-free disinfection effects induced by graphitic carbon nitride polymers under visible light illumination. *Chem. Commun.* **2014**, *50*, 4338–4340. [CrossRef]
197. Xu, J.; Wang, Z.; Zhu, Y. Enhanced Visible-Light-Driven Photocatalytic Disinfection Performance and Organic Pollutant Degradation Activity of Porous g-C$_3$N$_4$ Nanosheets. *ACS Appl. Mater. Interfaces* **2017**, *9*, 27727–27735. [CrossRef]
198. Li, Y.; Zhang, C.; Shuai, D.; Naraginti, S.; Wang, D.; Zhang, W. Visible-light-driven photocatalytic inactivation of MS$_2$ by metal-free g-C$_3$N$_4$: Virucidal performance and mechanism. *Water Res.* **2016**, *106*, 249–258. [CrossRef]

MDPI
St. Alban-Anlage 66
4052 Basel
Switzerland
Tel. +41 61 683 77 34
Fax +41 61 302 89 18
www.mdpi.com

International Journal of Molecular Sciences Editorial Office
E-mail: ijms@mdpi.com
www.mdpi.com/journal/ijms

www.ingramcontent.com/pod-product-compliance
Lightning Source LLC
LaVergne TN
LVHW070655100526
838202LV00013B/967